Ion Channel Drug Discovery

RSC Drug Discovery Series

Editor-in-Chief:
Professor David Thurston, *King's College, London, UK*

Series Editors:
Professor David Rotella, *Montclair State University, USA*
Professor Ana Martinez, *Medicinal Chemistry Institute-CSIC, Madrid, Spain*
Dr David Fox, *Vulpine Science and Learning, UK*

Advisor to the Board:
Professor Robin Ganellin, *University College London, UK*

Titles in the Series:

How to obtain future titles on publication:
A standing order plan is available for this series. A standing order will bring delivery of each new volume immediately on publication.

For further information please contact:
Book Sales Department, Royal Society of Chemistry, Thomas Graham House, Science Park, Milton Road, Cambridge, CB4 0WF, UK
Telephone: +44 (0)1223 420066, Fax: +44 (0)1223 420247
Email: booksales@rsc.org
Visit our website at www.rsc.org/books

Ion Channel Drug Discovery

Edited by

Brian Cox
Martin Gosling
Horsham Research Centre, Horsham, UK

THE QUEEN'S AWARDS
FOR ENTERPRISE:
INTERNATIONAL TRADE
2013

RSC Drug Discovery Series No. 39

Print ISBN: 978-1-84973-186-7
PDF eISBN: 978-1-84973-508-7
ISSN: 2041-3203

A catalogue record for this book is available from the British Library

Published by The Royal Society of Chemistry,
Thomas Graham House, Science Park, Milton Road,
Cambridge CB4 0WF, UK

Registered Charity Number 207890

For further information see our web site at www.rsc.org

Printed and bound by CPI Group (UK) Ltd, Croydon, CR0 4YY

This book is dedicated to the memory of our fathers, Raymond Cox (1927–2013) and Malcolm John Gosling (1947–2011), for their inspiration, love and support.

This book is also dedicated to the exceptional talented individuals of Ciba, Ciba-Giegy and Novartis who worked at the Horsham site (1937–2014) in their tireless pursuit of medicines to alleviate suffering.

Preface

Ion channels are located within the plasma membrane of nearly all cells and many intracellular organelles. These enigmatic proteins play a critical role in numerous fundamental physiological functions including generating and shaping action potentials, regulating cell volume and controlling epithelial secretion to name but a few. Ion channel modulators have historically proven their major therapeutic utility in numerous clinical paradigms including reduction of blood pressure, analgesia and local anaesthesia—however their potential as targets for novel and efficacious therapeutics is yet to be fully realized. The recent focus on the human genome has attributed an increasing number of diseases to genetic ion channel dysfunction (channelopathies). This emphasizes not only the important physiological and pathophysiological roles these proteins play but directs the drug discoverer to not only new targets but a clear understanding of potentially how they should be modulated.

Advances in our understanding of ion channel function combined with enabling technologies, such as automated electrophysiology, has fuelled interest and investment in ion channel modulation in both industrial and academic settings. The coming years promise to be exciting times as these new therapeutic opportunities are identified and realized.

This book builds on the platform created by the highly successful series of Royal Society of Chemistry 'Ion Channels as Therapeutic Targets' meetings in the last decade. These meetings, supported by both academic and industrial scientists, have covered themes including advances in screening technology, ion channel structure and modeling and up-to-date case histories of the discovery of modulators of a range of channels, both voltage-gated and non-voltage-gated channels.

Our goal was to provide a reference text of interest to a wide range of readers by eliciting contributions from highly respected academic and

RSC Drug Discovery Series No. 39
Ion Channel Drug Discovery
Edited by Brian Cox and Martin Gosling
© The Royal Society of Chemistry 2015
Published by the Royal Society of Chemistry, www.rsc.org

industrial researchers covering a combination of recent advances in the field, from technological and medicinal chemistry perspectives, as well as an introduction to the new 'ion channel drug discoverer'. We thank the many contributors to this book and hope you enjoy reading it as much as we did editing it.

Brian Cox and Martin Gosling
Horsham, UK

Contents

RSC Drug Discovery Series No. 39
Ion Channel Drug Discovery
Edited by Brian Cox and Martin Gosling
© The Royal Society of Chemistry 2015
Published by the Royal Society of Chemistry, www.rsc.org

**Chapter 11 The Therapeutic Potential of hERG1 K$^+$ Channels for
Treating Cancer and Cardiac Arrhythmias**   258
John Mitcheson and Annarosa Arcangeli

CHAPTER 1

Ion Channel Drug Discovery: a Historical Perspective

BRIAN COX

Global Discovery Chemistry, Novartis Institutes for Biomedical Research, Wimblehurst Road, Horsham, West Sussex, RH12 5AB, UK
Email: brian.cox@novartis.com

1.1 Introduction

Ion channels are membrane proteins that control the flow of ions across the cell membrane, are present in the membranes of all cells and make up one of the two traditional classes of ionophoric proteins along with ion transporters. Ion channels are responsible for maintaining resting membrane potential and all electrical signaling; they play a key role in the modulation of intracellular calcium levels crucial for the regulation of many cellular functions events and consequently a fundamental role in many physiological processes.

Ion channels are classified in a number of ways; by gating, *i.e.* what opens and closes the channels. Voltage-gated ion channels open or close depending on the voltage gradient across the plasma membrane, while ligand-gated ion channels open or close depending on binding of ligands to the channel. Further classifications are by the ion (or ions) that is (are) conducted *e.g.* sodium, calcium, potassium, proton, chloride or non-selective, or by the duration of the response to stimuli *e.g.* the transient receptor potential channels (TRP channels). Finally, for example in the potassium channel superfamily, they are classified by the number of pore loops contained in each channel forming subunit. The vast majority of channels form

RSC Drug Discovery Series No. 39
Ion Channel Drug Discovery
Edited by Brian Cox and Martin Gosling
© The Royal Society of Chemistry 2015
Published by the Royal Society of Chemistry, www.rsc.org

as tetramers of subunits that each contributes a single pore loop to the ion-conducting pore; however, there is a small family of two-pore-domain potassium channels (K2P), where channel subunits have two pore loops each; channels are formed as dimers of these subunits.[1]

In the same way as "kinome" is used to describe the protein kinase family of enzymes within the genome, the term "chanome" or "channelome" is often used to describe the >300 members of the ion channel family assembled from the ~500 annotated ion channel proteins as predicted in the human genome. With ion channels being such a large class of targets and involved in many key physiological processes it is surprising that less than twenty percent are currently commercially exploited.

When co-authoring an article on ion channels in 2005,[2] examination of the marketed ion channel modulators at that time highlighted that the majority of ion channels targeted up to that point had almost exclusively been ligand- or voltage-gated channels found in excitable tissues such as nerve and muscle. Since then there has been a number of drugs entering the market: retigabine (**1**), verenicline (**2**) and ivacaftor (**3**), and promisingly a considerable number of compounds in phase II/III clinical development with a growing number targeting channels in non-excitable tissues.[3] Table 1.1 lists a selection of the marketed ion channel modulators, their use and their specific ion channel target. Figure 1.1 illustrates the sustained commercial interest in the ion channel area as judged by the number of patent filings.

1   **2**

3

1.2 History of Ion Channel Drug Discovery

Classic examples of early ion channel modulator discovery are the voltage-gated sodium channel blockers used as local anesthetics and anticonvulsants and the L-type voltage-operated calcium channels blockers (L-VOCCs)

Table 1.1　Currently marketed ion channel modulators.

Ion channel target	Use	Drugs
L-type voltage-gated Ca^{2+} channel	Anti-hypertensive	Amlodipine, Nifedipine, Isradipine, Verapamil, Diltiazem, Nicardipine
L-type voltage-gated Ca^{2+} channel	Stroke	Nimodipine
L-type voltage-gated Ca^{2+} channel	Anti-arrhythmic	Verapamil, Diltiazem
N-type voltage-gated Ca^{2+} channel	Analgesic	Ziconitide
T-type voltage-gated Ca^{2+} channel	Anticonvulsant	Ethosuximide
KCNQ2 K_v 7.2	Anticonvulsant	Retigabine[a]
Cardiac voltage-gated Na^+ channel	Anti-arrhythmic	Procainamide, Quinidine, Lignocaine (aka Lidocaine)
Brain voltage-gated Na^+ channel	Anticonvulsant	Phenytoin, Lamotrigine, Carbamazepine
Voltage-gated Na^+ channel	Local anesthetics	Benzocaine, Lignocaine (aka Lidocaine), Procaine
Epithelial Na^+ channel	Diuretic	Amiloride
GABA Cl^- channel	Anticonvulsant	Diazepam
Nicotinic acetylcholine receptor	Neuromuscular blocker/ muscle relaxants	Atracurium
Nicotinic acetylcholine receptor	Smoking cessation	Varenicline[b]
$5HT_3$	Anti-emetic	Ondansetron, Granisetron
K_{ATP} channel	Diabetes	Tolbutamide, Glibenclamide, Gliclazide
CFTR channel	Cystic Fibrosis	Ivacaftor (VX-770)[c]

Approvals: [a]2010; [b]2006; [c]2012.

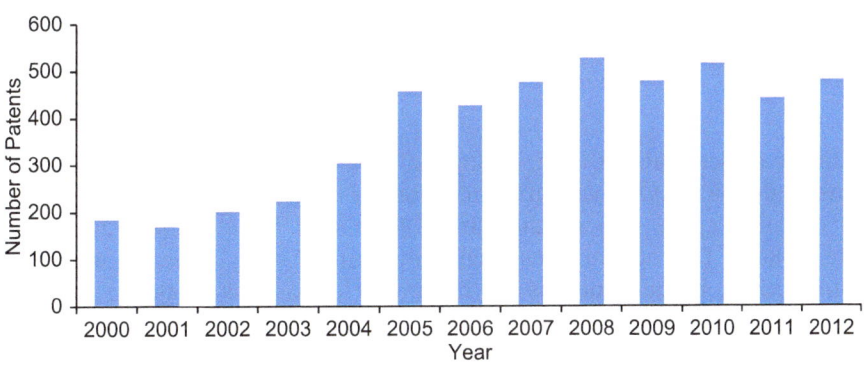

Figure 1.1　Patents filed listing "ion channel modulator" as mechanism of action (2000–2012).
Thomson Reuters Cortellis™.

which are still in clinical use many years after their first synthesis. Following an early understanding of the structural features of the natural product cocaine (**4**), the local anesthetic benzocaine (**5**) was synthesized in 1890, then amylocaine (**6**) in 1903 and procaine (**7**) in 1905. This was the start of a classic age of chemistry-driven drug discovery, optimizing compounds for efficacy and duration in animal models resulting in such compounds as procainamide (**8**) and lignocaine (lidocaine) (**9**).[4–6] It was not until the late 1950's that these compounds were shown to "inhibit" purported sodium channels which at the time were hypothesized to be integral in nerve impulse transmission.[7]

4   **5**

6   **7**

8   **9**

In the anticonvulsant field, phenytoin (**10**), synthesized as a phenobarbitone analogue in 1937, was introduced for the treatment of epilepsy followed by carbamazepine (**11**) in the early 1960's and lamotrigine (**12**) in the late 1980's. During this research compounds were selected primarily using animal models such as the maximal electroshock model in rodents. Later data from electrophysiological experiments on brain slices, in cloned channel systems and disposition studies has led to a better understanding of the mode of action of these compounds,[8] resulting in the derivation of anticonvulsant pharmacophore models for sodium channel blockers.[9,10]

10   11   12

Many of the anticonvulsant sodium channel blockers find off-label use in a variety of neuropathic pain conditions[11] *e.g.* compound **13** (CNV-1014802) is a compound from a distinct structural class, shown to be anticonvulsant[12] and is currently under clinical investigation for neuropathic pain. The compound, structurally related to a number of earlier published compounds (*e.g.* CO 102862 (**14**)[13] and safinamide (**15**)[14]) and recent compounds published by workers at Merck (*e.g.* **16**[15]) is described as a state-dependent sodium channel blocker that exhibits potency and selectivity against the Nav1.7 channel. This molecule was recently granted orphan-drug designation by the US FDA in July 2013 for the treatment of trigeminal neuralgia.[16] Even now we are still to fully understand the precise mode of action of these compounds on the nine members of the voltage-gated sodium channel family (Nav1.1-1.9) that contribute to the range of pharmacology exhibited (see chapter 5 for an extensive review of the Nav channel family).

13   14

15   16

The recently discovery of the novel anticonvulsant retigabine (**1**) followed a similar retrospective discovery route. Originally synthesized as an analogue of the opioid analgesic and muscle relaxant flupirtine in 1993 by researchers at ASTA Medica, it was found to be anticonvulsant. Mechanistically this was explained as the result of modulation of the neuronal "M current". Now with the benefit of molecular insight, retigabine has been shown to be an activator of Kv7.2 (KCNQ2/3). This is covered in detail in chapter 10.

The discovery of L-type calcium channel blockers and their utility in coronary disease occurred by chance in 1963, when it was reported that new compounds such as the phenylalkyamine (**17**, later named verapamil) mimicked the cardiac effect of simple calcium withdrawal, diminishing calcium-dependent high energy phosphate utilization, contractile force and oxygen requirement. In 1969, the term "calcium antagonist" was given a novel drug designation. In an extensive search for other calcium antagonists, a considerable number of substances that also met these criteria were identified; in 1975 the first dihydropyridines (*e.g.* nifedipine (**18**)) were discovered followed by many other members of this class in the following decades, including longer-acting compounds such as amlodipine (**19**). Also in 1975, a third class of L-type calcium channel blockers was discovered, the benzothiazepine class *e.g.* diltiazem (**20**).[17–19]

Ion channel targets in non-excitable tissues remained virtually unexplored in the early period of research apart from the discovery of the sulfonylurea class of anti-diabetic agents that modulate the ATP-sensitive potassium channels (KATP) in pancreatic cells. The sulfonylureas were synthetic offshoots from the extensive work on sulfonamide antibacterials which were in turn derived from azo dyes (the dye prontosil was found to be a metabolic pro-drug for sulfanilamide). Janbon and co-workers observed that using the sulfonamide isopropylthiadozole (**21**) to treat typhoid patients caused a high incidence of hypoglycemia and at the same time Loubatières working with the compound in dogs found it to have 'insulin-like' properties.[20] Compounds such as tolbutamide (**22**) entered the clinic for the treatment of type-II diabetes in 1956 but are now largely superseded by second-generation agents such as glibenclamide (glyburide) (**23**).[21] Almost 40 years after the discovery of the first compounds, the KATP channel in pancreatic β cells was identified as the molecular target for sulfonylureas,[22] and another decade passed before the molecular composition of the pancreatic KATP channel was identified, a complex of the pore-forming inwardly rectifying channel Kir 6.2 and sulfonylurea receptor type 1 (SUR1) regulatory subunit.[23]

In the 1970's and 80's ion channel research was driven by the pharma-cological discovery and classification of ion channel targets in tissue using a variety of synthetic tool molecules designed from endogenous ligands or pharmacologically active natural products. The discovery of membrane-permeable fluorometric indicator dyes allowed the measurement of intra-cellular calcium levels and membrane potential and the development of the patch-clamp technique allowed individual ion channel activity to be meas-ured for the first time. Data could therefore be provided that gave kinetic and biophysical information to aid the design of potent and functionally se-lective agents. The discovery stories associated with the 5HT$_3$ and the neuromuscular nicotinic acetylcholine receptors (nAChRs) have been well documented and resulted in a considerable number of therapeutically useful drugs, *e.g.* the antiemetic agent ondansetron (**24**) by workers at Glaxo (now GSK)[24,25] and the neuromuscular blocking agent atracurium (**25**) by aca-demic workers at Strathclyde University in collaboration with Wellcome (now GSK).[26] The latter was designed from consideration of other bi-quaternary compounds and D-tuborcurarine, an alkaloid from the South

American arrow poison curare, all used as neuroblocking agents at that point. The elimination of atracurium is not dependent on metabolic breakdown by esterases, unlike other agents, and was found to be under polymorphic control thus causing unpredictable duration of muscular relaxation between patients. Instead atracurium was designed to rely on pH dependent Hoffman elimination, which provided highly predictable onset and duration of action across patients.

24

25

Natural product research still remains an important source of ion channel modulators (see chapter 12) either as leads for medicinal chemistry programs or as drugs in their own right. The recently launched varenicline (2), a partial agonist of the α4β2 nicotinic acetylcholine receptor, utilized therapeutically in smoking cessation, was designed from (−)-cytisine (26) following observations that it was a partial agonist and antagonized the receptor response to its endogenous neurotransmitter, acetylcholine.[27]

2   **26**

The cone snail peptide ziconotide (**27**) entered clinical use in the last decade for the treatment of severe pain and is a highly potent blocker of the N-type voltage-gated calcium channel. It is delivered as an infusion into the cerebrospinal fluid using an intrathecal pump system.[28] Continuing with the larger molecule theme, antibodies represent a growing field of research for finding ion channel modulators; this subject is covered in chapter 13.

H-Cys-Lys-Gly-Lys-Gly-Ala-Lys-Cys-Ser-Arg-Leu-Met-Tyr-Asp-Cys-Cys-Thr-Gly-Ser-Cys-Arg-Ser-Gly-Lys-Cys-NH₂

27

In the 1990's advances in molecular pharmacology enabled the re-combinant expression of channel targets in heterologous cellular backgrounds that could be used in high throughput screening modes. Chapter 2 covers the impact and advances in high throughput screening in the ion channel field. VX-770 (ivacaftor) (**3**) was approved by the FDA in January 2012 for the treatment of cystic fibrosis in patients with the G551D mutation and probably represents the first marketed compound to be derived from a high throughput screen hit. This molecule stemmed from work carried out at Vertex using cell-based fluorescence membrane potential assays in high-throughput mode to identify potentiators and correctors of CFTR function. Early hits from the potentiator and corrector screens are exemplified by compounds VRT-422 (**28**) and VRT-532 (**29**), respectively.[29,30] Workers at Genzyme utilized a similar approach to identify hits such as **30**.[31]

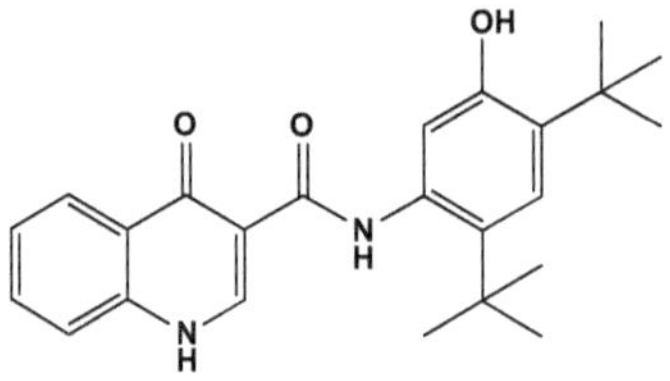

3

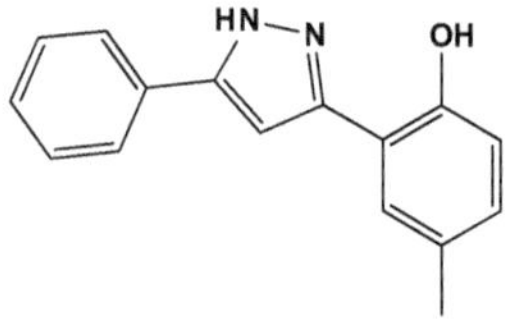

28

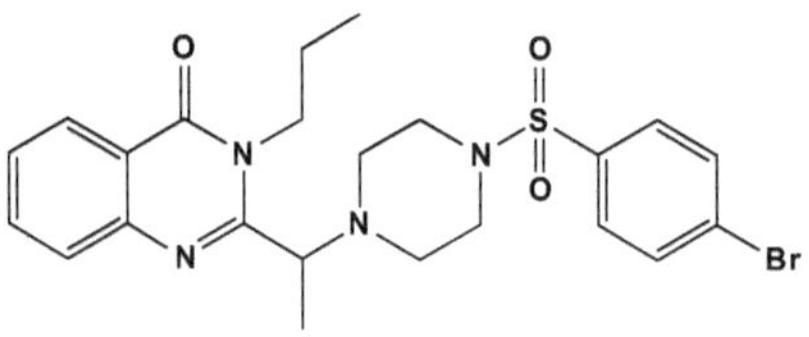

29

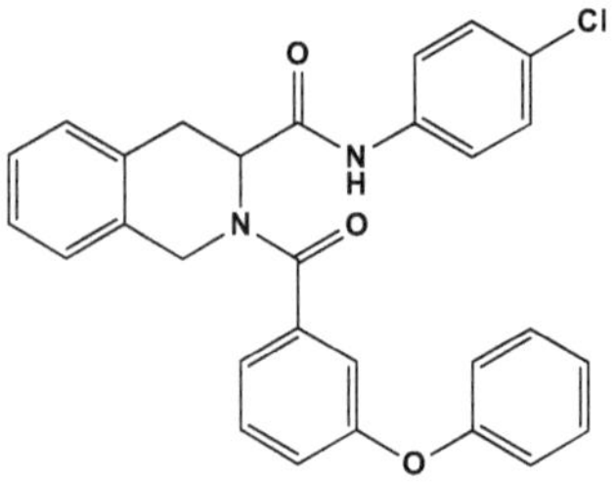

30

Another approach towards the treatment of cystic fibrosis is the blockade of the epithelial sodium channel (ENaC) with research focusing on topical administration into the lung of analogues of the potassium sparing diuretic amiloride (**31**), developed by Merck in the 1960's.[32] Amiloride itself has been studied in a number of clinical trials with conflicting results; the lack of efficacy of amiloride may be due to its short duration of action: it is cleared rapidly from the airways and its effect on lower airway potential difference lasts for only approximately 30 minutes.[33] Consequently, research efforts have focused on the design of longer acting analogues; this work is described in chapter 7. Recently, amiloride has also been shown to be a non-selective inhibitor of acid-sensing ion channels (ASICs) and exhibits a modest effect in rat pain models at high concentrations. Structural modification has provided more potent analogues, with particular focus on ASIC3 as the specific family member targeted for blocking chronic inflammatory pain.[34]

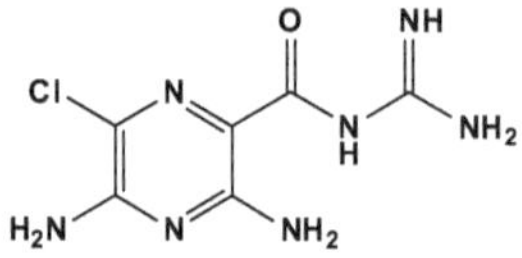

31

The ionotropic glutamate receptors (iGluRs) comprising of the AMPA, NMDA and kainate families have received much attention as targets for a variety of CNS disorders with a large number of compounds in clinical trial.[3] Perampanel (**32**), a non-competitive AMPA-type glutamate receptor antagonist, was recently reported to be well-tolerated and effective in reducing seizure frequency in partial onset epilepsy *versus* placebo.[35] Chapter 6 describes an approach to discovery of allosteric modulators of this target where co-crystallization of hits derived from screening with a ligand binding domain construct was used to aid structural based design.

32

Another family of ion channels that has received much interest over the past decade is that of the transient receptor potential channels (TRP channels) mentioned in the introduction. They are divided into two groups: group 1 containing the TRPC, TRPV, TRPM, TRPN and TRPA sub-groups and group 2 containing the TRPP and TRPML sub-groups. The fascinating prospect for modulation of these channels is that they mediate a variety of key bodily sensations such as pain, hot or cold, tastes, pressure, and aspects of vision.[36] The search for antagonists of TRPV1 as potential analgesics has been a major endeavor for many researchers and is covered in chapter 9.

One channel that has received more attention than almost any other is the hERG channel. HERG refers to the human Ether-à-go-go-Related Gene, alternately known as KCNH2, that codes for the potassium channel alpha subunit Kv11.1. This channel is involved in the mediation of the repolarizing current (I_{Kr}) in the cardiac action potential. HERG is actually a therapeutic target with compounds such as dofetilide (**33**), a class III antiarrhythmic agent, used for the maintenance of sinus rhythm.[37] Unfortunately the main attention has not been for positive reasons but as a target associated with off-target pharmacology (a so-called "anti-target") manifesting itself as life threatening drug-induced long-QT syndrome. Terfenadine (**34**) is perhaps the most well-known example, used widely as an antihistamine until reports of long-QT syndrome prompted its withdrawal. It was found to be a hERG blocker that under normal circumstances was rapidly metabolized by CYP3A4 on first-pass to an active metabolite that did not block hERG. However, many other drugs (*e.g.* erythromycin) and some foods (*e.g.* grapefruit juice) inhibit the function of CYP3A4 thus preventing the metabolism of terfenadine, leading to accumulation in the plasma and inevitable cardiac side-affects. The active metabolite of terfenadine, fexofenadine (**35**), is now marketed as an antihistamine in its own right and gave a structural insight into how to avoid hERG affects.[38] The structure–activity relationships of hERG modulation have been widely studied; a growing pharmacophoric understanding and careful pharmacological screening now aid drug design in avoiding hERG associated liabilities.[39,40] A full discussion on hERG is presented in chapter 11.

33

34

35

Since Rod Mackinnon and Peter Agre received the 2003 Nobel Prize for chemistry for discoveries concerning channels in cell membranes and for the discovery of water channels, respectively,[41] there has been an increasing numbers of ion channel crystal structures solved, with an expectation that many more structures will be determined in the coming years. The form and composition of native ion channels has also been under scrutiny and it is clear now that ion channels are not isolated ion-conducting pores but are integral components of structural and signaling complexes in cells. They do not just function in isolation but have closely associated accessory scaffolding sub-units and proteins, which are important for the functionality of the native channel, not only in terms of activation but also in terms of ensuring correct assembly, trafficking, insertion and retrieval.[42] Chapter 4 gives further insights into the study of ion channel structure.

Despite the availability of cellular systems containing cloned channel targets and the growing amount of information regarding ion channel structure and function, the pursuit of ion channel targets in high through-put screening mode has not progressed as well as other target classes, such as GPCRs and enzymes. Often screens use indirect assay readouts, such as ion-sensitive and potentiometric dyes, radiotracer flux and binding assays; although much higher throughput, all have well-documented limitations, such as sensitivity and a high potential for false positives.[43] Electro-physiology represents the highest fidelity option for ion channel screening, but was exceptionally low throughput, time consuming and required tech-nically skilled operators. However, the advent of automated electro-physiology offers the opportunity of running direct electrophysiology-based HTS campaigns; this is covered in chapter 3.

1.3 Conclusion

Ion channel research has come a long way in the last 100 or so years, pro-viding a wealth of novel and important medicines for the clinic. Continuing scientific and technological advances in the field of screening, structural-based design, the increasing chemical space of modulators available (low molecular weight, peptides, antibodies *etc.*) coupled with the availability of genetic information, raises the exciting prospect of taking us into a new more predictive and hopefully productive era of ion channel drug discovery.

Acknowledgement

Many thanks to Andrew Davis for carrying out the searches to provide the data expressed in Figure 1.1.

References

1. Z. Es-Salah-Lamoureux, D. F. Steele and D. Fedida, *Trends Pharmacol. Sci.*, 2010, **31**(12), 587.
2. S. Li, M. Gosling, C. T. Poll, J. Westwick and B. Cox, *DDT*, 2005, **10**(2), 129.
3. A. Wickenden, B. Priest and G. Erndemli, *Future Med. Chem.*, 2012, **4**(5), 661.
4. R. Sabatowski, D. Schaefer, S. M. Kasper, H. Brunsch and L. Radbruch, *Curr. Pharm. Des.*, 2004, **10**(7), 701.
5. J. Wojnar, *Chem. N. Z.*, 2007, **71**(1), 20.
6. W. O. Foye, in *Foye's Principles of Medicinal Chemistry*, ed. D. A. Williams and T. L. Lemke, Lippincott, Williams & Wilkins, Philadelphia, PA, 2002, p. 4.
7. S. Stolc, *Gen. Physiol. Biophys.*, 1988, 7, 177.

8. C. H. Large, M. Kalinichev, A. Lucas, C. Carignani, A. Bradford, N. Garbati, I. Sartori, N. E. Austin, A. Ruffo, D. N. C. Jones, G. Alvaro and K. D. Read, *Epilepsy Res.*, 2009, **85**, 96.

9. K. Unverferth, J. Engel, N. Höfgen, A. Rostock, R. Günther, H. J. Lankau, M. Menzer, A. Rolfs, J. Liebscher, B. Müller and H.-J. Hofmann, *J. Med. Chem.*, 1998, **41**(1), 63.

10. G. M. Lipkind and H. A. Fozzard, *Mol. Pharmacol.*, 2010, **78**(4), 631.

11. T. S. Jensen, *Eur. J. Pain*, 2002, **6**(Suppl. A), 61.

12. C. H. Large, S. Bison, I. Sartori, K. D. Read, A. Gozzi, D. Quarta, M. Antolini, E. Hollands, C. H. Gill, M. J. Gunthorpe, N. Idris, J. C. Neill and G. S. Alvaro, *J. Pharmacol. Exp. Ther.*, 2011, **338**(1), 100.

13. V. I. Ilyin, D. D. Hodges, E. R. Whittemore, R. B. Carter, S. X. Cai and R. M. Woodward, *Br. J. Pharmacol.*, 2005, **144**, 801.

14. *Drugs R D*, 2004, 5(6), 355.

15. S. Tyagarajan, P. K. Chakravarty, B. Zhou, B. Taylor, M. H. Fisher, M. J. Wyvratt, K. Lyons, T. Klatt, X. Li, S. Kumar, B. Williams, J. Felix, B. T. Priest, R. M. Brochu, V. Warren, M. Smith, M. Garcia, G. J. Kaczorowski, W. J. Martin, C. Abbadie, E. McGowan, N. Jochnowitz and W. H. Parsons, *Bioorg. Med. Chem. Lett.*, 2010, **20**(18), 5480.

16. http://www.convergencepharma.com/.

17. A. Fleckenstein, *Circ. Res.*, 1983, **52**, I3–I6.

18. M. Harrold, in *Foye's Principles of Medicinal Chemistry*, ed. T. L. Lemke and D. A. Williams, Lippincott, Williams & Wilkins, Philadelphia, PA, 2013, p. 768.

19. W. M. Yousef, A. H. Omar, M. D. Morsy, M. M. Abd El-Wahed and N. M. Ghanayem, *Int. J. Diabetes & Metabolism*, 2005, **13**(2), 76.

20. H. Kleinsorge, *Exp. & Clin. Endocrinol. & Diabetes*, 1998, **106**(2), 149.

21. S. W. Zito, in *Foye's Principles of Medicinal Chemistry*, ed. T. L. Lemke and D. A. Williams Lippincott Williams & Wilkins, Philadelphia, PA, 2013, p. 890.

22. N. C. Sturgess, R. Z. Kozlowski, C. A. Carrington, C. N. Hales and M. L. J. Ashford, *Br. J. Pharmacol.*, 1988, **95**(1), 83.

23. S. Seino, *Annu. Rev. Physiol.*, 1999, **61**, 337.

24. R. Barrett, D. Cavalla, I. H. Coates, C. Eldred, G. B. Ewan, N. Godfrey, D. C. Humber, P. C. North and A. W. Oxford, in *Trends Med. Chem. '90, Proc. Int. Symp. Med. Chem. 11ᵗʰ*, ed. S. Sarel, R. Mechoulam and I. Agranat, Blackwell, Oxford, 1992, p. 107.

25. Y. Chong and H. Choo, *Expert Opin. Investig. Drugs*, 2010, **19**(11), 1309.

26. R. D. Waigh, *Chem. Br.*, 1988, **24**(12), 1209.

27. J. W. Coe, P. R. Brooks, M. G. Vetelino, M. C. Wirtz, E. P. Arnold, J. Huang, S. B. Sands, T. I. Davis, L. A. Lebel, C. B. Fox, A. Shrikhande, J. H. Heym, E. Schaeffer, H. Rollema, Y. Lu, R. S. Mansbach, L. K. Chambers, C. C. Rovetti, D. W. Schulz, F. D. Tingley III and B. T. O'Neill, *J. Med. Chem.*, 2005, **48**, 3474.

28. H. E. Hannon and W. D. Atchison, *Mar. Drugs*, 2013, **11**, 680.

29. F. Van Goor, K. S. Straley, D. Cao, J. González, S. Hadida, A. Hazlewood, J. Joubran, T. Knapp, L. R. Makings, M. Miller, T. Neuberger, E. Olson, V. Panchenko, J. Rader, A. Singh, J. H. Stack, R. Tung, P. D. J. Grootenhuis and P. Negulescu, *Am. J. Physiol. Lung Cell Mol. Physiol.*, 2006, **290**, L1117.
30. F. Van Goor, S. Hadida, P. D. J. Grootenhuis, B. Burton, D. Cao, T. Neuberger, A. Turnbull, A. Singh, J. Joubran, A. Hazlewood, J. Zhou, J. McCartney, V. Arumugam, C. Decker, J. Yang, C. Young, E. R. Olson, J. J. Wine, R. A. Frizzell, M. Ashlock and P. Negulescu, *Proc. Natl. Acad. Sci. U. S. A.*, 2009, **106**(44), 18825.
31. B. H. Hirth, S. Qiao, L. M. Cuff, B. M. Cochran, M. J. Pregel, J. S. Gregory, S. F. Sneddon and J. L. Kane, Jr, *Bioorg. Med. Chem. Lett.*, 2005, **15**(8), 2087.
32. E. J. Cragoe, Jr, O. W. Woltersdorf, Jr, J. B. Bicking, S. F. Kwong and J. H. Jones, *J. Med. Chem.*, 1967, **10**(1), 66.
33. H. C. Rodgers and A. J. Knox, *Eur. Respir. J.*, 2001, **17**(6), 1314.
34. S. D. Kuduk, C. N. Di Marco, R. K. Chang, R. M. DiPardo, S. P. Cook, M. J. Cato, A. Jovanovska, M. O. Urban, M. Leitl, R. H. Spencer, S. A. Kane, M. T. Bilodeau, G. D. Hartman and M. G. Bock, *Bioorg. Med. Chem. Lett.*, 2009, **19**(9), 2514.
35. S. Rheims and P. Ryvlin, *Neuropsychiatr. Dis. Treat.*, 2013, **9**, 629.
36. *Transient Receptor Potential Channels. Advances in Experimental Medicine and Biology*, ed. M. S. Islam, Springer, Netherlands, 2011.
37. H. Roukoz and W. Saliba, *Expert Rev. Cardiovasc. Ther.*, 2007, **5**(1), 9.
38. I. Paakkari, *Toxicol. Lett.*, 2002, **127**(1–3), 279.
39. M. Recanatini, E. Poluzzi, M. Masetti, A. Cavalli and F. De Ponti, *Medicinal Research Reviews*, 2005, **25**(2), 133.
40. E. Raschi, V. Vasina, E. Poluzzi and F. De Ponti, *Pharmacol. Res.*, 2008, **57**(3), 181.
41. http://www.nobelprize.org/nobel_prizes/chemistry/laureates/.
42. C. Deutsch, *Neuron*, 2003, **40**(2), 265.
43. J. Xu, X. Wang, B. Ensign, M. Li, L. Wu, A. Guia and J. Xu, *DDT*, 2001, **6**(24), 1278.

CHAPTER 2

High-Throughput Screening

ALEXANDER BÖCKER, SABINE SCHAERTL AND
STEPHEN D. HESS*

Evotec AG, Essener Bogen 7, Hamburg, Germany 22419
*Email: Stephen.Hess@evotec.com

2.1 Introduction

Ion channels are a class of proteins that are attractive human, animal, and crop health targets. Ion channels can be considered high-gain (physiologically speaking), high-reward (therapeutic benefit) and high technology (difficult assay development) drug targets. Of the current 210 known human genes encoding ion channel alpha subunits in the human genome,[1] many have been functionally characterized to furnish an understanding of their biophysical and pharmacological properties, which then allows comparison to the congener channel in native tissues. Reviews covering the discovery, expression, function and pharmacological properties of ion channel families continue to be published (*c.f.* the recent review of Calcium-Activated Chloride channels[2]).

Ion channels are attractive therapeutic targets because they often have high control gain in physiological systems; *i.e.* opening or blocking many of the ion channels studied to date has a large effect on the function of the muscle, nerve, or cell containing the channels. For example, i.v. administration of lidocaine, a sodium channel blocker discovered in the 1940s,[3] can dampen and then stop action potential generation in the peripheral nerve of human subjects, and concomitant relief of pain symptoms in real time is reported by the subjects.[4] It is emphasized that this is in stark contrast to other popular drug targets, such as GPCRs and enzymes, where complex

RSC Drug Discovery Series No. 39
Ion Channel Drug Discovery
Edited by Brian Cox and Martin Gosling
© The Royal Society of Chemistry 2015
Published by the Royal Society of Chemistry, www.rsc.org

downstream signaling steps make it difficult or impossible to predict the overall gain of altering the function of these targets. The major issue confronting ion channel drug development teams is safety, as the same or closely-related ion channels are often widely distributed in the heart and CNS and activity at these channels needs to be minimal. Reviews that discuss ion channels as drug targets are available.[5,6]

It appears that every major pharmaceutical company and many biotechnology companies with drug discovery efforts have pursued at least one ion channel target in their portfolio following the appearance of the first papers clearly linking drug action to ion channels in the late 1940s (*i.e.*, xylocaine[3]). There are "blockbuster" drugs that target ion channels (*e.g.* DHPs for blood pressure, cardiac ion channel drugs), yet proportionately more effort has focused on other target classes such as GPCRs and kinases.[6] There are perhaps many reasons for this, but one is the historically poor performance of ion channel HTS efforts over the years. The majority of currently-prescribed ion channel drugs were most likely derived from classical pharmacology approaches. One success story is the CFTR potentiator ivacaftor (VX-770, Kalydeco) which was optimized from a lead scaffold discovered in an HTS of 228,000 compounds using a fluorescence-based membrane potential assay.[7] It is hoped that additional success stories are on the horizon.

To discover useful starting points for medicinal chemists to deliver efficacious, selective, and orally-available ion channel drug candidates a variety of methods to screen ever-larger collections of lead-like compounds, fragments, and antibodies have been used. Automated high-throughput patch clamp (HTPC) platforms, introduced commercially *circa* 2003 and now used worldwide by ion channel researchers, were expected to improve the odds and speed of finding good ion channel hits, and tracking hit-to-lead development, but today the platforms are mostly used to test for hERG liability and support medicinal chemistry activities after hits have been found using other approaches. There are notable exceptions (see later section on Automated Electrophysiology), but today almost all ion channel HTS are performed without using these platforms. That may soon change as the platforms evolve, and consumables costs decrease, but fluorescence-based assays remain the workhorse today.

Rather than providing a comprehensive survey of published ion channel HTS, we will instead briefly review the history of ion channel HTS techniques and then use examples from the authors' work with our collaborators to illustrate how the necessary approaches yield good results. We do not wish to ignore the excellent work performed by others, but instead use this contribution to describe in numerical detail our approach to assay development and full library HTS for selected examples in as much detail as possible. Confirmation of the fluorescence-based assay hits in electrophysiological assays is also presented. It is clear that ion channel HTS with carefully designed and pharmacologically relevant assays are now capable of providing suitably potent and selective starting points for medicinal chemistry programs that were not possible not so long ago. We predict an increased level of clinical success for this target class soon, as continual progress towards overcoming the difficult technical hurdles in making drugs is evident in the literature and in public disclosures. In sight are drugs that treat diseases that were previously poorly managed, or perhaps even cures for some diseases altogether.

2.2 Goal(S) of HTS Campaigns

The goal of an HTS effort is to identify a novel (that is patentable) chemical entity, active at the target screened, that has all of the properties necessary to be a safe and effective drug. Although in principle a drug molecule could be identified directly from an HTS, this probably has not happened very often. Instead, drug discovery teams use HTS campaigns to identify a handful of starting chemical scaffolds that can be progressed through years of medicinal chemistry effort to clinical development candidates that have acceptable potency and mechanism of action at the target, selectivity for the target, suitable physicochemical properties, an acceptable DMPK profile, can be dosed once a day orally, *etc*. Modern compound libraries now exceed millions of compounds, so that well-designed and executed HTS for ion channels usually identify suitable chemical starting points. But are these the best starting points? Would screening a different library give different and more valuable starting points? These questions are a source of anxiety to drug discovery team members, as there are no conclusive answers. Given the cost of all of the steps in drug discovery subsequent to the HTS step, it is crucial to design and execute the best library screen possible.

There are HTS efforts that seek alternative outcomes; *e.g.* screening a compound collection at a safety target, such as the ventricular potassium channel hERG, to gain an understanding of the overall liability of a collection for such a clear safety risk. Knowledge that a particular functional group causes a safety problem, or leads to intractable chemical series, has led to *in silico* screens to mark the members of the library that have the unwanted property. On the whole, most HTS campaigns examine large libraries, or substantial proportions of libraries focused into a collection with specified chemical properties, to identify novel chemical matter at the desired target. It is this type of HTS campaign that we will focus on.

2.3 HTS before and after the Cloning of Human Ion Channels

The first description of the cloning of an alpha subunit of an ion channel was published for the *Torpedo* nicotinic acetylcholine receptor by the Barnard lab in 1982.[8] A search of the PubMed database with the terms "Ion channel AND screen" yields a paper published in 1983 describing novel benzazepines as potential anxiolytics.[9] How were ion channel drugs discovered before human ion channel genes were cloned?

Before biochemical reconstitution or expression of cloned ion channels was achieved, screening strategies began with native tissues, cells, or membrane preparations known to express the ionic current of interest. Alternatively, analogues of a chemical structure with desired activity *in vivo* were directly screened in animal models such as pentylenetetrazol or electroshock maximal seizure using mice or rats,[10] where compounds targeting CNS ion channels or receptors were injected directly into the brain *via* the intracerebral ventricles. Active compounds would affect the required dose of pentylenetetrazol or alter the seizure threshold to the applied electroshock in this scenario. Of course a very limited number of compounds could be examined with these types of assays, so the breadth of chemical matter investigated was severely compromised.

The first functional expression of a mammalian ion channel, the rat brain sodium channel, was published from the Numa group in 1986.[11] As the efforts to clone and express individual ion channel subunits for human channels progressed rapidly through the late 1980s and 1990s, screening methodologies began to appear in the literature.[12–14] Multiple stories of ion channel HTS campaigns appear in the literature starting around the year 2000 (Figure 2.1), and a recent example for identifying openers of maxi-K channels is an excellent overview of a program that began with primary tissue assays and later utilized cloned human channels to identify and optimize selective tetrahydroquinoline maxi-K activators.[15]

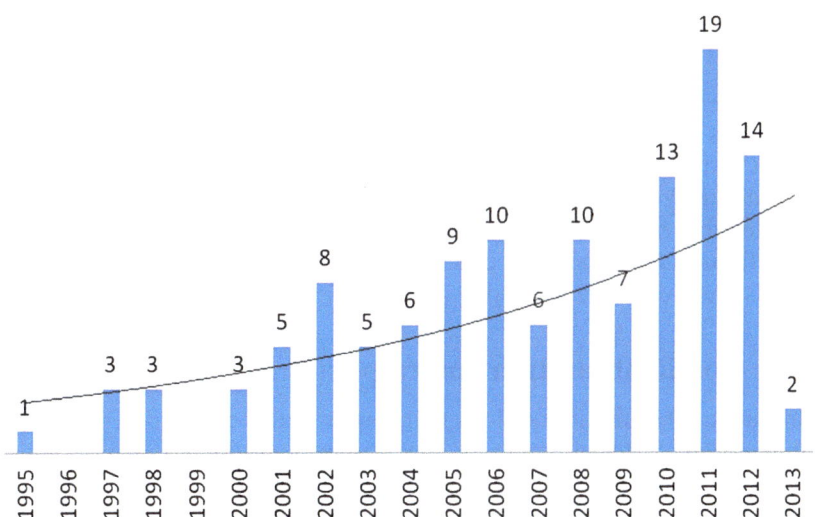

Figure 2.1 Number of publications per year using the search term "Ion Channel AND High Throughput Screen", 1995–present, from the Sciverse Scopus search tool. The solid line is a fit of an exponential function to the data.

2.4 A Brief History of Ion Channel HTS Methodologies

Here we will overview methods to study the interaction of test substances with ion channels. We include both methods that have not to date supported the compound throughput that most in the field accept is "High Through-put", *i.e.* millions of substances can be tested in a matter of a month or two, and those that have advanced to meet this level of throughput.

As mentioned above, before the cloning of alpha subunits for specific ion channels allowed HTS campaigns using recombinant cell lines, native tissues or membrane preparations were utilized. Studies of voltage-gated calcium channels in brain synaptosomes using ^{45}Ca uptake assays date back to at least the mid-to-late 1970s.[16,17] Although sensitive assays can in many cases be developed for various ion channels by using a labeled ion that fluxes through the channel of interest, the costs of radioligands used in the assay and disposal of the radioactive waste are usually incompatible with screening large libraries of compounds. Similar cost issues are true for radioligand displacement assays for ion channels. These were tried repeatedly in the 1980s and 1990s with limited success, as although high-affinity and specific peptide toxins were available (*e.g.*, batrachotoxin for voltage-gated sodium channels, omega CgTX for Cav2.2 channels, isradi-pine for Cav1.3, and others) the common problems are twofold: 1) hits in the assay can be agonists, antagonists, or positive or negative allosteric modulators, which dilutes the hit rate for the desired mechanism, and 2) it is well-recognized that many small molecules that are active in electro-physiological assays are missed by radioligand binding assays (*i.e.*, there is a high false negative rate). Nevertheless, radioligand binding to native tissues has allowed the discovery of novel ion channel compounds, in-cluding the use of ^{125}I-ChTx binding to bovine tracheal smooth muscle membranes as a tool to identify and characterize soyasaponin openers of maxi-K channels.[18] The commercial advent of fluorescent indicators useful for configuring ion channel assays and high-throughput readers (see below) led to a rapid decline in the use of radioligand displacement assays. Before discussion of the most commonly used fluorescent dyes, alternative methods should be mentioned.

2.4.1 Atomic Absorption Spectrophotometry

Quantitative measurement of ions using atomic absorption dates back to at least the 1960s. Measurement of ion flux as a result of channel activity should in principle be possible for Na^+, K^+, Ca^{2+} and Cl^- ions. A commercial platform using Rb^+ flux and a flame atomic absorption spectrometer developed by Aurora Biomed has been used by several groups to study hERG channels[19] in cardiac safety assays. Additional channels have been studied,[20–22] but to our knowledge this method is not routinely used to screen large compound libraries.

2.4.2 Isothermal Calorimetry

Isothermal calorimetry is a powerful tool for detecting both enthalpic and entropic contributions to the energetics of small molecule binding to immobilized proteins, including membrane proteins.[23] There are substantial challenges to using this platform with ion channels, including a requirement for large amounts of purified protein. For example, Krintel *et al.* used the technique to study the energetics of positive allosteric modulators to the solubilized ligand binding domain region of the rat GluA2 glutamate channel subunit.[24] This required enough purified soluble protein to reach a concentration of 45 μM in a 200 μL calorimeter cell, which is a high hurdle indeed. In addition, translation of this information to guide SAR chemistry decisions to predictable effects on function of the entire ion channel is an additional step that has not been technically feasible for most ion channels to date. Finally, the maximum throughput of approximately 30 compounds per day is not compatible with library screening in general. Thus, this method is not considered to be useful for high-throughput hit identification.

2.4.3 Fluorescence-based Assays

Papers describing fluorescence indicator-based assays on plate-based platforms capable of screening large numbers of compounds appeared in 2001.[25] There is a large volume of literature describing the use of fluorescence-based indicators for Ca^{2+}, membrane potential, and Cl^-. There are excellent reviews that cover this topic in detail; for example, see the review by Trivedi.[26] Because of the throughput that can be achieved, which allows screening of millions of compounds in a matter of weeks to a few months, the relatively low cost of consumables, and the ability to identify suitable chemical hits for starting medicinal chemistry campaigns, fluorescence-based assays remain the industry workhorse for ion channel HTS campaigns.

In general practice, ion channel assay developers first ask whether a direct ion indicator fluorescence-based approach is feasible. Many options are available to measure Ca^{2+} flux, including both single- and dual-wavelength indicators (*i.e.*, Fura-2[27] and Fluo indicators[28]). An excellent early review discusses practical approaches to using these indicators.[29] Successful assays to measure ion channels with these Ca^{2+} indicators include voltage-gated calcium channels, NMDA receptors, some of the calcium-permeable GluRs, some of the nAChRs, ICRAC, calcium-release proteins[30] and TRP channels.[31] For potassium channels, a thallium-sensitive dye has been used to configure assays for several inward rectifier channel members.[32,33] There are also modified fluorescent proteins that have been engineered to measure halides and allow assays for chloride channels.[34] A recent sodium indicator shows promise,[35] and there has been pent-up demand for a suitable direct indicator for sodium flux for years.

Where fluorescence-based direct flux tools are not available, membrane potential dyes have been developed. The first attempts used single-wavelength

bis-oxonols;[36] later quencher dyes were added to improve assay performance. Also, FRET approaches to measure membrane potential have been
used in assays for sodium channel targets[37] including assays where electrical
stimulation electrode arrays have been added to the fluorescence-reader
platform.[38,39] All membrane potential-based assays require additional
thought to understand how voltage-gated channels, or the activity of elec-
trogenic transporters present in the cell line or tissue being assayed, can
affect the results. Especially for ion channel opener assays, false-positive
rates can be unacceptably high and difficult to control.

2.4.4   Automated Electrophysiology

The current generation of automated patch-clamp platforms in widespread
use was first described in the literature around 2003.[40–42] A recent review
covers this history.[43] There are publications and public disclosures where
collections of compounds have been screened using these platforms (*e.g.*,
12,000 compounds at Nav1.7[44]), but for the most part, despite the high
quality data that can be obtained from these platforms the cost of the
consumable plates is apparently still a barrier to more widespread use in
HTS campaigns. This topic is covered in detail in Chapter 3 by J Dunlop in
this volume.

2.5   Strategies for Selection of Confirmation and Selectivity Assays

2.5.1   Ion Channel HTS – Experience

In the following we would like to address several points concerning assay
setup for ion channel HTS. Subsequently, results of several Ion channel HTS
campaigns will be presented followed by a chemical space analysis of the
obtained hits.

   As for any other HTS campaign, a proper assay setup is required for ion
channels to ensure a stable assay performance and reliable hit detection.
Examples of screening technologies have been outlined in the introduction
and will not be addressed further here. However, for ion channels, assay
setup goes beyond the mere fact of using the right assay technique (Ca-Flux,
Thallium-Flux, membrane potential, *etc.*). We would like to emphasize that
recombinant cell lines over-expressing the receptor are beneficial for
screening, and indeed expression levels can be chosen to facilitate searches
for agonist or antagonists.[45] Ideally host cell lines are used where endogenous expression of the target channel is not present. This approach, or
using an inducible expression system, can simplify the setup of a counter
assay in the native cell background and can also simplify assay validation for
the over-expressed target. However, this may be challenging for channels
with ubiquitous tissue expression. Especially in HEK cell lines, which allow
efficient over-expression of most target proteins, a range of endogenous

channels and especially potassium channels is expressed.[46,47] Proper set up of the cell line requires quantification of endogenous expression of the target by PCR, Western blotting, or ideally direct measurement of current. Stable and functional over-expression of the target channels has to be monitored over various passage numbers and confirmation of receptor pharmacology in patch-clamp experiments is required. Building stable cell lines expressing functional channels has hampered easy access to ion channel HTS over a number of years due the complex quaternary structure of many ion channels (pore forming alpha subunit(s) with multiple auxiliary subunits).

Depending on the expression level of the ion channel under investigation as well as the efficiency of the ion channel in transporting ions it may be possible to directly monitor the ion flux across the cellular membrane. This is possible for calcium and potassium channels, where the ion can be bound directly by a suitable fluorophore (calcium flux) or where the natural ion can at least be exchanged for an artificial ion which can be bound specifically (potassium replaced by thallium). If a direct flux technology provides only an insufficient signal or if no specific dye is available, membrane potential might be envisioned as a screening technology, which usually requires a smaller net ionic flux across the membrane and can be more sensitive. Each of these techniques comes with certain particularities which will be further outlined below. Once receptor pharmacology is confirmed in electro-physiology and in the HTS assay, cells may be cultured continuously for the HTS phase or cryopreserved stocks may be utilized, which are prepared as a large homogeneous batch and used directly in the assay without further cultivation. Cryopreserved cells can minimize day-to-day variance in cell-culture which is an advantage if target expression or activity is highly sensitive to cultivation conditions such as passage number, temperature, cell density or other factors. For setting up the optimal ion channel assay, careful consideration is needed with respect to the best activation mode. Some ion channels can only be activated indirectly *via* second messengers but most channels are activated directly by voltage or *via* a ligand, or *via* direct sensing of differences in temperature (TRPV1-4, TRPM8 and TRPA1), mechanical stress, osmolarity, pH differences (ASIC1-4) and other modalities.[48,49] Finding the right assay conditions under these circumstances is thus critical for running the HTS and requires optimization of cell number, buffer composition, pH, temperature, cell confluency, and many more factors.

Ion channel HTS can be run to detect either agonists or antagonists, or both. A typical assay scenario for calcium channel HTS is to administer compounds and directly measure the ion flux to detect potential agonists. Upon stimulation of the ion channel with an EC80 of a known direct or indirect agonist, antagonists can be detected by the decrease in calcium flux signal. Figure 2.2 shows an example of kinetic traces resulting from an ion channel calcium flux HTS for Target G in Table 2.1. In Figure 2.2(A) the signal after vehicle (assay buffer and DMSO screening concentration) or control compound (agonist and antagonist) addition is demonstrated and in Figure 2.2(B) the respective effect after application of a direct agonist at its

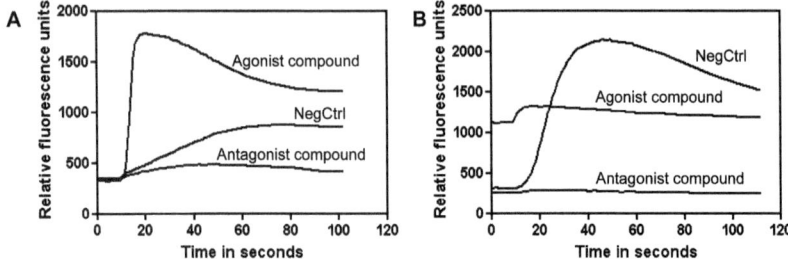

Figure 2.2   Kinetic traces resulting from a calcium flux assay for ion channel target G. (A) First step, application of compounds and controls to measure or detect agonism. (B) Second step, application of the selected activator after an incubation step to all wells to measure or detect antagonists. The x-axis represents the time in seconds and the y-axis corresponds to the measured relative fluorescence unit. Individual traces of the control with vehicle only (NegCtrl), a known agonist for the ion channel (Agonist compound) and a known antagonist (Antagonist compound) are indicated in the plot. Data were recorded using a FLIPR TETRA® system.

Table 2.1   Summary of 12 Ion channel HTS campaigns, n.d. = not determined.

Target	Mode/ Technology	Number of compounds	Mean Z'	Hit rate	Confirmation rate	Prioritized series/ singletons	Percentage confirmed in EPhys
A	Agonist/ Ca^{2+}-Flux	209,000	0.87	0.8%	68%	19/29	11%
B	Antagonist/ Ca^{2+}-Flux	205,000	0.72	0.7%	53%	31/74	57%
C	Agonist/ Ca^{2+}-Flux	254,000	0.72	1%	66%	28/92	n.d.
D	Antagonist/ Ca^{2+}-Flux	255,000	0.81	1.1%	72%	51/71	83%
E	Agonist/ Tl^{+}-Flux	252,000	0.69	0.5%	37%	5/16	44%
F	Antagonist/ Tl^{+}-Flux	254,000	0.81	1.5%	36%	46/143	89%
G	Antagonist/ Ca^{2+}-Flux	252,000	0.83	0.76%	15%	n.d.	n.d.
H	Antagonist/ MP	260,000	0.86	1.6%	33%	30/51	51%
I	Antagonist/ Ca^{2+}-Flux	258,000	0.64	1.4%	56%	30/28	n.d.
J	Antagonist/ YOPRO	257,000	0.83	1.8%	71%	67/240	99%
K	Agonist/MP	210,000	0.79	4.5%	57%	n.d.	6%
L	Antagonist/ MP	347,000	0.73	0.8%	70%	11/4	91%

EC80 is shown. The administration of the known agonist (referred to as positive control agonist) results in a kinetic response with approximately 2000 relative fluorescence units as maximum response. The data points of the kinetic trace before this agonist administration correspond to the baseline values of the measurement and are centered around 400 fluorescent units. An inactive compound or a negative control with DMSO translates only into a slight transient activation, an effect which is known for this target. In this particular case an antagonistic compound (or a known antagonist referred to as positive control antagonist) of the ion channel was also measured in the agonist mode, and this antagonist suppressed the slight transient activation caused by DMSO. In Figure 2.2(B), where an EC80 of a known agonist was added to each well, an inactive compound in the antagonist mode is indicated by the presence of an assay signal. As an indicator for inhibition the positive control antagonist blocks this signal. Interestingly, the positive control agonist from the first measurement shows an elevated baseline and no further activation at all. This suggests that the receptor is desensitized while the cell is still filled with calcium, an effect which has been described for various ion channels. Because of the absence of a clear signal increase in the second measurement (Figure 2.2(B)), the agonist thus appears as a functional antagonist but with an elevated baseline.

In the following we would like to address various characteristics when using different assay scenarios for detecting potential agonists and/or antagonists. Apart from setting the assay up so that agonists or antagonists can be clearly identified, it is also the goal during an HTS campaign to eliminate false positive hits or artifacts as early as possible. A false positive hit is caused by compounds which interfere with the assay in a target-independent way and generate a signal similar to a genuine hit. The source of false positives is strongly dependent on the assay set up, cellular background, target and assay readout. It includes both interference with the assay readout independent of the cells (*e.g.* autofluorescent or quenching compounds), but also a direct interaction of the compound with the cells but not with the relevant target (*e.g.* modulation of the membrane potential or causing an intracellular calcium increase by acting on other endogenous ion channels or GPCRs). It is usually not the goal to investigate the mechanism of a false positive hit, so this point is not addressed in great detail in the following discussion, but the early identification and exclusion of such artifacts is important because it saves resources needed for hit follow up of genuine hit series.

Calcium flux, agonist or antagonist mode, direct activation: A classical scenario would be a calcium flux assay with focus on potential agonists or antagonists and direct target stimulation. In such a scenario, calcium would be present in the assay buffer so that a response can be measured directly. As mentioned above, agonists are measured in a straightforward way but the appearance of false positives through activation of other surface receptors has to be carefully considered. In case of an antagonist scenario, a specific agonist (ideally at its EC80 concentration) is added in a second measurement after pre-incubation with the compound. As discussed above, at this level various effects can be monitored especially by analyzing the baseline, which can indicate autofluorescent or potential agonist compounds. Because several ion channel targets are prone to channel desensitization upon agonist stimulation, agonists would be detected as false positive antagonist hits in the final measurement in antagonist assays as no further activation is possible. Therefore, for both agonist and antagonist screens it would be most efficient to measure both compound addition and a final addition of a specific agonist in order to obtain the maximal information on the compounds. In addition, typical false positives in antagonist assays are compounds with cytotoxic properties that prevent activation of the channel in the second step through a target-independent mechanism.

Calcium flux, antagonist mode, indirect activation: A variant of such a scenario would be a calcium flux assay with focus on antagonists, but with indirect stimulation. This scenario might happen in the absence of any known direct specific activator of the ion channel, for example for an assay to find ICRAC signaling proteins or modulators.[50] In such a case it may be favorable to work in a low external calcium assay buffer especially if a result of the indirect stimulation is the generation of an intracellular calcium release signal. By re-application of calcium as a final step, the calcium signals originating from endogenous calcium stores and from influx through calcium channels can be separated, however, one cannot ascribe the effect of a compound to a direct effect on a calcium flux channel in this assay. Indirect activation might involve various upstream signaling partners and blocking the signaling of these partners would then translate into sources of false positives. Based on this it is recommended to measure the effect of the indirect channel activation separately which is possible in the aforementioned cases where intracellular calcium is released. Such effects are measurable and provide an elegant route to detect blockage of upstream channel partners. However, if the indirect activation does not provide a measureable readout, a secondary assay has to be applied on the hit confirmation or profiling level in order to exclude unspecific pathway inhibitors. This may lead to increased detection of false positives which can only be identified and eliminated after the primary screen. Consequently, we would like to suggest activating the channel directly wherever possible.

Thallium flux assays: For thallium flux assays, compounds are added, incubated, and in a final step a mixture of thallium and potassium is added to measure the signal increase. This scenario is used for both agonist and antagonist screens. For antagonist screens usually a specific activator will also be added before the final thallium addition in order to generate a suitable assay window if insufficient constitutive activity of the channel is present. Because the thallium and potassium indicator solution depolarizes the cells and opens voltage-dependent potassium channels, there is no activity-dependent signal that can be measured until the thallium and potassium solution is added. However, one should be aware that false positives will also be present. Autofluorescent compounds may be detected by monitoring the baseline before thallium addition, but it will not be possible in an antagonist screen to identify agonists which might desensitize the target channel. In any case, a suitable counter screen has to be applied which ideally uses the same readout technology so that crude readout artifacts can be identified in the same assay.

Membrane potential: Membrane potential assays are a special case where due to the high sensitivity careful selection of the assay buffer ion composition is needed to minimize contribution of other endogenous ion channels to the membrane potential. For an agonist membrane potential assay scenario there is a high probability for accumulation of false positives. To test specificity during the primary screen there may be the possibility to measure the agonist effect and then administer a specific antagonist of the channel under investigation. This could be done in the same measurement in a second addition step during the same plate readout so that the screening throughput is only slightly affected. For specific agonists, the antagonist may be able to reverse the membrane potential change while for unspecific compounds no effect is observed. Depending on the target, this strategy has the potential to reduce the false positive rate significantly. In an antagonist membrane potential assay scenario the measurement of compound addition as shown in Figure 2.2(A) will give valuable information on unusual compound behavior and is strongly recommended, even if not all sources of unspecific compound behavior in the second antagonist measurement can be entirely eliminated.

In conclusion, ion channel HTS assays can be carefully designed and run in a way where false positives are minimized and the confidence in any identified hit compound is increased. This setup highly depends on the assay technology and the desired mode of action. However, it will not be possible to exclude all potential sources of false positives. Therefore, it is necessary to set up appropriate counter screening scenarios, for example using the native cell line lacking the channel of interest or a cell line expressing a related ion channel. To monitor false positives caused by the use of a particular indicator technology, the applied assay principle should be identical to the primary assay. We are confident that combining all these measures ultimately leads to hit sets with a minimum number of false positives and will translate into the successful identification of ion channel ligands for subsequent hit-to-lead programs.

An aspect which has not been discussed so far is related to proper evaluation of the kinetic traces. Depending on the assay scenario and the mode of action, traces are evaluated by looking for the generation of signal (*e.g.* an agonist response in a calcium flux assay) or the absence of a signal (*e.g.* the compound blocks activation by an agonist in a calcium flux antagonist assay). An example of this has been outlined in Figure 2.2. We would like to point out that research is dedicated towards how to best evaluate kinetic traces.[51,52] In particular, the challenge is to understand the normal kinetics following activation of a particular ion channel so as to discriminate true actives from compounds that generate artifacts in the traces. We would like to illustrate this by using two examples. For simplicity it is assumed that the traces are evaluated based on the maximum or peak fluorescence level of the trace. A general methodology is that from the kinetic trace the individual baseline for each well is subtracted point by point in order to remove any background signal. All traces are then normalized to the baseline variation on the assay plate so that all measurements start with the same value and wells with high baseline signals are down-sized and *vice versa*. This is referred to as spatial uniformity correction and may be necessary to improve the assay statistics[51] (especially if uneven dye loading or cell density can be corrected). However, one should be aware of the pitfalls that such normalization can falsely transform compounds to hits, especially in antagonist measurements. One example can be seen in Figure 2.3(A) and (B) for ion channel antagonist assay B (Table 2.1) using calcium ion flux technology. In Figure 2.3 (A) the raw signals are present prior to global baseline normalization. The negative control results in a signal increase from 3000 RFU to 9000 RFU caused by the stimulation of an EC80 of a specific agonist. The positive antagonist control shows no such activation. Instead, a lower baseline is observed on the baseline as compared to the negative control. This may be due to antagonizing pre-activated ion channels prior to the agonist addition. In Figure 2.3(A) and (B) the kinetic trace of a compound is also demonstrated. This compound shows an elevated baseline and a maximum response up to the level of the negative control (close to the saturation of the instrument). This might be due to the autofluorescent properties of the compound. In Figure 2.3(B) spatial uniformity correction has been applied so that the different example traces are all normalized to the same baseline. For the positive antagonist control the original lower baseline effect is no longer evident. However, the antagonistic effect of the compound is still detected and only this extra information on the baseline is lost. The effect on the kinetics of compound action is more severe as it now

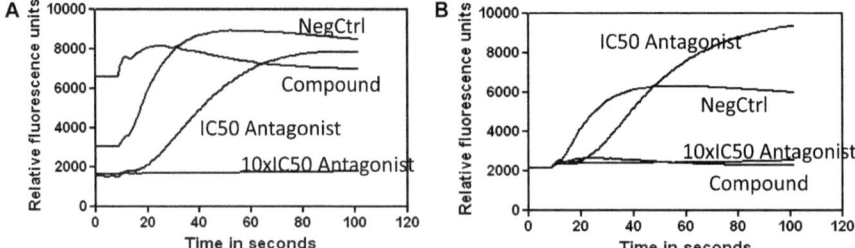

Figure 2.3   Kinetic traces resulting from a calcium flux assay in antagonist mode for an ion channel target B as shown in Table 2.1 (A) Antagonist measurement, (B) Antagonist measurement with spatial uniformity correction). The x-axis represents the time in seconds and the y-axis corresponds to the measured relative fluorescence units. Individual traces of the control with vehicle only (NegCtrl), a known antagonist with 10× its IC50 concentration (10xIC50 Antagonist), the antagonist at its IC50 concentration (IC50 Antagonist) and a compound are indicated in the plots. Data were recording using a FLIPR TETRA® system.

mimics an antagonist response. The compound would be falsely detected as an antagonist. The example shows that it is of crucial importance to carefully define how to set up the evaluation of the kinetic traces. In cases where a correction/normalization has to be applied to improve assay statistics, additional monitoring of baseline effects should be performed.

A comparable example is shown in Figure 2.3(A) and (B) for the IC50 control (*i.e.* the antagonist tested at its IC50 concentration to monitor sensitivity during the screen). The effect on the kinetics for this particular ion channel is that a time dependent elevation of fluorescence signal is observed. Evaluation of the kinetic trace based on the maximum response of the whole trace in Figure 2.3(A) would characterize the IC50 control as an inactive sample while evaluation up to 40 seconds would characterize the control compound correctly. It is thus important to be aware of such effects and to define up to which point the trace should be evaluated. This point can also be selected in such a way that the IC50 control on a particular plate generates a 50% response. This provides an elegant method to adjust for day-to-day differences in assay sensitivity. It should be noted that spatial uniformity correction for the IC50 control in Figure 2.3(B) projects the maximum signal markedly above the negative control. This again shows the need to carefully evaluate the effects of correction/normalization routines on the kinetic traces.

Our general process of running a high throughput screen is outlined in Figure 2.4. During HTS compounds are tested in singlicate mode at a fixed compound concentration. The appropriate compound test concentration is determined prior to the HTS by running a small compound subset of 2,500 compounds at different compound concentrations in replicate mode.[53] If possible the HTS concentration is then adjusted to obtain a hit rate in the range of 1%. However, other factors may have to be taken into consideration such as DMSO tolerance, the maximum achievable compound concentration, or expected false positive rates. Based on this, conditions can be changed so that the expected hit rates are deliberately higher or lower. In a third step, after completion of HTS, the obtained primary screening hits are tested again in replicate mode at the same compound concentration to confirm their effect. Subsequently, counter and selectivity assays can be added. In the final step, the confirmed hits are tested in concentration–response mode in replicates in the primary assay and in any available counter and selectivity assays. Next, the compounds are grouped into series and singletons using computational chemistry tools. Series are prioritized based on the obtained biological results as well as medicinal chemistry-related aspects. As a final step re-testing of compound material as powder is performed and selected representatives of the prioritized series/singletons are confirmed using electrophysiology measurements.

Table 2.1 gives an overview of 12 HTS assays performed internally using different technologies as well as different modes of action. On average more than 235,000 compounds per target were screened in singlicate at a fixed compound concentration. All campaigns provided good to excellent

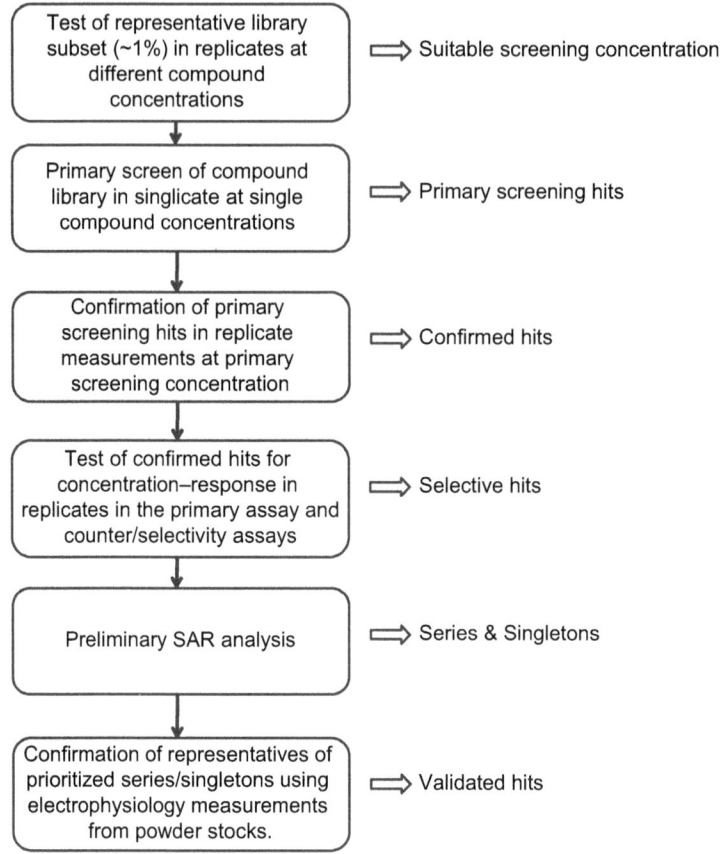

Figure 2.4   Schematic representation of the HTS work-flow applied in-house.

assay statistics as can be seen according to the obtained Z′ values.[54] Also, most campaigns resulted in "primary" hit rates (the percentage of hits obtained in the primary screen) in the range of 1%. However, there are exceptions like assay E with a low hit rate at the highest possible compound concentration. The data demonstrate that these hit rates highly depend on the target, the assay mode, the stimulus, and the screening concentration used. Also, as outlined above, these hit rates also include false positives, which are challenging if not impossible to eliminate at this early point in time.

Confirmation rates (the percentage of primary screening hits repeating in the confirmation measurement) are generally good (>50%), underlining the robust assay performance. Again, there are exceptions, such as antagonist assay G. In this screen the primary hits were not sufficiently potent. To not lose any structural information and to capture weak hit series, a higher than normal proportion of primary screening hits were selected with borderline activity from the primary screen. As a result of typical assay variability there is a high likelihood that a good fraction of these compounds will not confirm. This is the trade-off to obtain the benefit of minimal false negative rates using this approach.

A more interesting observation comes from the results of testing the confirmed hits in EPhys assays. The data in Table 2.1 show that all of these HTS campaigns translated into prioritized series and singletons. Obviously, there are more challenging cases where low hit rates in the primary screen and/or a low confirmation rate result in a low number of series/singletons. However, overall diverse sets of prioritized compounds can be expected from running a properly designed, executed and analyzed ion channel HTS. The hit compounds need to be validated by direct electrophysiological measurements to eliminate false positives which have not been detected with fluorescence-based counter- and selectivity assays. Consequently, the most interesting result is what proportion of the obtained prioritized screening library hits confirm upon re-testing as fresh powder and testing in electrophysiological assays. These results are present in the last column of Table 2.1 and range from 6% to 99%. There are several conclusions which can be drawn:

(1) There may be targets where it is not possible to set up an assay with direct flux technology. In these cases membrane potential technology may be the only possibility. As outlined above these assays have the potential to collect false positives and efforts have to be made to eliminate these false positives as early as possible. One can add antagonists for some of the "background" voltage-gated ion channels to minimize their contribution to the assay signal ion channels (*e.g.*, Bugianesi *et al.*[38]). However, there are practical limits to these approaches. This may explain the low confirmation rates for some of the examples in electrophysiological assays (assays A and K in Table 2.1).

(2) Agonist HTS scenarios are relatively simple to set up and measure. However, it is challenging to discriminate genuine agonists from false positives. Different sources of false positives exist in this context. One source is auto-fluorescent compounds which can be detected as hits in non-target-related assay scenarios. Another source is compounds having an unspecific agonist effect on the cell line under investigation. As this unspecific effect may be cell-line dependent it may be challenging to come up with an appropriate counter screen to triage such potential artifacts. As a consequence low confirmation rates in electrophysiology measurements have to be expected. This can be seen in Table 2.1 for the agonist screen A (appropriate counter assay was not available) compared to the agonist screen E (appropriate counter assay in wild type background was available). Moreover the agonist assay K represents an extreme case where due to time constraints no fluorescence-based counter assays were utilized and all hit compounds were directly moved into electrophysiology measurements. This can be seen as an extreme case.

(3) Taking the particulars on assay technology utilized and channel activation mode into account these data show that proper assay setup and
application of counter assays for HTS results in the successful identification of diverse sets of antagonists or agonists. These efforts have
a high likelihood of translating into high confirmation rates in subsequent electrophysiological measurements.

Another question is whether a correlation can be expected between the
data obtained in HTS and the results from hit confirmation using electrophysiological measurements. Figure 2.5 shows two examples where the
median inhibition of the hit confirmation in HTS is plotted against the respective confirmation measurement in electrophysiology. The example in
Figure 2.5(A) represents a good correlation while in Figure 2.5(B) this is not
the case. There are several reasons for this:

(1) The original HTS might have been performed from a DMSO stock of
the compound deck and the electrophysiological characterization was
performed with fresh powder stock. In this case the compounds might
differ in structure, concentration, and purity.
(2) The different technologies may provide different assay sensitivities.
Based on this the emphasis should be on obtaining the correct rank
ordering.
(3) The demonstrated correlations represent diverse sets of compounds
and not a within-series view. Compound properties like solubility,
stability, cell toxicity, the ability to cross membranes and many more
are unknown at this early point and may differ widely across the
chemical series. In turn, these chemical properties may impact the
fluorescence and electrophysiological measurements differently.

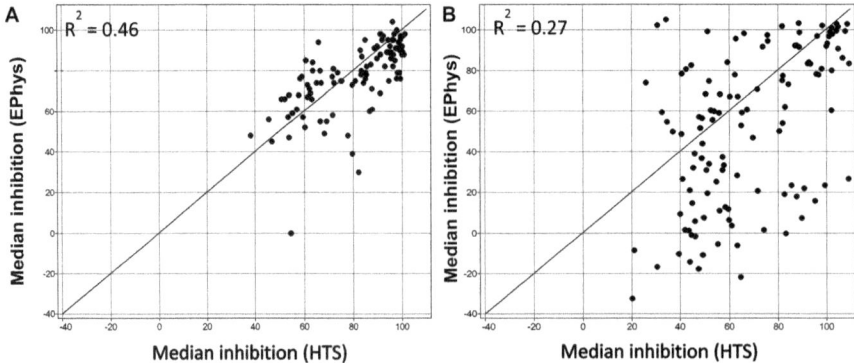

Figure 2.5   Plots of the median inhibition values obtained with fluorescence read-
out and electrophysiology read-out at a single concentration for two ion
channel targets. The x-axis represents the median inhibition from the
HTS and the y-axis corresponds to the inhibition using electrophysiology measurements. Each dot corresponds to a compound.

At this point we would like to point out that tool compounds (often drug substances or highly specific peptide blockers) which have been used for assay adaptation usually show a good correlation between both assay technologies. This might be due to the fact that one compares highly optimized, potent and specific compounds where the ability to accurately measure potency in different assays is good.

(4) It should be kept in mind that two highly different assay scenarios are compared. Flux or membrane potential assays are assay technologies where a signal is extracted from a population of cells. In electrophysiology data are recorded from single cells (or averaged from a few cells) given a certain voltage protocol. This allows controlling the ion channel state (open, closed, inactive) and gives a much more accurate representation of the interaction of the compound with the receptor. Another aspect is that incubation times are much longer for fluorescence-based assay technologies. During this period compounds may achieve a higher effective concentration at the binding site on the ion channel, or compound metabolism or toxic effects might occur and may contribute to unspecific signals.

What do we learn from all this? HTS is a technology which is well suited for the identification of diverse hit sets for ion channels. To eliminate ample sources of false positives and to confirm the compound mechanism on the target, these hits should be confirmed using electrophysiology. However, expecting a high correlation between the fluorescence and electrophysiology measurements at this early point may be premature or unrealistic. As a first step, the compound's *in vitro* ADME-T[55] and physicochemical properties should be investigated as well as a within series analysis of the correlation. If no correlation is evident between both measurements any potency values obtained in the HTS measurements should be taken with care as they may not be suitable for tracking SAR.

2.6   Library Composition

In the literature there is a constant debate about selecting the right compounds for screening. There is common agreement to focus on "drug-like" or "lead-like" compounds.[56–58] Such compounds have been defined by physicochemical properties comparable to known drugs and the absence of structural alerts (known toxicophores, frequent hitters and other undesirable structural features) and are assumed to have a higher likelihood of becoming a drug. The reason for this is related to endpoints not associated with the target like solubility, absorption, distribution, metabolism, excretion and toxicity.[55] In light of this, more focused collections have been proposed, *e.g.* for compounds that must pass the blood-brain barrier.[59] Also, fragment-based library design has been introduced. These compounds are much smaller compared to known drugs[60] and are thus tested at a higher compound concentration. However, in theory these compounds cover a much broader chemical space compared to drug-like compounds. This approach has been already successfully applied for ion channel screening.[61] Despite these general approaches another question is whether it is possible to design targeted libraries that ideally yield higher hit rates for a given target class. Successful applications are present in the literature for kinase enzymes[62] or aminergic GPCRs.[63] However for ion channels these approaches are challenging due to the high structural diversity of ion channel targets, the limited amount of crystal structure information and the absence of a general ion channel pharmacophore.[64,65] Consequently, ligand-based library design approaches have been proposed to build a selection of compounds with similarity to known ion channel binders.[64,65] For voltage-gated ion channels a conserved pore region has been suggested and used for defining a 3D structure of the binding pocket. This information was used for constructing a voltage-gated targeted ion channel library.[65] Whether or not these approaches offer an advantage over screening a diverse collection remains a matter of debate and may highly depend on the ion channel under investigation. Advantages of focused and thus smaller compound libraries clearly are a reduced effort and thus cost for primary screening. However, identifying hit compounds with sufficient novelty is potentially a down-side of this approach. In order to shed some light on this question we have investigated our screening library with respect to chemical diversity and the ability of creating targeted libraries for ion channel HTS using the following approach:

(1) Demonstrate that the used screening collection covers an appropriate bioactive chemical space, is not insufficiently diverse, nor does the distribution lie too far outside any known bioactive reference space.
(2) Compare the chemical space of the ion channel hits to the chemical space of the entire screening collection.
(3) Investigate if key physicochemical properties exist to guide the design of targeted libraries.

For this investigation our in-house screening library (300K drug-like compounds) was compared to the MDDR library,[66] a bioactivity reference catalogue with 190K bioactive compounds. The following physicochemical compound properties were calculated for all compounds using the software MOE:[67] number of hydrogen (H)-bond donors, number of H-bond acceptors, molecular weight, number of rotatable bonds, and logP (o/w). The following thresholds were set: number of H-bond donors > 5, number of H-bond acceptors > 10, molecular weight > 500, number of rotatable bonds > 10, and logP (o/w) > 5. It was then counted how often a compound exceeded these thresholds. Compounds with <2 counts were categorized as drug-like. In addition, a set of >130 substructure filters was also developed. This set was designed, refined and extended by in-house medicinal chemists to exclude known toxicophores, frequent hitters and compounds with other undesirable structural features. Compounds containing such substructures were also marked as non-drug-like in the individual sets. For this study we focused on one descriptor set, MOE physicochemical descriptors[67] (N = 156) and Unity fingerprints[68] capturing structural features of the compounds. Both libraries were compared using principle component analysis (PCA) and self-organizing maps (SOM). PCA is a projection technique which projects a given data set onto a pre-defined number of dimensions (scoring vectors) by explaining the variance in the data. In the first round the data set is projected onto the first scoring vector. A loading vector is created containing the weighting (influence) of the original descriptors. The remaining unexplained variance forms the new data set for the next iteration. This process is applied until a predefined number of dimensions is reached.[69] In this study the number of dimensions was set to two. The PCA approaches applied in this study were able to explain at least 70% of the variance in the data set with these two dimensions.

The SOM technique on the other hand is a non-linear clustering method, allowing visualizing a high dimensional data set as a two-dimensional (or more) map containing n clusters (or neurons). Employing Kohonen's algorithm a "topology" preserving projection of the high dimensional space is obtained (*i.e.* molecules that are close to each other in the original high-dimensional space are also close on the SOM).[70] In the present study so-called toroidal (border-less) emergent SOMs were calculated using the ESOM algorithm with 500×820 neurons and 10 training cycles.[71] We have extensively analyzed the resulting maps and compared our screening collection to the 190K bioactive reference set. We were able to show that the screening library covers to a high degree the structural and physicochemical property space of the "drug-like" section of the bioactive reference collection. This is evident irrespective of the way compounds are described and of the projection methodology. We believe this to be an important prerequisite for the subsequent analysis.

For this study we have projected the identified ion channel ligands from the HTS assays above onto the PCA plots and SOMs. We emphasized hits previously confirmed in electrophysiological assays. Figure 2.6(A) shows a SOM plot obtained with physicochemical descriptors. Grey dots correspond to compounds from the screening deck while black dots represent ion channel hits. White regions correspond to non-drug-like bioactive property space. It is evident that the ion channel HTS hits are highly diverse and cover the drug-like physicochemical property space to a large extent. The same conclusion is obtained when analyzing the first two dimensions of a PCA plot in combination with structural fingerprints. This is exemplified in Figure 2.6(B). These studies show that screening a diverse screening collection for an ion channel target will result in a highly diverse set of hit compounds. Our studies did not reveal particular chemical spaces populated with ion channel hits. This finding is also supported when analyzing the distribution of ion channel ligands from the MDDR on the different plots (data not shown) and underlines the high structural diversity of the ion channel proteins. We would like to conclude that there might be the possibility of generating targeted libraries for specific ion channels or for ion

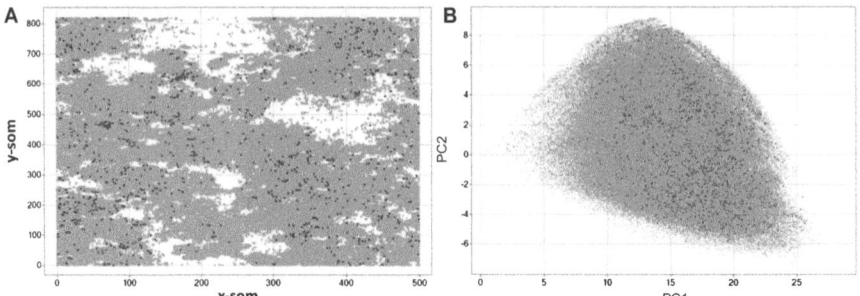

Figure 2.6   (A) SOM maps obtained for the in-house screening deck and the MDDR library in combination with MOE 2D physicochemical descriptors. Each dot corresponds to a drug-like compound from the in-house screening deck. Grey dots represent screening compounds which were inactive in ion channel HTS assays while black dots represent ion channel HTS hits. White regions correspond to non-drug-like bioactive property space. (B) Plot of the first two principle components obtained for the in-house screening deck and the MDDR library in combination with Unity fingerprints. Grey dots represent screening compounds which were inactive in ion channel HTS assays while black dots represent ion channel HTS hits.

channel (sub)families. However, such approaches will be challenging for ion channel targets in total.

Are there specific key physicochemical properties which could be used for designing ion channel libraries? We have selected the same set of compounds as described above. As a control we have added the information if a compound was a HTS GPCR hit from our database. The reason for this was that GPCR HTS assays were all cell-based assays, and GPCRs, like ion channels, are membrane proteins. Moreover, several of these assays were also utilizing Ca-Flux kinetic read-out as screening technology. Various key physicochemical properties were studied. Figure 2.7 represents a plot of molecular weight *versus* logD at pH 7.4.[72] Grey dots represent non-hits from the ion channel and GPCR campaigns. Black dots in Figure 2.7(A) exemplify ion channel HTS hits and black dots in Figure 2.7(B) correspond to GPCR hits. Interestingly, a paucity of low molecular weight ion channel HTS hits with low lipophilicity was observed. This observation is not obvious for GPCR HTS hits. Consequently, logD and molecular weight may serve as triaging filters for HTS starting points for ion channel targeted screening libraries. We would like to point out that this finding is not supported when analyzing ion channel hits from the MDDR. Thus our finding might suggest that the observation is a particularity of the ion flux assays under investigation. However, we rather believe that this is related to the fact that in the MDDR, compounds which have progressed through hit-to-lead and lead optimization efforts are usually reported. Throughout this process physico-chemical properties might have been optimized in a certain direction (*e.g.* to lower TPSA in order to increase blood-brain barrier penetration).

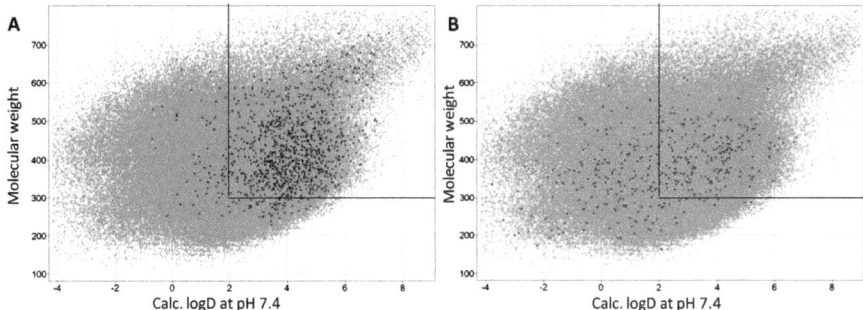

Figure 2.7   (A) Ion channel HTS, plot of calculated logD at pH 7.4 (x-axis) *versus* molecular weight (y-axis). Each dot corresponds to a compound from the screening deck. Grey dots represent screening compounds which were inactive in ion channel HTS assays and black dots represent ion channel HTS hits. (B) GPCR HTS, plot of calculated logD at pH 7.4 (x-axis) *versus* molecular weight (y-axis). Each dot corresponds to a compound from the screening deck. Grey dots represent screening compounds which were inactive in GPCR HTS assays while black circles correspond to compounds detected as GPCR HTS hits. Triaging filters for molecular weight > 300 and logD > 2 are shown as black lines in the plots.

2.7   Conclusions and Future Perspective

High-throughput screening is a highly automated technology where large compound collections are screened to find high quality chemistry starting points for a target of interest. Ion flux and membrane potential assay technologies have been developed for running HTS for ion channel targets in fluorescence mode. Our conclusion is that these technologies are well suited for finding such starting points. However, careful assay design is required to minimize collection of potential false positives. This includes optimal set up of the assay and proper evaluation of the generated kinetic read-out, as well as designing the most relevant counter assay strategies. Using this approach will result in a high proportion of high quality starting points confirming in electrophysiological measurements.

It can be expected that due to the large number of ion channel targets, the paucity of structural information for this class, and the relatively undefined pharmacophore binding sites for most ion channels more needs to be learned before chemists and biologists can more efficiently pursue drug discovery on existing and new ion channel drug targets. A disciplined approach, beginning with a well-thought-through assay design and well-executed HTS and subsequent utilizing of appropriate selectivity and counterscreening *in vitro* assays, is required and should increase the ability to define and optimize molecules entering *in vivo* testing and progressing to clinical candidacy. The tools and knowledge are available at present, and advances in high-throughput automated patch clamp platforms (especially if consumables costs reach a value point that is more widely appealing) should soon drive us to more and more "shots on goal" and successful new ion channel drugs.

Acknowledgement

We thank all of our colleagues at Evotec that performed the work we describe in this chapter. We also thank Dr Thomas Hesterkamp and Dr Andreas Scheel for their comments and suggestions on the manuscript.

References

1. J. L. Sharman, H. E. Benson, A. J. Pawson, V. Lukito, C. P. Mpamhanga, V. Bombail, A. P. Davenport, J. A. Peters, M. Spedding and A. J. Harmar, NC-IUPHAR, *Nucleic Acids Res.*, 2013, **41** (database issue).
2. F. Huang, X. Wong and L. Y. Jan, *Pharmacol. Rev.*, 2012, **64**(1), 1.
3. L. E. Tammelin and N. Lofgren, *Acta Chem. Scand.*, 1947, **1**(10), 871.
4. A. M. Taha and A. M. Abd-Elmaksoud, *Br. J. Anaesth.*, 2013, **110**(6), 1040.
5. G. J. Kaczorowski, O. B. McManus, B. T. Priest and M. L. Garcia, *J. Gen. Physiol.*, 2008, **131**(5), 399.
6. S. K. Bagal, A. D. Brown, P. J. Cox, K. Omoto, R. M. Owen, D. C. Pryde, B. Sidders, S. E. Skerratt, E. B. Stevens, R. I. Storer and N. A. Swain, *J. Med. Chem.*, 2013, **56**(3), 593.
7. F. Van Goor, S. Hadida, P. D. Grootenhuis, B. Burton, D. Cao, T. Neuberger, A. Turnbull, A. Singh, J. Joubran, A. Hazlewood, J. Zhou, J. McCartney, V. Arumugam, C. Decker, J. Yang, C. Young, E. R. Olson, J. J. Wine, R. A. Frizzell, M. Ashlock and P. Negulescu, *Proc. Natl. Acad. Sci. U. S. A.*, 2009, **106**(44), 18825.
8. K. Sumikawa, M. Houghton, J. C. Smith, L. Bell, B. M. Richards and E. A. Barnard, *Nucleic Acids Res.*, 1982, **10**(19), 5809.
9. E. J. Trybulski, L. Benjamin, S. Vitone, A. Walser and R. I. FryerI, *J. Med. Chem.*, 1983, **26**(3), 367.
10. L. S. Goodman, E. A. Swinyard, W. C. Brown, D. O. Schiffman, M. S. Grewal and E. L. Bliss, *J. Pharmacol. Exp. Ther.*, 1953, **108**(4), 428.
11. M. Noda, T. Ikeda, H. Suzuki, H. Takeshima, T. Takahashi, M. Kuno and S. Numa, *Nature*, 1986, **322**(6082), 826.
12. N. Roessler, D. Englert and K. Neumann, *J. Recept. Res.*, 1993, **13**(1–4), 135.
13. G. J. Doucette, M. M. Logan, J. S. Ramsdell and F. M. Van Dolah, *Toxicon.*, 1997, **35**(5), 625.
14. M. A. Varney, C. Jachec, C. Deal, S. D. Hess, L. P. Daggett, R. Skvoretz, M. Urcan, J. H. Morrison, T. Moran, E. C. Johnson and G. J. Veliçelebi, *Pharmacol. Exp. Ther.*, 1996, **279**(1), 367.
15. C. G. Ponte, O. B. McManus, W. A. Schmalhofer, D. M. Shen, G. Dai, A. Stevenson, S. Sur, T. Shah, L. Kiss, M. Shu, J. B. Doherty, R. Nargund, G. J. Kaczorowski, G. Suarez-Kurtz and M. L. Garcia, *Mol. Pharmacol.*, 2012, **81**(4), 567.

16. M. P. Blaustein, *J. Physiol.*, 1975, **247**(3), 617.

17. D. A. Nachshen and M. P. Blaustein, *Biophys. J.*, 1979, **26**(2), 329.

18. O. B. McManus, G. H. Harris, K. M. Giangiacomo, P. Feigenbaum, J. P. Reuben, M. E. Addy, J. F. Burka, G. J. Kaczorowski and M. L. Garcia, *Biochemistry*, 1993, **32**(24), 6128.

19. K. W. Chaudhary, J. M. O'Neal, Z. L. Mo, B. Fermini, R. H. Gallavan and A. Bahinski, *Assay Drug Dev. Technol.*, 2006, **4**(1), 73.

20. S. Gill, R. Gill, Y. Xie, D. Wicks and D. Liang, *Assay Drug Dev. Technol.*, 2006, **4**(1), 65.

21. S. Gill, R. Gill, D. Wicks and D. Liang, *Assay Drug Dev. Technol.*, 2007, **5**(3), 373.

22. S. Trivedi, J. Liu, R. Liu and R. Bostwick, *Expert Opin. Drug Discov.*, 2010, **5**(10), 995.

23. M. W. Freyer and E. A. Lewis, *Methods Cell Biol.*, 2008, **84**, 79.

24. C. Krintel, K. Frydenvang, L. Olsen, M. T. Kristensen, O. de Barrios, P. Naur, P. Francotte, B. Pirotte, M. Gajhede and J. S. Kastrup, *Biochem. J.*, 2012, **441**(1), 173.

25. K. L. Whiteaker, J. P. Sullivan and M. Gopalakrishnan, *Curr. Protoc. Pharmacol.*, 2001, Chapter 9: Unit 9.2.

26. S. Trivedi, J. Liu, R. Liu and R. Bostwick, *Expert Opin. Drug Discov.*, 2010, **5**(10), 995.

27. G. Grynkiewicz, M. Poenie and R. Y. Tsien, *J. Biol. Chem.*, 1985, **260**(6), 3440.

28. A. Minta, J. P. Kao and R. Y. Tsien, *J. Biol. Chem.*, 1989, **264**(14), 8171.

29. P. A. Negulescu and T. E. Machen, *Methods Enzymol.*, 1990, **192**, 38.

30. F. Striggow and B. E. Ehrlich, *J. Gen. Physiol.*, 1996, **108**(2), 115.

31. N. T. Blair, J. S. Kaczmarek and D. E. Clapham, *J. Gen. Physiol.*, 2009, **133**(5), 525.

32. C. D. Weaver, D. Harden, S. I. Dworetzky, B. Robertson and R. J. Knox, *J. Biomol. Screen.*, 2004, **9**(8), 671.

33. R. Raphemot, C. D. Weaver and J. S. Denton, *J. Vis. Exp.*, 2013, **71**(pii), 4209.

34. L. V. Galietta, S. Jayaraman and A. S. Verkman, *Am. J. Physiol. Cell. Physiol.*, 2001, **281**(5), C1734.

35. C. M. Lamy and J. Y. Chatton, *NeuroImage*, 2011, **58**(2), 572.

36. A. Waggoner, *J. Membr. Biol.*, 1976, **27**(4), 317.

37. J. P. Felix, B. S. Williams, B. T. Priest, R. M. Brochu, I. E. Dick, V. A. Warren, L. Yan, R. S. Slaughter, G. J. Kaczorowski, M. M. Smith and M. L. Garcia, *Assay Drug Dev. Technol.*, 2004, **2**(3), 260.

38. R. M. Bugianesi, P. R. Augustine, K. Azer, C. Dufresne, J. Herrington, G. S. Kath, O. B. McManus, C. S. Napolitano, A. Rush, J. Sachs, N. Simpson, M. K. Wismer, G. J. Kaczorowski and R. S. Slaughter, *Assay Drug Dev. Technol.*, 2006, **4**(1), 21.

39. C. J. Huang, A. Harootunian, M. P. Maher, C. Quan, C. D. Raj, K. McCormack, R. Numann, P. A. Negulescu and J. E. González, *Nat. Biotechnol.*, 2006, **24**(4), 439.

40. L. Kiss, P. B. Bennett, V. N. Uebele, K. S. Koblan, S. A. Kane, B. Neagle and K. Schroeder, *Assay Drug Dev. Technol.*, 2003, **1**(1 Pt 2), 1.
41. C. C. Shieh, *Drug Discov. Today*, 2004, **9**(13), 551.
42. H. J. Neubert, *Anal. Chem.*, 2004, **76**(17), 327A.
43. C. Farre and N. Fertig, *Expert Opin. Drug Discov.*, 2012, **7**(6), 515.
44. N. Castle, D. Printzenhoff, S. Zellmer, B. Antonio, A. Wickenden and C. Silvia, *Comb. Chem. High Throughput Screen.*, 2009, **12**(1), 107.
45. C. G. Ponte, O. B. McManus, W. A. Schmalhofer, D. M. Shen, G. Dai, A. Stevenson, S. Sur, T. Shah, L. Kiss, M. Shu, J. B. Doherty, R. Nargund, G. J. Kaczorowski, G. Suarez-Kurtz and M. L. Garcia, *Mol. Pharmacol.*, 2012, **81**(4), 567.
46. S. Berjukow, F. Döring, M. Froschmayr, M. Grabner, H. Glossmann and S. Hering, *Br. J. Pharmacol.*, 1996, **118**(3), 748.
47. P. Thomas and T. G. Smart, *J. Pharmacol. Toxicol. Methods.*, 2005, **51**(3), 187.
48. D. E. Clapham, L. W. Runnels and C. Strübing, *Nat. Rev. Neurosci.*, 2001, **2**(6), 387.
49. T. W. Sherwood, E. N. Frey and C. C. Askwith, *Am. J. Physiol. Cell. Physiol.*, 2012, **303**(7), C699.
50. S. L. Zhang, A. V. Yeromin, X. H. Zhang, Y. Yu, O. Safrina, A. Penna, J. Roos, K. A. Stauderman and M. D. Cahalan, *Proc. Natl. Acad. Sci. U. S. A.*, 2006, **103**(24), 9357.
51. FLIPR® Tetra User Guide, 0112-0109G, © 2010 Molecular Devices, Inc., Sunnyvale, California, USA.
52. P. Gribbon, C. Chambers, K. Palo, J. Kupper, J. Mueller and A. Sewing, *J. Biomol. Screen.*, 2006, **11**(5), 511.
53. P. E. Ilouga and T. Hesterkamp, *J. Biomol. Screen.*, 2012, **17**(6), 705.
54. J. H. Zhang, T. D. Chung and K. R. Oldenburg, *J. Biomol. Screen.*, 1999, **4**, 67.
55. H. van de Waterbeemd and E. Gifford, *Nat. Rev. Drug Discov.*, 2003, **2**(3), 192.
56. C. A. Lipinski, *J. Pharmacol. Toxicol. Methods*, 2000, **44**(1), 235.
57. T. I. Opera, C. L. Waller and G. R. Marshall, *Drug Des. Discov.*, 1994, **12**(1), 29.
58. S. Bickerton, S. Jiwpanich and S. Thayumanavan, *Mol. Pharm.*, 2012, **9**(12), 3569.
59. T. T. Wager, R. Y. Chandrasekaran, X. Hou, M. D. Troutman, P. R. Verhoest, A. Villalobos and Y. Will, *ACS Chem. Neurosci.*, 2010, **1**(6), 420.
60. M. Congreve, R. Carr, C. Murray and H. Jhoti, *Drug Discov. Today.*, 2003, **8**(19), 876.
61. F. Giordanetto, L. Knerr, N. Selmi, A. Llinàs, A. Lindqvist, Q. D. Wang, P. Ståhlberg, F. Thorstensson, V. Ullah, K. Nilsson, G. O'Mahony, G. Högberg, E. Lindhardt, A. Strand and G. Duker, *Bioorg. Med. Chem. Lett.*, 2011, **21**(18), 5557.

62. C. J. Harris, R. D. Hill, D. W. Sheppard, M. J. Slater and P. F. Stouten, *Comb. Chem. High Throughput Screen.*, 2011, **14**(6), 521.
63. R. Heilker, M. Wolff, C. S. Tautermann and M. Bieler, *Drug Discov. Today.*, 2009, **14**(5–6), 231.
64. N. Y. Mok and R. Brenk, *J. Chem. Inf. Model.*, 2011, **51**(10), 2449.
65. C. J. Harris, R. D. Hill, D. W. Sheppard, M. J. Slater and P. F. Stouten, *Comb. Chem. High Throughput Screen.*, 2011, **14**(6), 521.
66. M D L Drug Data Report, Version December 2009; Symyx Technologies Inc.: Santa Clara, CA, 2009.
67. Molecular Operating EnVironment (MOE), version 2008.10, Chemical Computing Group Inc.: Montreal, Canada, 2006.
68. SYBYL-X, Version 1.2; Tripos International: St. Louis, USA, 2010.
69. M. Otto, *Chemometrics. Statistics and Computer Application in Analytical Chemistry*, Wiley-VCH, Weinheim, Germany, 1998.
70. T. Kohonen, *Biol. Cybern.*, 1982, **43**, 59.
71. A. Ultsch and F. Moerchen, *ESOM-Maps: Tools for Clustering, Visualization, and Classification with Emergent SOM, Technical Report*, Dept. of Mathematics and Computer Science, University of Marburg, Germany, 2005.
72. MarvinView version 5.9.0, Chemaxon Ltd., Budapest, Hungary, 2012.

Automated Electrophysiology in Ion Channel Drug Discovery

JOHN DUNLOP

Neuroscience Innovative Medicine Unit, AstraZeneca, 141 Portland St, Cambridge, MA 02139, USA
Email: john.dunlop@azneuro.com

3.1 Ion Channels as Targets for Therapeutic Intervention

Ion channels are integral membrane proteins that control the passage of ions across cellular membranes with diverse ion channel families responding to and activated by varied stimuli including changes in membrane voltage, exposure to ligands, changes in temperature and mechanical force. As such ion channels play a critical role in many physiological processes and underlie the key property of cellular excitability in cardiac, muscle and neuronal tissues. Consequently, ion channels have also been implicated in many pathophysiological processes, *e.g.*, neuronal hyperexcitability associated with neuropathic pain or epilepsy, and as such many ion channels have been the focus of therapeutic drug development efforts. That this target class has been a fruitful source of drug candidates is evidenced by the fact that the site of pharmacological action of a number of approved drugs can be attributed to their activity on specific ion channels including voltage gated Na^+, K^+, Ca^{2+} and Cl^- channels, ligand-gated glutamate, GABA and nicotinic receptors and receptors for ryanodine, IP3 and transient receptor

RSC Drug Discovery Series No. 39
Ion Channel Drug Discovery
Edited by Brian Cox and Martin Gosling
© The Royal Society of Chemistry 2015
Published by the Royal Society of Chemistry, www.rsc.org

potential channels controlling cellular Ca^{2+} homeostasis.[1] Of the FDA approved drugs in 2011–2012, four agents have their primary activity at ion channels: Potiga/retigabine, a KCNQ2/3 channel opener for partial onset seizures; Horizant/gabapentin enacarbil, an alpha2/delta calcium channel modulator prodrug for restless leg syndrome; Kalydeco/ivacaftor, a CFTR potentiator for cystic fibrosis, and Fycompa/perampanel, an AMPA receptor antagonist for partial onset seizures. Despite such apparent success in developing ion channel targeted therapeutics, it is generally well accepted that these proteins have been a notoriously difficult class of targets to find new drugs for. This is based on the historical absence of high throughput automated screening platforms that could recapitulate the high fidelity and data quality of manual patch clamp electrophysiology—the highly labor-intensive but gold standard technique for studying ion channels—in a scale amenable to supporting the compound throughput that has become inherent in the modern drug discovery paradigm for other target classes such as GPCRs and kinases. Consequently, historical approaches to ion channel targeted drug discovery leaned heavily on non-electrophysiological and indirect measures of ion channel function, especially in the early screening and lead optimization phase, reserving the manual patch clamp studies for confirmation of only the most interesting candidate molecules. The last ten years has seen the introduction and rapid evolution of multiple automated electrophysiology systems to support ion channel screening offering much promise for the future generation of ion channel targeted therapeutics. The evolution, implementation and practical examples of the utility of automated electrophysiology platforms is discussed here and several reviews on this topic would serve as excellent supplementary reading.[2–5]

3.2 Non-electrophysiological Approaches to Ion Channel Screening

Prior to the introduction of automated screening platforms for electrophysiological assessment of drug activity at ion channels, non-electrophysiological approaches were the mainstay supporting drug discovery efforts. In particular, radioligand binding assays, ion flux measurements and optical approaches have all been utilized successfully in early screening approaches toward the identification of ion channel modulators. Ion flux measurements using radioactive ion tracers such as ^{86}Rb to assess potassium channel function have been particularly useful, and with the trend away from the use of radioactivity in high throughput screening this has seen a significant evolution itself to atomic absorption spectroscopy methodology for non-radiolabelled assessment of ion channel flux.[6–10] Optical methods such as those using fluorescent indicator dyes for calcium or membrane potential on platforms such as the fluorometric imaging plate reader (FLIPR) have revolutionized high-throughput screening (HTS) approaches to ion channel screening enabling truly scalable ion channel screening at industrialized levels. This has not come without its own limitations as the optical approaches represent indirect measurements of ion channel activity/function and consequently are subject to false positives that are intrinsic to the technology itself, *e.g.*, fluorescent interference, or the need to use non-physiological indirect methods to activate voltage-gated ion channels. Nevertheless, there is still an important role for the non-electrophysiological approaches especially in the earliest stages of drug screening where throughput considerations outweigh considerations of how close the assay format gets to a direct measurement of ion channel function.

3.3   Patch Clamp Automation (R)Evolution

The successful introduction of automated electrophysiology platforms to support ion channel targeted drug discovery efforts can be considered as both a revolution and an evolution, first by revolutionizing the way we can prosecute ion channels as drug targets and to recognize that this new technology itself is constantly evolving and improving. The need for such technology has been driven by a number of factors including the extremely low-throughput nature of conventional manual whole-cell patch clamp, the associated scaling of the drug discovery process where it became typical to screen in excess of one million compounds as an initial HTS campaign, and the need to support higher-throughput assessment of compound activity/action at cardiac ion channels, specifically the hERG (human *ether-a-go-go*) channel involved in repolarization of the cardiac action potential. Clearly, the latter has been a significant driver of this technology introduction and evolution with the removal of certain drugs from the market being attributed to cardiac arrhythmias, or *torsade de points*, and subsequent association of this effect with the ability of drugs to block the hERG channel requiring an early derisking strategy for this activity very early in the drug discovery process.

   The use of manual whole-cell patch clamp electrophysiology remains the gold-standard for studying ion channel activity and modulation by compounds allowing for detailed mechanistic and biophysical understanding of compound action at a level of resolution unsurpassed by the automated platforms. That being said, the labor intensive nature of this approach and the need for highly trained and skilled electrophysiologists to conduct these studies has limited its utility to evaluation of small numbers of compounds, typically those that are considered high quality leads with the potential to elevate to clinical candidate nomination. Similarly, with the requirement to introduce hERG screening earlier into the drug discovery process, manual patch clamp throughput only permitted evaluation of compounds late in their discovery cycle, and a positive result at this stage negated years of significant investment and often significant delays in the discovery of next generation molecules. With that framework in mind the objective in the early development of automated electrophysiology platforms was to replace traditional glass patch electrodes with a planar array of recording interfaces miniaturized on the surface of either a silicon, polymer or glass substrate to permit higher throughput parallel simultaneous recordings, in an automated manner by non-specialized operators. It is hard to dispute that this goal has largely been achieved by what is now commonly known as the chip-based planar patch clamp technology and multiple platforms have now been successfully introduced, including IonWorks, PatchXpress, QPatch, SynchroPatch, IonFlux and Dynaflow, affirming the revolution. To speak of the evolution of these platforms we have seen in the last ten years from the introduction of the IonWorkHT, a platform using the perforated patch clamp technique, and originally relying on a single hole in each well of the recording chip to enable cell recording, its evolution to the IonWorks Quattro, using 64 holes/well in the so called population patch clamp mode and enabling higher success rates and throughput, to the IonWorks Barracuda enabling even higher throughput with 384 simultaneous recordings. Similarly, we have witnessed the evolution of the Port-A-Patch to Patchliner to SyncroPatch; the QPatch to QPatch HT to QPatch HTX; the IonFlux 16 to IonFlux HT, each driving higher throughput while maintaining the same high quality cell recordings.

The subsequent discussion will focus on the impact of these new technologies in the drug discovery process and some specific examples of how they have been utilized, more than on a discussion of the relative strengths and weaknesses of each platform, as these will largely depend on the specific application of the user, such as but not limited to throughput requirements, ligand *versus* voltage gated channel applications, enterprise scale of the ion channel targeted effort, and even practical considerations such as available budget and degree of access to complementary technologies.

3.4 Impact of Automated Ion Channel Screening Technology

There are clearly a number of ways that the introduction of automated ion channel screening platforms have impacted the drug discovery process including: (i) improved screening cascades and ability to make decisions on compounds, (ii) enablement of early cardiac/hERG screening, (iii) facilitating the generation of the highest quality cell reagents, and (iv) allowing for an earlier readout on compound mechanism of action. Each of these will be discussed in some more detail.

(i) Automated electrophysiology platforms for ion channels are most effective when their use is fully integrated with other technologies for ion channel screening and with consideration of their different screening capacity, specific read-out and a clear recognition of the value and limitations of the data they are delivering to you. Such an integrated flow-scheme is illustrated in Figure 3.1 where primary

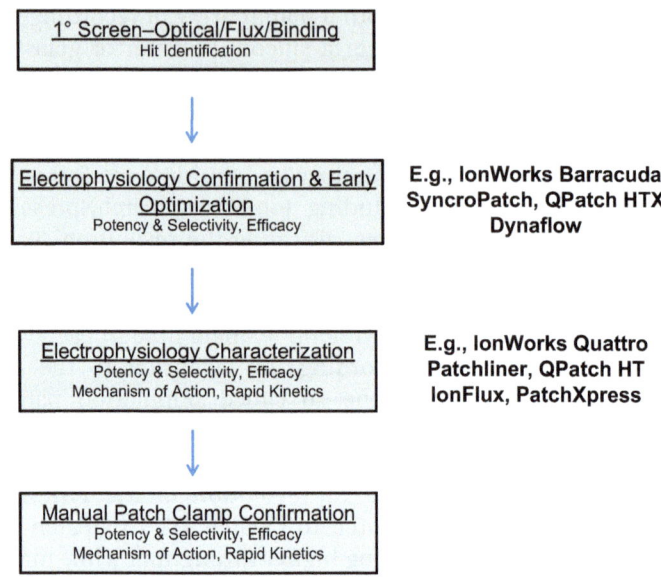

Figure 3.1 Illustrative example of an ion channel targeted drug discovery flow scheme. Following the initial HTS using a non-electrophysiological assay read-out in this example, automated electrophysiology platforms can be selected for integration into the workstream using the highest throughput platforms early on to drive the capacity for confirming larger numbers of initial hits and subsequent support of early structure–activity relationship generation and lead optimization. As the chemotypes mature in activity, the lower throughout platforms can be incorporated but utilized to gain additional mechanistic detail beyond primary potency and selectivity. Definitive manual patch clamp confirmation, if desired at all, can be reserved for only the highest quality compounds with the potential to be elevated to candidate status.

screening at scale can be supported by non-electrophysiological approaches such as optical or ion flux methodologies followed by lead confirmation and optimization on the higher throughout automated electrophysiology platforms. More detailed mechanistic screening can subsequently be supported by automated electrophysiology and finally the compounds of highest interest can be rigorously profiled using manual patch clamp technology. Each of the technologies is incorporated into the screening scheme according to scale of compound throughput and the limitations of the earliest employed non-direct screening methods can be minimized to a large extent by the ability to take larger numbers of compounds into confirmatory electrophysiology studies with increasing complexity as you flow through the scheme. Imagine trying to do an analog of this scheme prior to the automation (r)evolution where the heavy burden was placed on the manual patch clamp operator.

The extent to which automated electrophysiology platforms has evolved is impressive with the earliest platforms supporting screening in the order of hundreds of compounds per week to the present day where the highest throughput platforms can now accommodate screening well in excess of 10,000 compounds per week. It is conceivable that one can now choose to eliminate the early use of indirect screening methods for ion channels at the point of the primary screen by accepting not to screen the typical full HTS set in excess of 1×10^6 compounds and selecting compound subsets representing the chemical diversity of the larger collection, or ion channel focused libraries.

(ii) Without any doubt the most significant driver toward the implementation of automated electrophysiology in drug discovery has been to support higher throughput screening of compound activity on the cardiac hERG channel as a de-risking strategy to minimize the likelihood for subsequent effects on action potential QT interval and cardiac arrhythmias. This is substantiated by the vast majority of publications that have appeared describing the use of high throughput electrophysiology platforms focusing on hERG screening,[11-18] and more recently expanded to other cardiac ion channels such as Nav1.5, Cav1.2 and KCNQ1.[19-22] As with drug discovery efforts targeted toward specific ion channels, early hERG screening approaches largely utilized binding assays (*e.g.*, [³H]dofetilide) or ion flux (⁸⁶Rb) measurements to determine activity on hERG channels. A major limitation of the binding assays in particular was that this did not report functional effects on the hERG channel and was prone to false negatives.

The heightened emphasis on hERG channel screening that resulted from the removal of certain drugs from the market due to their effects on cardiac arrhythmias and association of this signal with their ability to inhibit the hERG channel placed an initial heavy burden on the manual patch clamp operator. Subsequently, the early adoption and implementation of automated electrophysiology platforms by groups handling heavy hERG channel screening burdens has been instrumental in driving their widespread utility today for both hERG screening and ion channel targeted drug discovery. An example of the impact of the automated platforms is shown in Figure 3.2 where in a single experimental run of approximately one hour duration on an IonWorksHT platform, three compounds were tested in concentration–response mode alongside the positive control quinidine allowing for the estimation of their potencies for hERG inhibition. An experiment of this scope would have taken several weeks to accomplish using manual patch clamp electrophysiology and would typically have occurred in two phases, firstly testing the compounds at a single high concentration to select compounds such as 1 and 2 with demonstrated inhibitory activity for subsequent concentration–response screening. Enabling hERG screening with the higher throughput electrophysiology platforms has allowed this to become part of the iterative chemistry design cycle by considering both potency at the primary target of interest, in concert with the

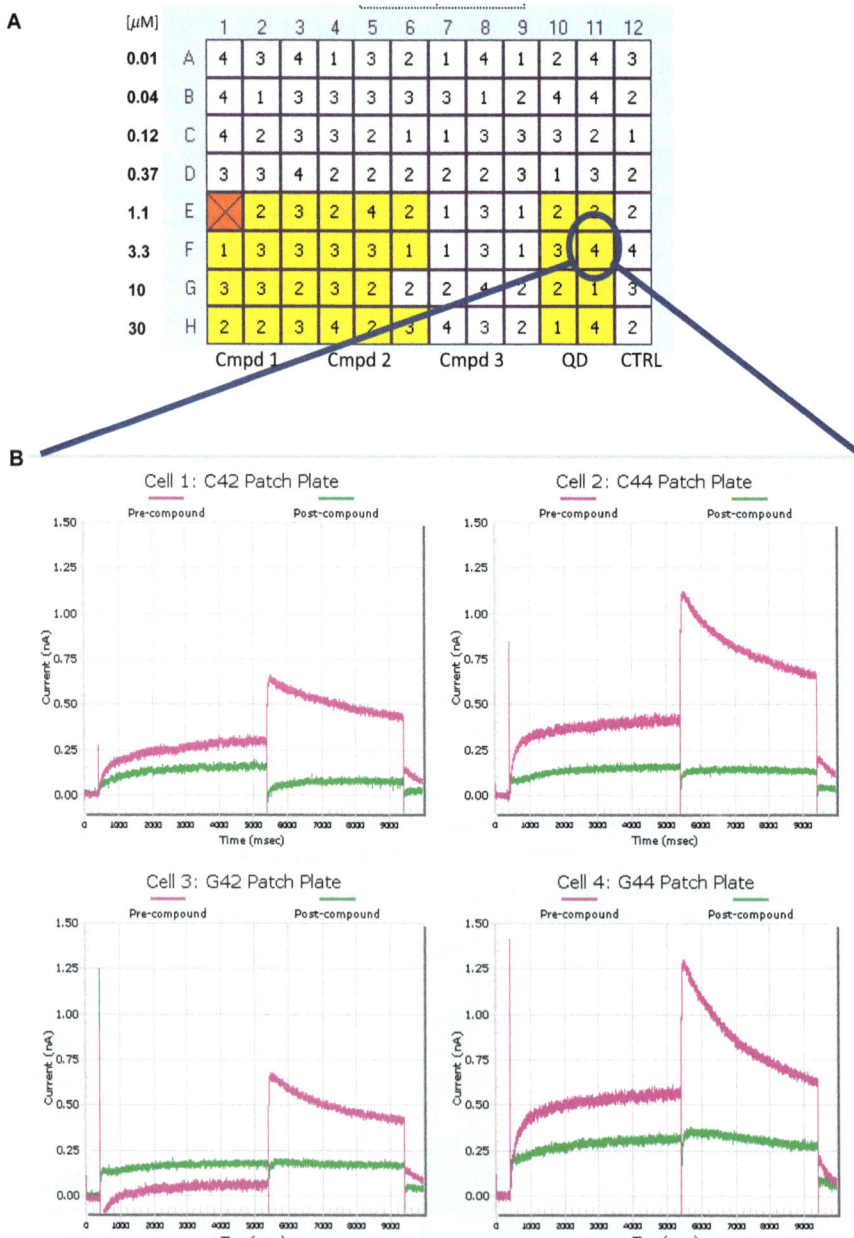

hERG channel activity, allowing for a much earlier de-risking of this signal than was ever possible with the manual patch clamp approach.

(iii) The success of automated electrophysiology is highly impacted by the quality of the cells utilized, something that cannot be emphasized strongly enough, and also of relevance is that the cell line that is utilized for manual patch clamp or for optical screening approaches is often not suitable for automated electrophysiology applications. In this regard the high throughput nature of these platforms can be exploited to generate the highest quality cell reagent by allowing for the screening of multiple clonal cell lines in the cell selection and optimization phase on the platform to be ultimately used for auto-mated screening. This can be illustrated by our experience of tran-sitioning the screening of the potassium channel KCNQ2/3 expressed in Chinese hamster ovary cells from manual patch clamp to the IonWorks platform as shown in Figure 3.3. Initial attempts of clonal expansion of this cell line and testing on IonWorks were un-successful. This was likely attributed to the known fact that in manual patch clamp experiments with these cells only a sub-population of the cells yielded robust KCNQ2/3 current of the type shown in Figure 3.3 and further that the trained eye of the manual patch clamp operator could select these cells with high success rate based on their morphology. Subsequently, we took advantage of this knowledge to hand pick cells based on morphology *via* micropipette for clonal expansion and subsequent screening for activity on IonWorks testing 12 different subclones simultaneously. This ap-proach resulted in the successful generation of a clonal cell line that gave robust KCNQ2/3 currents on the IonWorks platform and with 75% overall success rate.[23]

(iv) Although the optical and binding approaches to ion channel screening have been instrumental in ion channel drug discovery ef-forts one of their major limitations in comparison to electro-physiological approaches is related to the ability to gain detailed

---

Figure 3.2   (A) 96-well compound plate lay-out scheme for screening three test compounds (Cmpd) and the positive control quinidine (QD) on hERG currents using IonWorksHT. Each well of the 96-well plate maps to four different recording wells of an IonWorks patch plate as illustrated in B for the selected well. The number in the well is the number of successfully completed IonWorks recordings from a possible total of four. Wells are color coded as yellow according to a pre-specified software setting to indicate a calculated inhibition greater than 50% allowing for an immediate qualitative assessment of compound activ-ity. A red well indicates four out of four recordings failed to complete. (B) The hERG tail currents for the four recording wells highlighted from A. in the absence (magenta) and presence (green) of quinidine illustrating a complete block of hERG currents at 3.3 μM QD.

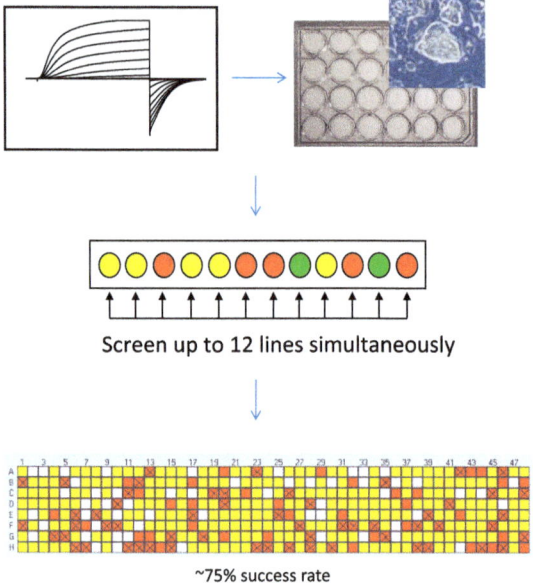

Screen up to 12 lines simultaneously

~75% success rate

Figure 3.3   Using the power of automated electrophysiology to optimize the cell reagent. Stable CHO cells expressing KCNQ2/3 gave robust currents using whole-cell manual patch clamp electrophysiology as indicated by the traces. By selecting cells for clonal expansion with a typical morphology associated with robust recordings multiple sub-clones were selected for evaluation for KCNQ2/3 currents on IonWorks with ultimately one clone chosen for further work based on an overall 75% success rate.

mechanistic insight into compound action. This can be best illustrated by considering properties such as voltage- and use-dependence of compound action on ion channels, properties that cannot be assessed with the non-electrophysiological screening assays in a meaningful manner. Illustrative examples are shown in Figure 3.4 demonstrating properties of voltage and use-dependence in the inhibitory action of compounds on the cardiac sodium channel Nav1.5 revealed *via* screening on the IonWorks platform. The voltage-dependence of Nav1.5 block by ambroxol is demonstrated by measuring inhibitory potency at two different holding potentials resulting in a rightward shift in compound potency at more negative holding potential. The use-dependent Nav1.5 block by propafenone is demonstrated by measuring the inhibition of channel activity as a function of the number of stimulation pulses at the same holding potential. For any given concentration of propafenone, the magnitude of inhibition increases with the number of stimulation pulses (essentially how much the channel is 'used').

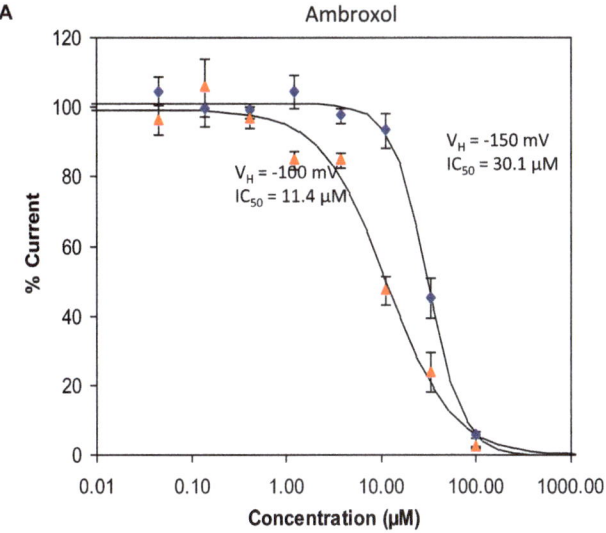

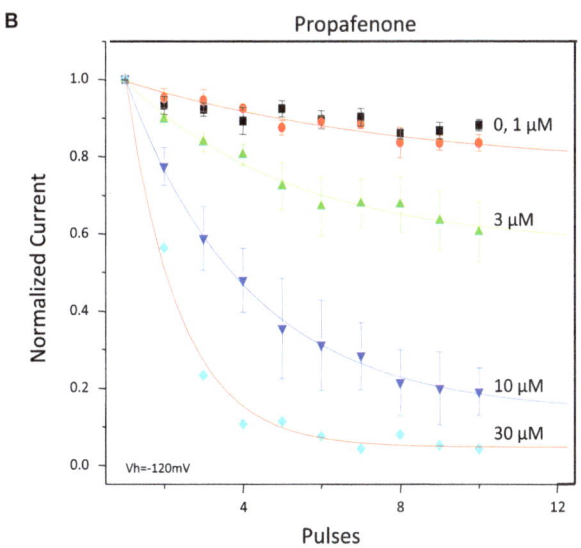

Figure 3.4   Using automated electrophysiology to gain mechanistic insight into compound action. (A) Voltage-dependent block of the cardiac sodium channel Nav1.5 with ambroxol determined using IonWorks recordings. (B) Use-dependent block of the cardiac sodium channel Nav1.5 with propafenone determined using IonWorks recordings.

3.5   Drug Discovery Applications

It is clear that automated electrophysiology has become firmly established in the drug discovery process for ion channels with multiple applications

A

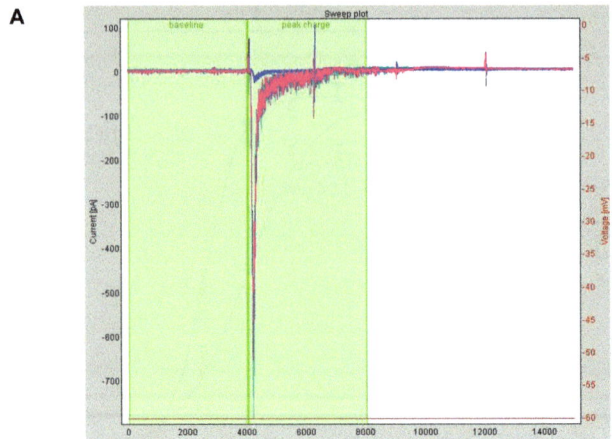

B

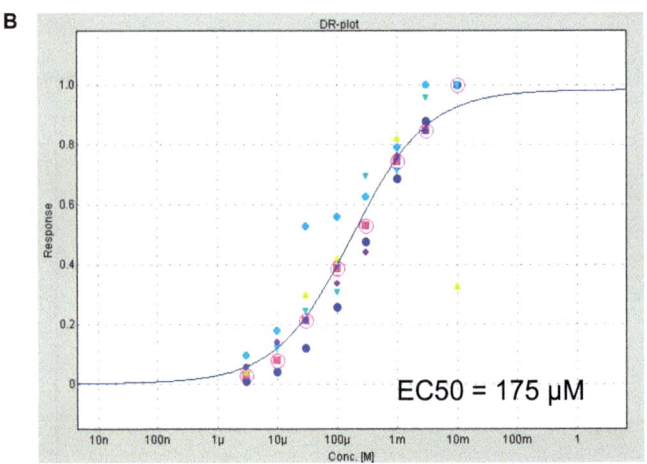

EC50 = 175 µM

C

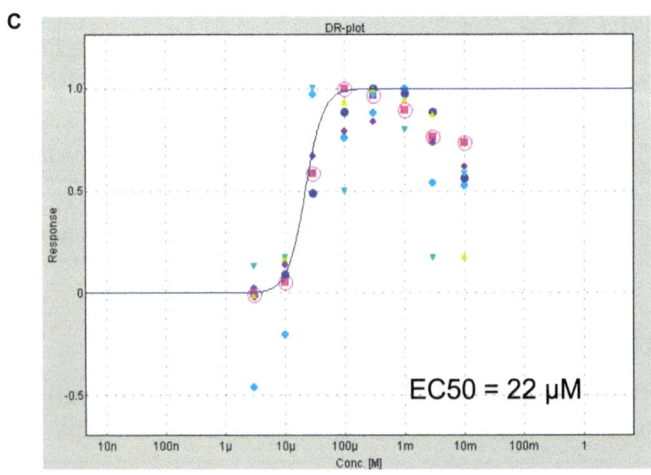

EC50 = 22 µM

now validated and presented including several examples of hERG screening to support de-risking of potential for QT prolongation as discussed, in addition to a diverse array of ion channels targeted for possible therapeutic utility including KCNQ2/3,[23] Nav1.7,[24,25] Nav1.3,[25] HCN1,[26] Kv1.3,[27] Kv2.1,[28] α3β4 and α4β2 nicotinic receptors[29] and P/Q-type Ca^{2+} channels.[30]

As an illustrative example, one of the channels that has received much attention from the drug discovery perspective is the alpha7 subtype of the nicotinic acetylcholine receptor (nAChR), implicated in a number of studies to be a therapeutic target for cognitive impairment associated with schizophrenia and Alzheimer's disease. As part of a strategy to improve screening throughput for the characterization of novel alpha7 nAChR agonists and positive allosteric modulators we developed both optical assays using calcium mobilization on the FLIPR platform and an automated electrophysiology solution, in this case using the QPatch. A hallmark characteristic of the alpha7 nAChR receptor is its rapid desensitization in the msec timescale following agonist application, and we were very interested to determine if such rapid kinetics could be resolved on high throughput automation platforms. For both the PatchXpress (unpublished) and the QPatch, agonist-evoked currents were readily resolved and concentration–response curves on the QPatch recapitulated the pharmacological profile for acetylcholine (Figure 3.5) and other agonists that were previously characterized using manual whole-cell patch clamp electrophysiology.[31]

3.6　Concluding Remarks

The last ten years has witnessed an explosion in the availability and evolution of diverse platforms for automated electrophysiology and these have now been fully integrated into the drug discovery process both for ion channel targeted drug discovery and in support of cardiac ion channel screening. These platforms have facilitated unprecedented screening capacity for ion channels allowing for a much more comprehensive interrogation of compound libraries and enhanced structure–activity relationship campaigns compared to historical efforts relying on labor-intensive and manual screening. These advances should greatly facilitate ion channel targeted drug discovery efforts now and in the future.

Figure 3.5　Measuring acetylcholine-stimulated alpha7 nAChR currents using the QPatch. (A) Representative current traces following application of acetylcholine to GH4C1 cells expressing the rat alpha7 nAChR confirming the ability to resolve rapid channel kinetics using automated electrophysiology. Estimated EC50 values for the activation of alpha7 nAChRs expressed in GH4C1 cells calculated using (B) peak current or (C) area-under-the-curve were determined as 175 and 22 µM, respectively, essentially identical to values obtained using manual patch clamp electrophysiology.

References

1. P. Imming, C. Sinning and A. Meyer, *Nat. Rev. Drug Discov.*, 2006, **5**, 821–834.
2. J. Dunlop, M. Bowlby, R. Peri, D. Vasilyev and R. Arias, *Nat. Rev. Drug Discov.*, 2008, 7, 358–368.
3. Q. Lu and F. An, *Comb. Chem. High Throughput Screen.*, 2008, **11**, 185–194.
4. B. T. Priest, A. M. Swenson and O. B. McManus, *Curr. Pharm. Design*, 2007, **13**, 2325–2337.
5. T. J. Dale, C. Townsend, E. C. Hollands and D. J. Trezise, *Mol. Biosys.*, 2007, **3**, 714–722.
6. G. C. Terstappen, *Anal. Biochem.*, 1999, **272**, 149–155.
7. S. Rezazadeh, J. C. Hesketh and D. Fedida, *J. Biomol. Screen.*, 2004, **9**, 588–597.
8. C. W. Scott, D. E. Wilkins and S. Trivedi, *Anal. Biochem.*, 2003, **319**, 251–257.
9. S. Gill, R. Gill and D. Wicks, *Assay Drug Dev. Tech.*, 2007, **5**, 373–380.
10. F. Jow, E. Tseng and T. Maddox, *Assay Drug Dev. Tech.*, 2006, **4**, 443–450.
11. A. E. Dubin, N. Nasser and J. Rohrbacher, *J. Biomol. Screen.*, 2005, **10**, 168–181.
12. H. Guthrie, F. S. Livingston and U. Gubler, *J. Biomol. Screen.*, 2005, **10**, 832–840.
13. L. Guo and H. Guthrie, *J. Pharmacol. Toxicol. Methods*, 2005, **52**, 123–135.
14. S. Sorota, X. S. Zhang and M. Margulis, *Assay Drug Dev. Tech.*, 2005, **3**, 47–57.
15. M. H. Bridgland-Taylor, A. C. Hargreaves and A. Easter, *J. Pharmacol. Toxicol. Methods*, 2006, **54**, 189–199.
16. J. Q. Ly, G. Shyy and D. L. Misner, *Clin. Lab. Med.*, 2007, **27**(1), 201–208.
17. T. R. Bridal, M. Margulis, X. Wang, M. Donio and S. Sorota, *Assay Drug Dev. Tech.*, 2010, **8**, 755–765.
18. R. Mannikko, G. Overend, C. Perrey, C. L. Gavaghan, J.-P. Valentin, M. Armstrong and C. E. Pollard, *Br. J. Pharmacol.*, 2010, **159**, 102–114.
19. X. Cao, Y. T. Lee, M. Holmqvist, Y. Lin, Y. Ni, D. Mikhailov, H. Zhang, C. Hogan, L. Zhuo, Q. Lu, M. E. Digan, L. Urban and G. Erdemli, *Assay Drug Dev. Tech.*, 2010, **8**, 766–780.
20. A. R. Harmer, N. Abi-Gerges, A. Easter, A. Woods, C. L. Lawrence, B. G. Small, J.-P. Valentin and C. E. Pollard, *J. Pharmacol. Toxicol. Methods*, 2008, **57**, 30–41.
21. B. M. Heath, Y. Cui, S. Worton, B. Lawton, G. Ward, E. Ballini, C. P. A. Doe, C. Ellis, B. A. Patel and N. C. McMahon, *J. Pharmacol. Toxicol. Methods*, 2011, **64**, 258–268.
22. G. J. Kaczorowski, M. L. Garcia, J. Bode, S. D. Hess and U. A. Patel, *Front. Pharmacol.*, 2011, **2**, 1–11.
23. F. Jow, R. Shen and P. Chanda, *J. Biomol. Screen.*, 2007, **12**, 1059–1067.
24. S. Trivedi, K. Dekermendjian, R. Julien, J. Huang, P.-E. Lund, J. Krupp, R. Kronqvist, O. Larsson and R. Bostwick, *Assay Drug Dev. Tech.*, 2008, **6**, 167–179.

25. N. Castle, D. Printzenhoff, S. Zellmer, B. Antonio, A. Wickenden and C. Silvia, *Comb. Chem. High Throughput Screen.*, 2009, **12**, 107–122.
26. D. Vasilyev, Q. J. Shan, Y. T. Lee, V. Soloveva, S. P. Nawoschik, E. J. Kaftan, J. Dunlop, S. Mayer and M. R. Bowlby, *J. Biomol. Screen.*, 2009, **14**, 1119–1128.
27. K. Liu, M. Samuel, J. Tillett, J. K. Hennan, B. Mekonnen, V. Soloveva, R. K. Harrison, J. W. Paslay and J. Larocque, *J. Biomol. Screen.*, 2010, **15**, 185–195.
28. J. Herrington, K. Solly, K. S. Ratliff, N. Li, Y.-P. Zhuo, A. Howard, L. Kiss, M. L. Garcia, O. B. McManus, Q. Deng, R. Desai, Y. Xiong and G. J. Kaczorowski, *Mol. Pharmacol.*, 2011, **80**, 959–964.
29. J. D. Graef, L. C. Benson, S. S. Sidach, H. Wei, P. M. Lippiello, M. Bencherif and N. B. Federov, *J. Biomol. Screen.*, 2013, **18**, 116–127.
30. D. Hermann, M. Mezler, A. M. Swensen, C. Bruehl, A. Obergruber, K. Wicke, H. Schoemaker, G. Gross, A. Draguhn and V. Nimmrich, *Comb. Chem. High Throughput Screen*, 2013, **16**, 233–243.
31. S. Friis, C. Mathes, M. Sunesen, M. R. Bowlby and J. Dunlop, *J. Neurosci. Methods*, 2009, **177**, 143–148.

CHAPTER 4

Structural Understanding of Ion Channels in Atomic Detail

PHILLIP J. STANSFELD

Department of Biochemistry, University of Oxford, South Parks Road, Oxford, OX1 3QU, UK
Email: phillip.stansfeld@bioch.ox.ac.uk

4.1 Introduction

All molecular cells are encased in an approximately 40 Å thick hydrophobic sheath of lipid molecules, compartmentalising all the chemical components essential for life. This cell membrane acts as a barrier to larger polar molecules and charged atoms, called ions. The theory of permeable pores within the membrane can be traced back to the middle of the 19th century. A century later it was shown that nerve impulses are propagated by the movement of Na^+ and K^+ ions across cell membranes.[1–5] Yet the concept of ion channel proteins as an explanation for this phenomenon remained controversial until the 1970s, with the alternative hypothesis suggesting that charged ions could be transferred through the membrane by molecular carriers.[6]

Pioneers in the field of electrophysiology quickly identified the existence of ion channels and elucidated their functional roles.[7–10] Early structural assumptions proposed that ion channels were selective aqueous pores formed by intrinsic membrane proteins which could open and close to control ionic passage and could be blocked by molecules physically occluding the permeation pathway.[11,12] The development of molecular biology tools led to the cloning of the genes that encoded individual ion channels,

RSC Drug Discovery Series No. 39
Ion Channel Drug Discovery
Edited by Brian Cox and Martin Gosling
© The Royal Society of Chemistry 2015
Published by the Royal Society of Chemistry, www.rsc.org

providing the amino-acid sequence for these proteins.[13] This permitted studies to determine the topology and stoichiometry, whilst also identifying key structural features involved in gating, ligand binding and ion permeation. At the same time structural biologists developed advanced techniques for protein structure determination, permitting the resolution of the first membrane protein crystal structure in 1984.[14]

The identification of ion channel genes within prokaryotic genomes provided new targets for structural characterisation.[15] In 1998 a groundbreaking discovery was made with the first X-ray structure determined for a potassium (K^+) selective ion channel, called KcsA.[16] Since then numerous ion channel structures have followed. This chapter will discuss the developments that have made this possible, outline the key features of the ion channels resolved to date and consider how these and future developments will aid drug discovery within the field.

4.2   Methods for Resolving Membrane Protein Structures

Integral membrane proteins are notoriously difficult targets for structural studies. Despite comprising approximately 25% of most known genomes, only 1% of the structures in the Protein Data Bank (PDB) belong to this class of protein.[17] Of these structures only 340 or so are unique and many are bacterial in origin.[18] This deficiency is gradually being reconciled by improvements to the methodologies involved. This section will discuss the recent progress within the field, considering improvements that have aided the elucidation of both ion channel and other membrane protein structures.

Eukaryotic membrane proteins were originally deemed too complicated to be solved by X-ray crystallography or NMR, due to difficulties in expressing the proteins in *Escherichia coli*, their post-translational modifications and their overall complexity. For this reason the majority of the early targets for high-resolution structural studies were bacterial homologues. As the number of libraries for bacterial genomes increased, proteins were identified that shared homology with their mammalian counterparts. As a result, approximately three-quarters of the ion channels resolved to date are bacterial in origin.

The number of eukaryotic structures is exponentially increasing due to methods that enhance their expression, purification, crystallisation and structural determination. The most successful systems for expressing eukaryote membrane protein include the yeasts, *Pichia pastoris* and *Saccharomyces cerevisiae*, insect, *Spodoptera frugiperda* (sf9), and mammalian (*e.g.* HEK293S) cell lines.[19]

One of the most difficult aspects of membrane protein structural biology is mimicking the lipid bilayer environment. In order to produce high-resolution crystal structures, proteins are extracted from their native membrane and reconstituted into detergent micelles that mimic the hydrophobic core of the bilayer. There is no definitive rule for which detergent should be used for a given membrane protein, therefore proteins generally undergo detergent screens to detect the most suitable bilayer mimic. This usually involves an initial first pass with a gentle detergent, such as dodecyl maltoside (DDM), before screening other suitable candidate molecules for solubilisation and purification.

Lipidic cubic phase (LCP) has also provided a successful medium, with the single acyl chain, monoolein, typically the lipid used. This technique has primarily been used for elucidation of members of the GPCR family and for membrane proteins with minimal soluble domains. The complex of the β-adrenoceptor in association with its G-protein is one notable exception to this rule. In this instance a shorter chain version of monoolein was used to increase the volume of the soluble region.

Ideally the likelihood of crystallisation is assessed at an early stage, by testing a variety of constructs for their monodispersity/aggregation and stability. One of the most successful approaches is Fluorescent Size-Exclusion Chromatography (FSEC).[20,21] This method covalently fuses GFP to different constructs of a membrane protein permitting their expression, localisation, monodispersity and molecular mass to be appraised. Subsequent truncations or mutations can then be made to increase the likelihood of structural determination. Mutations may also be made to remove glycosylation sites, lock the structure in a particular orientation through disulphide bridges or enhance the thermal stability of a protein.[22]

Other strategies have spliced together bacterial and eukaryotic sequences to produce a chimeric protein, usually with the transmembrane portion bacterial and the soluble region eukaryotic.[23,24] Similar approaches have used chimeric constructs of T4-Lysozyme, covalently linked to the membrane protein of interest.[25] This improves crystal packing by presenting an increased surface area and allowing protein–protein contacts to form. An equivalent strategy is to raise antibodies against the membrane protein, to allow recognition and tight binding to the non-transmembrane regions of the target protein. These antibodies are usually either smaller fragments of antibodies (FABs or Fvs) or nanobodies; antibodies from sharks or camelids, which lack the light chain portion of the antibody.

4.3   Potassium Channels and their Relatives

Ever since Hodgkin and Huxley proposed the existence of carriers for K^+ and Na^+ across the squid giant axon,[1–5] structural and functional studies have sought to resolve the properties and characteristics of the proteins involved. Almost half a century later, a K^+ channel became the first ion channel to be resolved to high resolution.

In 1995, a bacterial protein, called KcsA, was revealed to share strong sequence similarity with mammalian voltage-gated (Kv) channels for the pore domain, yet lacked the additional mechanistic parts such as a voltage-sensor (VS) and T1-domain. It was therefore considered a sensible target for

X-ray crystallography.[26] Even for this relatively simple target, the C-terminal domain was cleaved, while the N-terminus was disordered. An initial structure was resolved to 3.2 Å in 1998, with a higher resolution structure resolved three years later, using FAB fragments, which were targeted to the extracellular potion of the channel.[16,27] This improved crystal packing and increased the resolution to 2.0 Å. Further bacterial structures followed, with structures appearing for homologues of inwardly rectifying (Kir; KirBac1.1),[28] calcium-activated (KCa; MthK)[29] and voltage-gated (Kv; KvAP)[30] K[+] channels.

4.3.1 Conserved Pore Architecture

The determination of the structures of KcsA revealed for the first time the 3-dimensional architecture of the pore domain of an ion channel. This domain is conserved across all K[+] channels, with similar domains also present in voltage-sensitive Na[+] (Nav) and Ca[2+] (Cav) channels, and glutamate receptors (GluR). A single monomer consists of an outer transmembrane helix, a half-helix (pore helix), a highly conserved coiled domain, called the selectivity filter, and an inner pore-lining transmembrane helix, which forms the channel gate. Four subunits oligomerise to create the pore architecture, which comprises the selectivity filter for K[+], a central water-filled cavity and a hydrophobic gate formed by the crossing of the four inner helices (Figure 4.1).

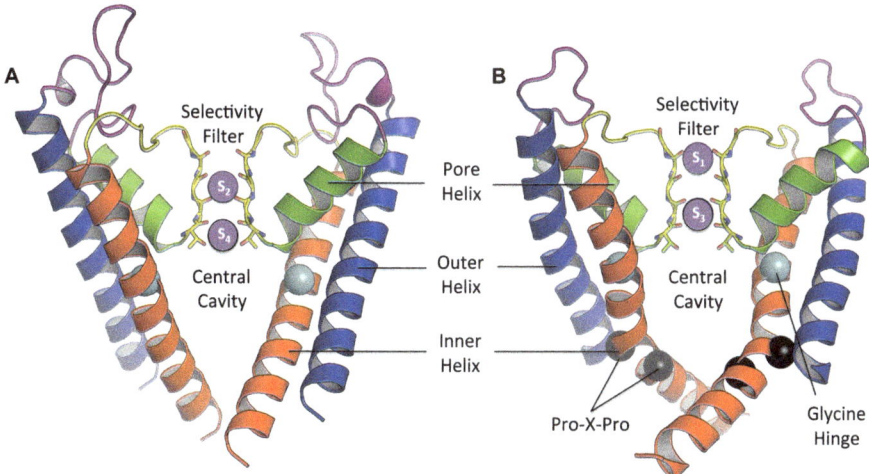

Figure 4.1 The architecture of a K[+] channel pore domain, illustrating the differences between open and closed states. The outer (blue), pore (green) and inner (red) helices are shown, along with the selectivity filter (yellow). Only two subunits are shown for clarity. (A) Closed state structure of KcsA. (B) Open state structure of KvChim. The Pro-X-Pro residues are shown as black spheres. The conserved glycine hinge is shown as cyan spheres.

Selective ion permeation across the membrane is fundamental to preserving the cell membrane potential. The selectivity filter in K^+ channels is formed by the well-conserved TVGYG sequence, which is structured as a ladder of exposed carbonyl oxygens that define four ion-binding sites, termed S_1 to S_4. Ions bind to the filter in either S_1:S_3 or S_2:S_4 configuration, with water molecules binding to the unoccupied sites (Figure 4.2(A)).[31,32] In order to enter the narrow selectivity filter, a K^+ ion must lose its surrounding water, with the carbonyl oxygens mimicking the first hydration shell in each of the four sites. Further sites are proposed above the filter, S_0, and in the central cavity S_{cav}. A K^+ ion is not fully dehydrated in either of these sites.

The ionic radius of a K^+ ion is perfectly suited for the dimensions of the selectivity filter, whereas a Na^+ ion is too small to be fully accommodated in its dehydrated state, as the carbonyl oxygens are unable to approach each other due to electrostatic repulsion. Indeed it was shown that even in low K^+ concentrations Na^+ does not enter the KcsA selectivity filter, with K^+ ions found only in sites S_1 and S_4 and a collapse of sites S_2 and S_3.[27] This is expected to be the non-conducting conformation or 'C-type inactivated' state of the selectivity filter.[27] This is not true for all K^+ channels, with another channel, MthK, interacting in a completely different fashion. For this structure a high concentration of Na^+, in the absence of K^+, still results in an active conformation of the selectivity filter.[33] The one noticeable difference is that the S_1 site is shifted towards the extracellular side, with the Na^+ ion being coordinated in a planar site, interacting with the four backbone carbonyl oxygens between S_1 and S_0. A similar binding site was proposed for Li^+ and by analogy, Na^+, for the carbonyl oxygens between S_3 and S_4.[34] Indeed molecular simulations have suggested that in the absence of K^+, which otherwise obstructs Na^+ passage, Na^+ can bind in plane between all four sites.[35] This is of special interest in hypokalaemia, when the concentration of K^+ drops to below a blood concentration of 3 mM. In this instance it has been shown that TWIK-1 channels can conduct an inward flux of Na^+ that could potentially lead to cardiac arrest.[36]

The identification of a bacterial non-selective K^+ channel homologue provided a means to further test the hypotheses of selectivity. This channel, called NaK, has the unique selectivity filter sequence of GDG rather than the more standard GYG[37] (Figure 4.2(B)). The structure of this channel illustrates a novel conformation for the selectivity filter, with sites S_1 and S_2 merged into a single site. This appears to lower the energetic cost for Na^+ passage through the filter as there are only two remaining sites, with only one of these solely coordinated by carbonyl ligands. It is possible to transform NaK into a K^+ selective channel by incorporating mutations into the selectivity filter (NaK2K).[38] Similar studies have been used to mutate the selectivity filter of NaK to that of a cyclic-nucleotide gated (CNG) channel to explain why these channels are non-selective, despite retaining a similar fold to K^+ channels.[39] The crystal structure of this construct suggests that the S_1 site is absent in CNG channels (Figure 4.2(C)). Analogous studies of the K^+-selective MthK channel suggested it is possible to make a channel non-selective by mutating the selectivity filter

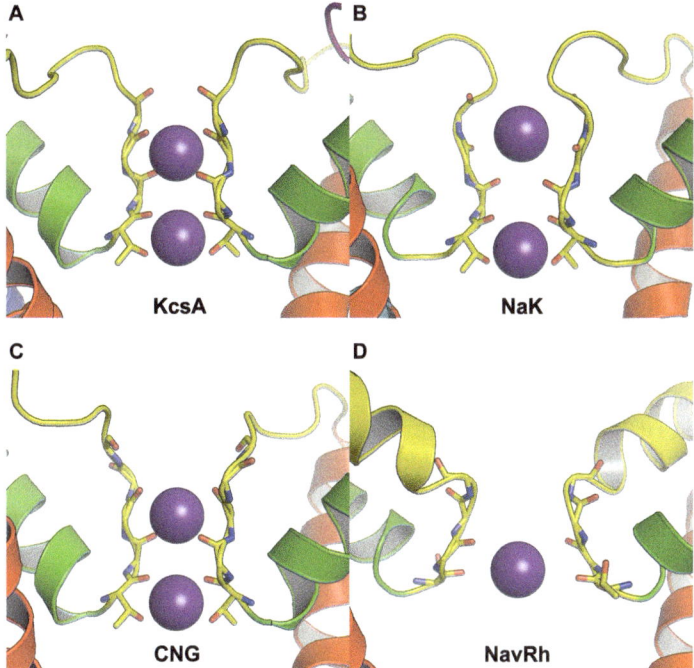

Figure 4.2   Selectivity in the pore-loop superfamily of channels. (A) KcsA. K^+ selective. (B) NaK. K^+ and Na^+ selective. Sites S_1 and S_2 are combined, with sites S_3 and S_4 the same as in a K+ selective channel. (C) NaK-CNG mimic. K^+ and Na^+ selective. There is no S_1 site in this structure. S_{2-4} are the same as in a K^+ selective channel. (D) NavRh. Na^+ selective. The structure is resolved with Ca^{2+} in the S_{in} site, which is analogous to S_3. There is also a second pore half-helix (P2) in the bacterial Nav structures.

threonine to alanine to remove the S4 site.[40a] Evaluation of both studies suggests that all four sites are required for K^+ selectivity.[40a] Similar strategies have been used to transform both the selectivity filter and central cavity of the bacterial NaK and KcsA channels to that of hERG, a channel that is serendipitously blocked by a number of pharmaceutical compounds.[40b] (See Chapter 11 in this book by Mitcheson).

The primary activation gate of K^+ channels is formed by the inner helices. The first structures of KcsA were resolved in the closed state, with the inner helix bundle constricting the channel pore. A few years after the determination of KcsA structure a second channel, MthK, was crystallised in the open state.[29,41] This channel was solved with its native activatory ligand Ca^{2+} bound to the large regulator of K^+ conductance (RCK) C-terminal domains, thereby suggesting the structure was captured in the activated state. Although only the backbone of the transmembrane portion of the channel was resolved there was sufficient detail to indicate that the inner transmembrane helices were splayed apart *via* a bend at a highly conserved glycine residue.

The linkers between the transmembrane and C-terminal domains were not resolved, so it was not certain how the calcium binding induces the opening of the channel.

Since then, eukaryotic structures of RCK domains from BK channels have appeared.[42–44] These have been crystallised in what are believed to be the open and closed states of the C-terminal gating ring. Although the transmembrane domain and linkers are absent in these structures it is possible to propose how widening of the gating ring can pull on the inner transmembrane helices to induce pore opening. Structures for the cytosolic domains of a number of other K^+ channels and their relatives have been resolved, including the T1-domain and beta-subunits from Kv channels, the cytoplasmic domains of Kir channels,[45,46] the PAS domain from hERG[47] and the cyclic nucleotide binding domains (CNBD) from both the bacterial MlotiK[48] and eukaryotic HCN and KCNH channels.[49,50]

Recent developments have led to the elucidation of transmembrane structures for eukaryotic proteins from the three major families of K channels: inward rectifiers (Kir), voltage-gated (Kv) and two-pore (K2P) (Figure 4.3). The next sections will cover these structures in detail and outline the mechanisms of gating used by these channels.

4.3.2   Kir Channels

The Kir channel family was initially characterised by the structure of a bacterial channel, KirBac1.1,[28] a set of cytoplasmic domain structures of eukaryotic proteins,[45,46] a chimera between bacterial and eukaryotic Kir channels (Kirchim)[23] and subsequent homology models based on these coordinates. More recently structures have appeared for eukaryotic Kir2.2 and Kir3.2 channels,[51,52] along with further KirBac proteins in multiple conformational states.[53] This section will describe the overall architecture of a Kir channel, detail the key structural features and outline the mechanisms proposed for channel activation.

The initial structure of the cytoplasmic domain of Kir3.1 (GIRK1) revealed that the four cytoplasmic subunits associated as a tetramer with four-fold rotational symmetry. They contain a central cytoplasmic pore, which extends the total length of the channel by a further 60 Å from the membrane.[45] Acidic residues line this pore, providing a negatively charged funnel for directing both K^+ ions and the native blockers of Kir channels, Mg^{2+} and polyamines, thereby governing the rectification properties of these channels. At the apex of these structures, closest to the membrane, a constriction point, termed the G-loop, is apparent in this largely β-stranded domain. This region is believed to form a second gate, in addition to the barrier formed at the transmembrane domain. The KirBac1.1 structure contains a cytoplasmic domain largely similar to its eukaryotic cousins, whilst its transmembrane domain is very similar to the closed state structure of KcsA. The structure also contains an interfacial (slide/M0) helix, formed by the N-terminal domain.[28]

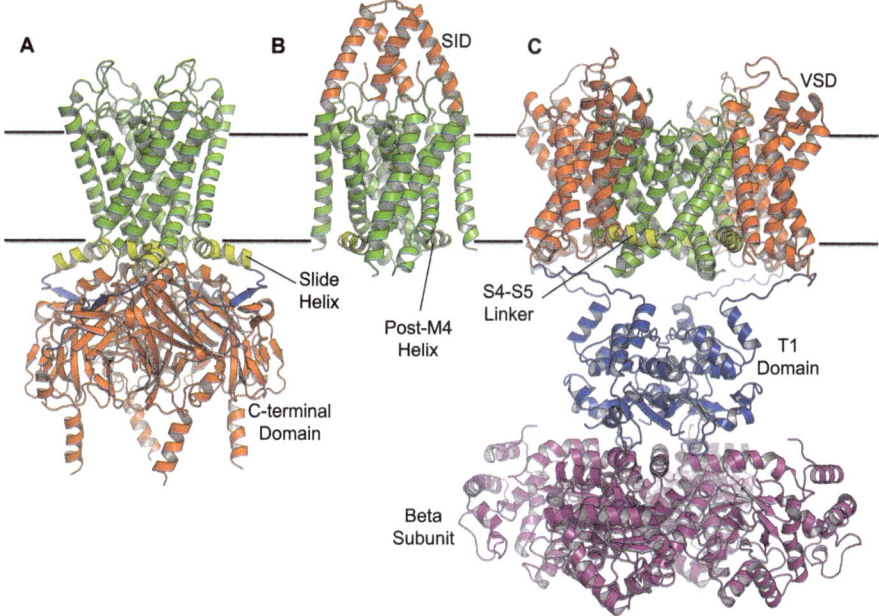

Figure 4.3 Structures of the eukaryotic K$^+$ channel superfamily. In each case the pore domain is coloured green. In the three channels an interfacial helix is believed to be important in gating (yellow). (A) Inward Rectifier (Kir) channel—cKir2.2. (B) Two-pore (K2P) channel—hK2P1.1. The extracellular cap or self-interacting domain (SID) (red) is found on the extracellular side, with both monomers often covalently linked by a disulphide bridge. (C) Voltage-gated (Kv) channel—rKv1.2/2.1 (KvChim). The voltage-sensitive domain (VSD) (red) sits within the bilayer. The N-terminal Tetramerisation (T1) domain (blue) is found on the intracellular side, forming a complex with the beta-subunit (purple).

The elucidation of the cKir2.2 structure confirmed three key structural features that had previously been proposed through biochemical and electrophysiological experiments.[54] The first was a highly conserved aspartate residue within the central cavity of all strongly rectifying Kir channels. This is suggested to be required for the coordination of polyamines and Mg^{2+} ions.[55,56] This residue has previously been described as the rectification controller and is not conserved in voltage-gated channels. A second structural feature is a salt-bridge behind the selectivity filter that is believed to be required for stabilisation of the conductive conformation of the filter.[57] A similar interaction is suggested for KcsA between two acidic residues, which form a carboxyl-carboxylate linkage through protonation of a glutamate in the pore helix.[58] The third feature is a disulphide bridge between the turret (M1-P linker) and the post-selectivity filter loop. The intra-subunit cysteine linkage and a well-defined network of H-bonds results in a highly ordered extracellular mouth of the channel that is structurally distinct from the other channel members.[59] This is believed to reduce the susceptibility of the channel to being blocked by certain toxins. Subtle differences in this region may make this region tractable to designing suitable compounds for subtype-specific block of Kir channels.

The most recent Kir channel structures reveal the coordinates of the binding site for PIP_2, the endogenous activator of Kir channels.[51,52] The binding site was previously identified by a number of mutagenesis studies and predicted structurally by molecular simulation.[60] The X-ray structures reveal that the principal residues are basic in nature and are located at the N-terminal end of the TM1 helix and in the C-linker between TM and C-terminal domains. There is also a tight network of H-bonds that connect the C-linker to the slide helix of the adjacent subunit. This set of structures concurs with the previous Kirchim structure and a multi-state set of KirBac structures that there is a rotation and upward motion of the C-terminal domain towards the membrane upon channel activation. In eukaryotic proteins this appears to be driven by the presence of the PIP_2 molecules, as they induce the folding of a helix within the otherwise coiled C-linker. The structural rearrangement pulls the C-terminal domain upwards and positions the G-loop gate in close proximity to the transmembrane gate.

Despite the presence of PIP_2, neither of the two wild-type Kir channels is resolved in the predicted open state. In an attempt to capture an open state channel, a constitutively active mutant of G-protein activated Kir3.2 was crystallised. The R201A mutant channel activates in the absence of G-protein, thereby decoupling the channel from this activation pathway. In the absence of PIP_2 the wild-type channel is closed at both gates. The R201A mutant appears to induce opening of the G-loop gate in the absence of G-protein.[52] When PIP_2 is then added the channel opens at both G-loop and transmembrane gates. In this conformation both gates are in close proximity, while the inner transmembrane helices are splayed apart, focused at the glycine-hinge.

4.3.3   K2P Channels

K2P channels are responsible for the background or leak currents of K^+ ions that stabilise the negative electrical (resting) potential of the cell. There are 15 K2P channels in humans, which are separated into six different subtypes, depending on the sequence identity, modulation and function: TWIK, TREK, TASK, TALK, THIK and TRESK. K2P channels are modulated by pH, mechanical stress, temperature, anaesthetics and antidepressants.[61]

Recently two structures of members of the K2P channel family have been crystallised.[62,63] Unlike the majority of K^+ channels, which are tetrameric, K2P channels are comprised of only two subunits, with a single subunit consisting of two selectivity filter sequences. The structures of TRAAK and TWIK were both resolved with the helix-bundle in the open conformation. Both structures share a pore domain structure that is similar to that previously observed. They also contain a novel turret region on the extracellular side of the channel, formed by the linker between the M1 and P1 helices (extracellular cap or self-interacting domain [SID]). As previously suggested by modelling studies this region is composed of a helix-turn-helix structure,[64] which interacts with the adjacent subunit on the turn, either through H-bonds or frequently through a disulphide bridge. This region is predicted to reduce the susceptibility of the channel to extracellular toxins and tetra-alkylammonium compounds, providing side windows for the passage of ions. There are also unique regions at the ends of the inner helices, with M2 connected directly to M3 and M4 followed by an interfacial helix in the TWIK structure. The arrangement of the inner helices presents side fenestrations that connect the hydrophobic core of the membrane to the central cavity and provides a potential passageway for hydrophobic molecules. Despite the dimeric nature of the channel the selectivity filter contains four-fold rotational symmetry, suitable for the octahedral coordination of a K^+ ion.

4.3.4 Kv Channels

Kv channels have a similar pore domain to that previously described for other members of the family, formed by the fifth and sixth TM (S5-S6). The remaining four TM helices (S1-S4) comprise the voltage-sensor domain (VSD) and move in response to changes in membrane potential. The N-terminal domain in most Kv channels is responsible for the tetramerisation of the channel (T1-domain) and often associates with a beta-subunit. Structures for both units and their assembly were first resolved components for Kv channels.[65–67] The transmembrane domain was resolved a few years later, for the KvAP channel from archaebacteria, as either the full TM structure (S1–S6) or as an isolated VSD (S1–S4).[30] While the VSD structure appeared to be well ordered, the VSD in the full channel structure appeared unusually distorted, in an orientation that would not correctly span the membrane. Initially the FAB fragments were believed to induce the misfolding of the structure. This has latterly been shown to be incorrect, with the protein suggested to require the membrane environment to retain its stability.[68]

A stable structure was subsequently resolved for Kv1.2 from rat.[69] This structure presented the full transmembrane structure in complex with both its T1-domain and cytosolic beta-subunit. As there is no membrane potential applied during X-ray crystallography this channel is resolved in the open state, with the S6 helices in the pore domain splayed apart and VSD, in particular the S3–S4 'paddle', in the up-state. In addition to the glycine hinge, the S6 helices contain a well-conserved Pro-X-Pro motif that is also responsible for channel gating. This region tightly packs against an interfacial helix between the pore and voltage-sensing domains; called the S4–S5 linker. This helix is pulled or pushed by the motions of the S4 helix in response to changes in the membrane potential. The motions of S4 are controlled by the conserved basic residues every fourth residue on the helix, which are shielded from the hydrophobic core of the membrane either by snorkelling towards the phospholipid headgroups or through interactions with the acidic residues on the other three TM helices in the VS domain. The latter contacts were not fully resolved until the determination of a Kv1.2/2.1 chimeric structure (KvChim).[24] A hydrophobic seal at the centre of the VS domain also appears to be crucial to the gating mechanism and also to prevent proton leakage.[70] A highly conserved phenylalanine residue in S2, which is found in Kv, Nav, Cav, Hv proton channels and voltage-sensitive phosphatases (VSP), forms the main component of the gating charge transfer centre. A rigid cyclic ring is believed to be required at this site to control the motions of the S4 helix in response to changes in voltage. The first four S4 basic residues in the activated channel are extracellular to this residue, with the fifth positive charge (K5) found below the ring structure, interacting with two highly conserved acidic residues on S2 and S3. The accompanying electrophysiological studies suggest that the K5 lysine tightly binds at this site to stabilise the open conformation of the channel. When the first S4 arginine (R1) in Shaker is mutated to lysine the closed state appears to be favoured. It is therefore proposed that R1 in the closed state binds to the same site as K5 in the open state. Molecular simulation studies have investigated the transition of the voltage-sensor from the activated to resting state. The dynamics suggest that the S4 helix undergoes a rotation and change in secondary structure from α- to 3_{10}-helix in order to negotiate the cyclic ring of F233, by thinning and elongating the helix, whilst also retaining the charge–charge interactions with the other helices in the voltage-sensor.[71] Other studies have attempted to generate a consensus model for the resting state of the VSD based on the vast amount of experimental data.[72]

4.3.5 Voltage-gated Sodium (Nav) Channels

The overall architecture of Kv channels is conserved in Nav channels, although the eukaryotic versions of these proteins are comprised of a single subunit, containing four six-transmembrane pseudodomains. There are tetrameric relatives of these channels within the bacterial genome. One such channel from *Arcobacter butzleri*, termed NavAb, has been solved (see also Chapter 5 by Catterall) (Figure 4.2(D)).[73]

Although the VS domain of this structure is in the activated conformation the pore domain remains locked closed, seemingly due to disulphide bridges between the inner S6 helices in the mutated construct used for the initial crystallography. Interestingly the wild-type structure and a structure of a homologous channel, NavRh, have also been resolved with the pore closed, despite the activated conformation of the voltage-sensor.[74,75] Indeed, the VSD of the NavRh structure is even more activated than the wild-type NavAb structure, with the innermost S4 basic residue (analogous to K5) lying above, rather than below, the gating charge transfer centre discussed in the previous section.[75] The mechanism by which the pore is uncoupled from the VSD in these structures, to allow this 'inactivated' state, remains uncertain.

These discrepancies may be reconciled by structures of the pore in its open conformation. An open state structure is expected in the near future for a simplified pore-domain construct of the NachBac channel, from *Bacillus halodurans*.[76]

The comparison of these structures should allow an enhanced understanding of how gating occurs in Na$^+$ channels and by homology allow a better grasp of the function and drug interactions of mammalian Nav, Cav and Transient Receptor Potential (TRP) channels.

4.4   Ligand Gated Ion Channels

This section will consider the three primary superfamilies of ionotropic ligand-gated receptors for which we have high-resolution structural images (Figure 4.4).

4.4.1   Glutamate Receptors

Glutamate is the primary excitatory neurotransmitter within the mammalian central nervous system, which acts on either metabotropic (G-protein coupled) or ionotropic (ion channel) receptors (iGluRs). The ionotropic members of this family contain ligand-binding domains (LBD) for glutamate that are coupled to membrane embedded cation-selective channels. The activity of these ion channels is central to normal nervous system function and as a result a number of diseases including Alzheimer's, Huntington's and Parkinson's, disorders such as schizophrenia, and the damage to neurons

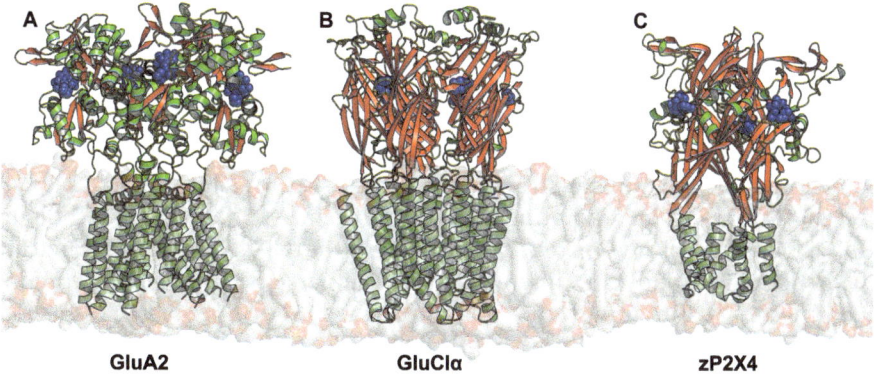

GluA2   GluClα   zP2X4

Figure 4.4   Ligand-gated ion channels. Ligands are shown as blue spheres and the lipid bilayer as a grey surface. (A) The Glutamate receptor, rGluA2, in the closed state. The competitive antagonist, ZK200775, is bound at the glutamate-binding site. (B) The Cys-loop receptor, GluClα, in its activated conformation, with glutamate bound to the receptor. The ATD is not shown to allow better comparison with the other two structures. (C) The Ionotropic Purinergic receptor, zP2X4, in its open state. Adenosine triphosphate (ATP) is bound to the ectodomain.

that accompanies stroke and traumatic brain injury result from malfunction of these ion channels.[77] Glutamate-gated ion channels can be split into four subtypes, principally named by the original ligand used to identify the channel subtype: AMPA (GluA1–GluA4), NMDA (GluN1, GluN2A–GluN2D, GluN3A, and GluN3B) and Kainite (GluK1–GluK5). The fourth members are the δ receptor subunits (GluD1 and GluD2).[78] Despite their specificity for different agonists all subtypes are physiologically activated by glutamate.

The discovery of a K^+ selective glutamate receptor (GluR0) in prokaryotes suggested that the eukaryotic glutamate channels possess a similar pore domain to K^+ channels despite their loss of K^+ selectivity.[79] The finding also revealed that GluR channels are inverted with respect to K^+ channels, with the selectivity filter and half-helix on the intracellular side and gate on the extracellular face of the membrane, and are therefore also members of the Pore-loop superfamily. The unearthing of this bacterial ancestor a year after the resolution of the KcsA channel provided hope that a structure may also be forthcoming for a full glutamate receptor.

In the meantime structures appeared for the ligand-binding domain of the glutamate family by using the ingenious trick of removing the transmembrane (TM) domain and introducing a short linker in its place (see Chapter 6 by Ward).[80] This allowed the crystallisation of only the LBD, with the amino-terminal domain (ATD) also cleaved in the constructs used for crystallography. This protocol allows the resolution of a variety of different compounds within the LBD of glutamate receptors, permitting the structural pharmacology of ligands bound to different members in this superfamily of receptors.

Since the initial structure was resolved in 1998, over 100 structures have appeared for the LBDs of glutamate receptors. A number of structures have also been solved for the ATD.[81–83] The LBD structure can be split into two separate domains—D1 and D2. In the dimeric structure the D1 domain forms the principal contacts between subunits. The binding of glutamate induces a clamshell like motion of the two domains, with D2 clamping down on the relatively static D1 domain.

The first near-complete structure of a GluR was resolved for the GluA2 receptor from rat, incorporating the ATD, LBD and TM domains.[84] Although the channel is formed of four subunits only the TM domain is tetrameric with both ATD and LBD interacting in a dimers-of-dimers configuration (Figure 4.4(A)). Intriguingly the subunit pairing in the dimer-of-dimers is not the same for both domains, *e.g.* while subunits A and B interact in the LBD, subunit B interacts with C in the ATD. The nature of this entwined assembly is of especial interest in heteromeric channels, such as the NMDA family, which are comprised of two GluN1 subunits, which bind glycine, and two GluN2, which bind glutamate. The structural study also provided evidence that these subunits are arranged GluN1-N2-N1-N2, with N1 found at the A and C subunit positions, while N2 is at B and D.[84]

The channel was resolved in its closed conformation, with an antagonist (CNQX) bound to the LBD. For this structure the first three helices (M1 to M3) are comparable to the structure of a closed K channel—outer (M1), pore half-helix (M2) and inner (M3). The fourth helix (M4) sits on the outside of the TM bundle. The region that is expected to form the selectivity filter of the channel is not resolved in the structure.

Each extracellular loop connecting the TM helices is interwoven in the LBD, with the ATD located above the LBD. The binding of glutamate and clamshell motions of the LBD are expected to pull on the linkers to the TM domains and thereby induce a conformational rearrangement of the M3 helices that open the channel pore. After a period of activation these receptors undergo desensitisation, whereby the LBD relax to a conformation that relinquishes the tension imposed on the TM helices and induces channel closure. Structures of the LBD have been resolved for the desensitised state, but none have yet been forthcoming for the full channel complex.[85]

Future structural studies should hopefully resolve the full channel construct at a higher resolution and in different conformational states, ideally with different ligands bound. Trapping the channel in the open state will likely require a non-desensitizing mutant channel or use of compounds that block desensitisation or promote channel opening.

4.4.2 Cys-loop Receptors

The Cys-loop superfamily includes receptors for acetylcholine (ACh), serotonin (5-HT), glycine and gamma-aminobutyric acid (GABA) that are responsible for fast excitatory/cationic (5-HT$_3$ and AchR) and inhibitory/anionic (GABA$_A$ and GlyR) transmission in the peripheral and central nervous systems.[86] These channels have long been targets for compounds used in anaesthesia, or even further back as the molecular target for the paralyzing poison, curare, which was coated onto arrows or blowgun darts by the indigenous people of South America.

For many years, the electric organ of the electric eel has provided an enticing prospect for understanding the electrical properties of molecular cells. Within this organ there is a dense population of Ach receptors, which has been utilised for both electrophysiological and structural studies. Indeed the AchR from this organ was the first membrane protein to be isolated.[87] Whilst it also provided one of the earliest images of an ion channel structure, albeit to 17 Å resolution,[88] and then subsequently refined to 9 Å.[89] Further work from the same lab resulted in higher resolution images using electron diffraction methods, culminating in a 4 Å structure of the receptor, initially for the transmembrane domain[90] and latterly the entire complex.[91]

This was not before an X-ray structure appeared for a homologous ligand-binding domain (AchBP) from the snail, *Lymnaea stagnalis*.[92] This soluble protein lacks the TM domain and is found in cholinergic synapses, secreted by glial cells to moderate the level of neurotransmitters in the synaptic cleft.[93] This structure bears strong homology to the ectodomain of the full Ach receptor, being principally comprised of long β-strands.

The first X-ray structures to contain both transmembrane and extra-cellular domains were for two cation-selective bacterial homologues, from *Erwinia chrysanthemi* (ELIC) and *Gloebacter violaceous* (GLIC), in their resting/desensitised and activated states, respectively.[94–96] By comparing the two structures it is possible to propose how the binding of ligand, 30 Å from the TM domain, induces pore gating in Cys-loop receptors. In the ELIC structure the second TM helices form a constriction point towards the extracellular side of the membrane, preventing ionic flux. This constriction is removed in GLIC, as the M2 helices tilt outwards through a quaternary twist to increase the diameter of the extracellular side of the pore. In turn this reduces the diameter of the intracellular side. In simplest terms the binding of a ligand, protons in the case of GLIC, to the ectodomain appears to induce a conformational rearrangement of the first two β-strands (β1-β2) in a subunit, which couples to the loop connecting TM2 to TM3 and induces the outward motions of both the TM helices to open the channel.[95,96] In these studies GLIC was identified as being proton sensitive. ELIC, on the other hand, was initially described as an orphan receptor, with uncertainty surrounding the nature of its activatory ligand. ELIC has since been shown to activate in response to the binding primary amines that include the neurotransmitter GABA, but not tertiary amines such as Ach, which act as antagonists.[97–99]

Due to the initial absence of the ligand for ELIC, much of the subsequent structural studies have been performed on GLIC. These studies include the identification of binding sites for the general anaesthetics propofol and desflurane, towards the centre of a single subunit's four-helix TM bundle, closely packing against the pore-lining second TM helix (M2).[100] The study proposes that the anaesthetics act by binding to the closed state of the channel and preventing the conformational switch to the open state. A similar binding site has also been proposed for alcohols, with an F to A mutation inducing alcohol sensitivity upon GLIC.[101] In both cases lipids are likely to influence both accessibility and stability of the compounds in their binding sites. Further structural studies of GLIC have used cross-linking mutations to promote channel closure, with a variety of crystal structures revealing a locally closed state conformation of this channel, through re-arrangement of the M2 helices.[102]

More recently a structure was resolved for the anionic glutamate-gated chloride channel α (GluClα) from *Caenorhabditis elegans*, which bears closest homology to the GlyR subset of channels (Figure 4.4(B)).[103] Along with glu-tamate bound to the ligand binding domain, this set of structures depict the interactions made between the receptor and both an allosteric activator, ivermectin, which is used to treat river blindness and a channel blocker, picrotoxin. Ivermectin is found to bind between subunits on the extracellular side of the TM domain, in close proximity to the anaesthetic binding site and with extensive exposure to the lipid bilayer. Picrotoxin, on the other hand, binds to the cytosolic side of the TM pore, close to the selectivity filter of the channel. Picrotoxin has also been shown to block the bacterial homologues.[104] Lidocaine and tetraalkylammonium compounds have all been show to bind to GLIC in a similar fashion.[105]

Ion selectivity in these channels principally derives from the net charge at the N-terminal end of TM2. Comparable to the mechanism used in chloride channels (CLC) the anion selectivity stems from the positive helix dipole at the intracellular mouth of the pore. In cation channels, five glutamate residues, one from each subunit, counteract the net dipole charge to induce cation selectivity. Comparison of the GLIC and GluCl structures also reveals that the funnel within the ectodomain is electropositive in the anion channels and electronegative in cation family members.[103]

The determined structures provide a means of extrapolating by homology the interactions of drugs with mammalian members of the Cys-loop family. Of especial interest is the α7 nicotinic receptor, for which a number of molecular models have appeared.[106]

4.4.3 Ionotropic Purinergic Receptors (P2X) and their Relatives

Purinergic ionotropic (P2X) channels provide a third major class of ligand-gated ion channels. This subset of receptors is found exclusively in eukaryotes, and therefore has no tractable bacterial homologues for structural studies. They are involved in a wide number of physiological processes, including nociception, synaptic transmission, smooth muscle contraction, taste and inflammation and therefore are a potential therapeutic target for cardiovascular, inflammatory and neuronal disease. There are seven subtypes of P2X$_{(1-7)}$ receptor, which vary in their localisation, expression, pharmacology and activity.[107]

X-ray crystallography studies of the P2X family have yielded high-resolution structural data. In 2009 two structures appeared for the zebrafish P2X4 channel in its closed, apo state.[108] This structure is trimeric, with each subunit comprised of a large ectodomain interconnecting two transmembrane helices, the latter of which (TM2) lines the pore of the channel. The architecture of a single subunit is said to resemble the shape of a dolphin, with the TM domains acting as the tail/fluke, a set of long β-strands comprising the body and extending toward the head domain, which contains a number of disulphide bridges previously reported.[109] Other structural features extend from the body and are annotated based on the dolphin analogy as left and right flippers and the dorsal fin.

More recently a zP2X4 construct with ATP bound has revealed the structure of the channel in its activated state (Figure 4.4(C)).[110] The same study also produced a higher resolution structure of the apo state that fixes a residue register error for a small portion of the original structure. By comparing the two states observed in the crystal structures it is possible to draw a mechanism for the ATP-driven gating of the channel. Upon binding of ATP to an inter-subunit cleft the left flipper is pushed out of the site by the phosphate groups, while the head and dorsal fin domains collapse around the ligand-binding site. In response there are rigid body motions of the lower body domain, separating the three subunits and widening the extracellular vestibule above the TM domain to a diameter of ~ 10 Å. These motions are directly coupled to the TM helices, ratcheting them apart in an iris-like motion to open the channel pore.

Unlike the Cys-loop superfamily, the pore of the channel does not extend through the ectodomain;[111] instead large inter-subunit fenestrations are found at the membrane interface. These side portals extend to a large acidic central vestibule that feeds into the TM pore. The TM pore in the X-ray structure is predominantly lined by hydrophobic residues, except for a highly conserved aspartate residue at the C-terminal end of TM2, which is likely to mediate the channel's cation selectivity.

Acid/amiloride sensitive ion channels (ASIC) share a similar topology to P2X channels, despite low sequence homology. These channels are also of therapeutic interest as antagonists of these channels can reduce ischemic cell death after stroke. A number of crystal structures have been resolved for the closed and desensitised states of this protein.[112,113] This protein bears close homology to the epithelial Na^+ channel (ENaC), which is involved in mucociliary clearance within the lung and is of especial therapeutic interest for diseases such as cystic fibrosis. As there is not an X-ray structure of either channel, homology models have been generated for ENaC, using ASIC as a template.[114]

4.5 Ion Channels with the Transporter Scaffold

There are a number of ion channels that are effectively broken transporters or have used transporter function to modulate channel activity. This section will briefly describe three specific examples of ion channels that bear close homology to known transporter structures.

4.5.1 Chloride Channels (CLC)

The first example is the chloride channel/transporter (CLC) family, which were first identified in 1979[115] and later isolated in 1990.[116] There are nine human members of this family, with many involved in diseases, such as Bartter's syndrome and Dent's, Thomsen and Becker diseases. A structural understanding of this family of proteins is somewhat ambiguous as the majority of functional data is based on the subset of eukaryotic voltage-sensitive *channel* members (CLC-0, CLC-1 and CLC-2), while all the high-resolution structures are for bacterial H^+/Cl^- coupled *transporters*.[117] Despite the uncertainty surrounding the ion conduction mechanism, the overall topology is conserved across the CLC family, with the channels essentially acting as broken transporters.[118] All are formed of a two-fold symmetrical homodimer, with each monomer containing a single ion pathway. The structures of two bacterial homologues, ecCLC and stCLC, were resolved in 2002.[119] Like K^+ channels a half-helix dipole is important for forming the ionic pathway. In CLC channels and transporters this is inverted, with respect to K^+ channels, with the N-terminal ends coordinating the permeation path. The first sets of structures were initially deemed to function as a channel until later studies revealed that they were in fact H^+/Cl^- antiporters.[120]

The first eukaryotic structure of a CLC transporter (CmCLC) discovered that a single glutamate residue, which has long been known to be functionally and structurally important, may act as the sole obstruction along the pore pathway.[121] This residue has previously been referred to as the gating or extracellular glutamate (Glu_{ex}) and is essential for both the coupling between H^+ and Cl^- in CLC transporters and for voltage-dependent gating in CLC channels. Mutating the Glu_{ex} to glutamine opens the extracellular mouth of the transporter and is suggested to represent the conformation of the transporter when Glu_{ex} is protonated.[122] This conformation in the CmCLC structure allows a continuous ionic pathway across the membrane that is more a hallmark of an ion channel.[121] The CLC transporters retain their function, as it is believed that it is energetically more favourable for Glu_{ex} to lose its proton than for more than two Cl^- ions to enter the pore pathway. The energetic barrier for Cl^- ion entry is expected to be lower or even absent in voltage-sensitive CLC channels, retaining the open ion pathway and allowing permeation rates to reach 10^{7-8} ions/s. This matter may remain a cause of debate until the structure of a CLC channel is resolved.

4.5.2 Cystic Fibrosis Transmembrane Regulator (CFTR)

A second chloride channel that is essentially a broken transporter is the Cystic Fibrosis Transmembrane Regulator (CFTR), which is homologous to the ATP-binding Cassette (ABC) family of transporters. Genetic mutations to this channel induce Cystic Fibrosis (CF) by either preventing channel opening, or in more severe cases preventing correct channel folding and expression at the plasma membrane. This reduces chloride ion influx and thereby inhibits mucociliary clearance.

There are currently a number of X-ray structures for the cytosolic ABC region of the channel,[123] including the defective ΔF508 construct[124–126] and other F508 mutants.[127] Electron microscopy (EM) maps have also been solved for the whole channel,[128] with the bacterial ABC transporter, Sav1866, modelled into the density.[129] The Sav1866 structure has also been used to construct homology models of CFTR in its open, nucleotide-bound state.[130,131]

4.5.3 ATP-Sensitive K^+ Channel (KATP)

Finally, an ABC transporter called the sulfonylurea receptor (SUR) is found to associate with a Kir channel, Kir6.2, to form the ATP-Sensitive K^+ Channel (KATP) complex.[132,133] Mutations to either component of the complex can induce another genetic disorder, called neonatal diabetes (ND), in which the channel remains open despite the presence of ATP, which is responsible for channel closure. Consequently, this prevents the cascade of events that leads to insulin secretion.[134] This channel has long been a target for sulfonylurea drugs, which were originally identified for treatment of Type II diabetes. These drugs have recently been repurposed to treat ND, due to their ability to induce closure of the mutant channels and thereby stimulate insulin secretion.[135] Structural, molecular modelling and electrophysiological studies have elucidated a binding site for ATP, the small molecule that induces channel closure,[136] and proposed an oligomeric structure for the KATP complex.[137] Despite this recent progress, a structural image of the binding site for sulfonylurea drugs remains elusive and a full structural understanding of many of the ND mutations remains incomplete.

4.6 Conclusions

The ion channels discussed in this chapter are not the only channels for which we have high-resolution structural information. There are also structures of a homologue of the plant SLAC1 anion channel (TehA),[138] mechanosensitive ion channels (MscL, MscS),[139–142] proton (M2) channels from the influenza virus[143–145] and light-gated channelrhodposin (ChR) cation channels.[146] Indeed we now have representative high-resolution structures for the majority of the ion channel superfamilies. The major absentees are for the superfamilies of ion channels that permit calcium release from intracellular organelles, which include the inositol-triphosphate (IP$_3$R) and ryanodine receptors (RyR), and the Ca^{2+}-release activated Ca^{2+} (CRAC) channels. There are structures for the IP$_3$R and RyR soluble domains, but none that include the transmembrane domain.[147–149] In a number of cases higher homology between the representative structure and channel of interest would be preferred. Furthermore, a range of structures captured in a variety of states, or with other ligands bound, would further aid the understanding of an ion channel's molecular function. A comprehensive list of currently resolved ion channel structures can be found online at http://blanco.biomol.uci.edu/mpstruc/listAll/list (Table 4.1).

The recent exponential expansion in the number and nature of ion channel structures has indicated that structural biology lies excitingly on the verge of providing further structures for not only known drug targets, but also future ion channel candidates. Until then the resolution of ion channels with novel folds and structures with more significant homology will allow better understanding of disease-related and drug-targeted proteins. The past 15 years of high-resolution ion channel structures have enlivened an already exciting field. Hopefully the next 15 years will be just as enlightening as more novel structures emerge to support our understanding of the molecular complexity of ion channel diseases.

Table 4.1 High resolution structures for the TM domains of Ion channels.

Name	Species	Key Points	PDB	Ref
Potassium Channel Pore Domain				
KcsA	*Streptomyces lividans*	Closed 3.2 Å	1BL8	16
		Closed 2.0 Å	1K4C	27
		Inactivated	1K4D	
		Open	3F5W	150
MthK	*Methanothermobacter*	Open	1LNQ	29
	thermautotrophicus		3LDC	33
NaK	*Bacillus cereus*	Closed	2AHY	37
		Mutations	2Q67	151
		Open	3E86	152
		NaK2K Mutations	3OUF	38
		CNG Mutations	3K0D	39
Voltage-Sensitive Potassium Channels				
KvAP	*Aeropyrum pernix*	S1-S6	1ORQ	30
		S1-S4	1ORS	
		S1-S4	2A0L	68
		NMR S1-S4	2KYH	153
Kv1.2	*Rattus norvegicus*	Open State	2A79	69
KvChim	*Rattus norvegicus*	Chimera	2R9R	24
		F233W Mutant	3LNM	70
MlotiK1	*Mesorhizobium loti*	Closed State (not voltage-sensitive)	3BEH	154
Two-pore Potassium Channels				
TWIK-1	*Homo sapiens*	Open State	3UKM	62
TRAAK	*Homo sapiens*	Open State	3UM7	63
Inward Rectifier Potassium Channels				
KirBac1.1	*Burkholderia pseudomallei*	Closed State	1P7B	28
KirBac3.1	*Magnetospirillum*	Multi-state	2WLK	53
	magnetotacticum	S129R Mutant	3ZRS	155
KirChim	*Mus musculus/Burkholderia xenovornas*	Closed State	2QKS	23
Kir2.2	*Gallus*	Closed State	3JYC	54
		Closed with PIP_2	3SPI	51
Kir3.2	*Mus musculus*	Closed with PIP_2	3SYA	52
		R201 Mutant Open with PIP_2	3SYQ	
Sodium Channels				
NavAb	*Arcobacter butzleri*	I217C Mutant	3RVY	73
		Wild-Type	4EKW	74
NavRh	*Alphaproteobacterium HIMB114*	Wild-Type	4DXW	75
Ionotropic Glutamate Receptors				
GluA2	*Rattus norvegicus*	Closed State	3KG2	84
Cys-Loop Receptors				
Nicotinic Receptor (EM)	*Torpedo marmorata*	TM Domain	1OED	90
		Full Structure	2BG9	91
ELIC	*Erwinia chrysanthemi*	Closed State	2VL0	94
		with Acetylcholine	3RQW	97

Table 4.1   (*Continued*)

Name	Species	Key Points	PDB	Ref
GLIC	*Gloebacter violaceus*	Open State	3EHZ	95
		Open State	3EAM	96
		with Lidocaine	2XQ3	105
		with anaesthetics	3P50	100
GluClα	*Caenorhabditis elegans*	With ivermectin and glutamate	3RIF	103
		With picrotoxin	3RI5	
Ionotropic Purinergic Receptors and their Relatives				
P2X4	*Danio rerio*	Closed State	3H9V	108
		Closed State	4DW0	110
		with ATP Open State	4DW1	
ASIC2	*Gallus gallus*	Closed State	2QTS	112
		Desensitised	3HGC	113
Other Ion Channels				
M2 proton channel	*Influenza A*	X-ray Structure	3BKD	144
		Solution NMR	2RLF	143
		Solid State NMR	2L0J	145
TehA	*Haemophilus influenzae*		3M71	138
CLC H^+/Cl^- Transporter	*Escherichia coli*	Wild-type	1KPK	119
		Glu_{Ex} Mutants	1OTU	122
		Monomer	3NMO	156
	Cyanidioschyzon merolae	Eukaryotic	3ORG	121
Channelrhodopsin	*Chlamydomonas reinhardtii*	Closed State	3UG9	146
MscL	*Mycobacterium tuberculosis*	Closed State	2OAR	139
	Staphylococcus aureus	Intermediate State	3HZQ	142
MscS	*Escherichia coli*	Closed State	2OAU	140
		Open State	2VV5	141

References

1. A. L. Hodgkin and A. F. Huxley, *J. Physiol.*, 1952, **117**, 500–544.
2. A. L. Hodgkin and A. F. Huxley, *J. Physiol.*, 1952, **116**, 497–506.
3. A. L. Hodgkin and A. F. Huxley, *J. Physiol.*, 1952, **116**, 473–496.
4. A. L. Hodgkin and A. F. Huxley, *J. Physiol.*, 1952, **116**, 449–472.
5. A. L. Hodgkin and R. D. Keynes, *J. Physiol.*, 1955, **128**, 61–88.
6. B. Hille, C. M. Armstrong and R. MacKinnon, *Nat. Med.*, 1999, **5**, 1105–1109.
7. B. Hille, *Prog. Biophys. Mol. Biol.*, 1970, **21**, 1–32.
8. F. Bezanilla and C. M. Armstrong, *J. Gen. Physiol.*, 1972, **60**, 588–608.
9. E. Neher and B. Sakmann, *Nature*, 1976, **260**, 799–802.
10. B. Hille and W. Schwarz, *J. Gen. Physiol.*, 1978, **72**, 409–442.
11. C. M. Armstrong, *J Gen Physiol.*, 1971, **58**, 413–437.
12. C. M. Armstrong and B. Hille, *J. Gen. Physiol.*, 1972, **59**, 388–400.
13. B. L. Tempel, Y. N. Jan and L. Y. Jan, *Nature*, 1988, **332**, 837–839.
14. J. Deisenhofer, O. Epp, K. Miki, R. Huber and H. Michel, *Nature*, 1985, **318**, 618–624.

15. R. Milkman, *Proc. Natl. Acad. Sci. U. S. A.*, 1994, **91**, 3510–3514.
16. D. A. Doyle, J. M. Cabral, R. A. Pfuetzner, A. Kuo, J. M. Gulbis, S. L. Cohen, B. T. Cahit and R. MacKinnon, *Science*, 1998, **280**, 69–77.
17. H. M. Berman, J. Westbrook, Z. Feng, G. Gilliland, T. N. Bhat, H. Weissig, I. N. Shindyalov and P. E. Bourne, *Nucl. Acids Res.*, 2000, **28**, 235–242.
18. S. H. White, *Nature*, 2009, **459**, 344–346.
19. E. P. Carpenter, K. Beis, A. D. Cameron and S. Iwata, *Curr. Opin. Struct. Biol.*, 2008, **18**, 581–586.
20. T. Kawate and E. Gouaux, *Structure*, 2006, **14**, 673–681.
21. D. Drew, S. Newstead, Y. Sonoda, H. Kim, G. von Heijne and S. Iwata, *Nat. Protoc.*, 2008, **3**, 784–798.
22. T. Warne, M. J. Serrano-Vega, C. G. Tate and G. F. Schertler, *Protein Expression Purif.*, 2009, **65**, 204–213.
23. M. Nishida, M. Cadene, B. T. Chait and R. MacKinnon, *Embo J.*, 2007, **26**, 4005–4015.
24. S. B. Long, X. Tao, E. B. Campbell and R. MacKinnon, *Nature*, 2007, **450**, 376–382.
25. V. Cherezov, D. M. Rosenbaum, M. A. Hanson, S. G. Rasmussen, F. S. Thian, T. S. Kobilka, H. J. Choi, P. Kuhn, W. I. Weis, B. K. Kobilka and R. C. Stevens, *Science*, 2007, **318**, 1258–1265.
26. H. Schrempf, O. Schmidt, R. Kummerlen, S. Hinnah, D. Muller, M. Betzler, T. Steinkamp and R. Wagner, *Embo J.*, 1995, **14**, 5170–5178.
27. Y. Zhou, J. H. Morais-Cabral, A. Kaufman and R. MacKinnon, *Nature*, 2001, **414**, 43–48.
28. A. Kuo, J. M. Gulbis, J. F. Antcliff, T. Rahman, E. D. Lowe, J. Zimmer, J. Cuthbertson, F. M. Ashcroft, T. Ezaki and D. A. Doyle, *Science*, 2003, **300**, 1922–1926.
29. Y. Jiang, A. Lee, J. Chen, M. Cadene, B. T. Chait and R. MacKinnon, *Nature*, 2002, **417**, 515–522.
30. Y. Jiang, A. Lee, J. Chen, V. Ruta, M. Cadene, B. T. Chait and R. Mackinnon, *Nature*, 2003, **423**, 33–41.
31. J. Åqvist and V. Luzhkov, *Nature*, 2000, **404**, 881–884.
32. J. H. Morais-Cabral, Y. Zhou and R. MacKinnon, *Nature*, 2001, **414**, 37–42.
33. S. Ye, Y. Li and Y. Jiang, *Nat. Struct. Mol. Biol.*, 2010, **17**, 1019–1023.
34. A. N. Thompson, I. Kim, T. D. Panosian, T. M. Iverson, T. W. Allen and C. M. Nimigean, *Nat. Struct. Mol. Biol.*, 2009, **16**, 1317–1324.
35. I. H. Shrivastava, D. P. Tieleman, P. C. Biggin and M. S. P. Sansom, *Biophys. J.*, 2002, **83**, 633–645.
36. L. Ma, X. Zhang and H. Chen, *Sci Signal*, 2011, **4**, ra37.
37. N. Shi, S. Ye, A. Alam, L. Chen and Y. Jiang, *Nature*, 2006, **440**, 570–574.
38. D. B. Sauer, W. Zeng, S. Raghunathan and Y. Jiang, *Proc. Natl. Acad. Sci. U. S. A.*, 2011, **108**, 16634–16639.
39. M. G. Derebe, W. Zeng, Y. Li, A. Alam and Y. Jiang, *Proc. Natl. Acad. Sci. U. S. A.*, 2011, **108**, 592–597.

40. (a) M. G. Derebe, D. B. Sauer, W. Zeng, A. Alam, N. Shi and Y. Jiang, *Proc. Natl. Acad. Sci. U. S. A.*, 2011, **108**, 598–602; (b) J. F. Cordero-Morales, V. Jogini, V. Vásquez, R. W. Bourdeau, H. Yu, B. Roux, M. Tristani-Firouzi and E. Perozo, Engineering the hERG1 Selectivity Filter into the NaK Pore Domain (Poster), *55th Annual Meeting of the Biophysical Society*, Baltimore, MD, 2011.
41. Y. Jiang, A. Lee, J. Chen, M. Cadene, B. T. Chait and R. MacKinnon, *Nature*, 2002, **417**, 523–526.
42. Y. Wu, Y. Yang, S. Ye and Y. Jiang, *Nature*, 2010, **466**, 393–397.
43. P. Yuan, M. D. Leonetti, A. R. Pico, Y. Hsiung and R. MacKinnon, *Science*, 2010, **329**, 182–186.
44. P. Yuan, M. D. Leonetti, Y. Hsiung and R. MacKinnon, *Nature*, 2012, **481**, 94–97.
45. M. Nishida and R. MacKinnon, *Cell*, 2002, **111**, 957–965.
46. S. Pegan, C. Arrabit, W. Zhou, W. Kwiatkowski, A. Collins, P. A. Slesinger and S. Choe, *Nat. Neurosci.*, 2005, **8**, 279–287.
47. J. H. M. Cabral, A. Lee, S. L. Cohen, B. T. Chait, M. Li and R. MacKinnon, *Cell*, 1998, **95**, 649–655.
48. G. M. Clayton, W. R. Silverman, L. Heginbotham and J. H. Morais-Cabral, *Cell*, 2004, **119**, 615–627.
49. M. J. Marques-Carvalho and J. H. Morais-Cabral, *Acta Crystallogr., Sect. F: Struct. Biol. Cryst. Commun.*, 2012, **68**, 337–339.
50. T. I. Brelidze, A. E. Carlson, B. Sankaran and W. N. Zagotta, *Nature*, 2012, **481**, 530–533.
51. S. B. Hansen, X. Tao and R. MacKinnon, *Nature*, 2011, **477**, 495–498.
52. M. R. Whorton and R. MacKinnon, *Cell*, 2011, **147**, 199–208.
53. O. B. Clarke, A. T. Caputo, A. P. Hill, J. I. Vandenberg, B. J. Smith and J. M. Gulbis, *Cell*, 2010, **141**, 1018–1029.
54. X. Tao, J. L. Avalos, J. Chen and R. MacKinnon, *Science*, 2009, **326**, 1668–1674.
55. B. A. Wible, M. Taglialatela, E. Ficker and A. M. Brown, *Nature*, 1994, **371**, 246–249.
56. E. Ficker, M. Taglialatela, B. A. Wible, C. M. Henley and A. M. Brown, *Science*, 1994, **266**, 1068–1072.
57. J. Yang, M. Yu, Y. N. Jan and L. Y. Jan, *Proc. Natl. Acad. Sci. U. S. A.*, 1997, **94**, 1568–1572.
58. D. Bucher, L. Guidoni and U. Rothlisberger, *Biophys. J.*, 2007, **93**, 2315–2324.
59. M. L. Leyland, C. Dart, P. J. Spencer, M. J. Sutcliffe and P. R. Stanfield, *Pflugers Arch.*, 1999, **438**, 778–781.
60. P. J. Stansfeld, R. J. Hopkinson, F. M. Ashcroft and M. S. P. Sansom, *Biochem.*, 2009, **48**, 10926–10933.
61. K. A. Ketchum, W. J. Joiner, A. J. Sellers, L. K. Kaczmarek and S. A. Goldstein, *Nature*, 1995, **376**, 690–695.
62. A. N. Miller and S. B. Long, *Science*, 2012, **335**, 432–436.
63. S. G. Brohawn, J. del Marmol and R. MacKinnon, *Science*, 2012, **335**, 436–441.

64. K. H. Yuill, P. J. Stansfeld, I. Ashmole, M. J. Sutcliffe and P. R. Stanfield, *Pflugers Arch.*, 2007, **455**, 333–348.
65. A. Kreusch, P. J. Pfaffinger, C. F. Stevens and S. Choe, *Nature*, 1998, **392**, 945–948.
66. J. M. Gulbis, S. Mann and R. MacKinnon, *Cell*, 1999, **97**, 943–952.
67. J. M. Gulbis, M. Zhou, S. Mann and R. MacKinnon, *Science*, 2000, **289**, 123–127.
68. S. Y. Lee, A. Lee, J. Chen and R. MacKinnon, *Proc. Natl. Acad. Sci. U. S. A.*, 2005, **102**, 15441–15446.
69. S. B. Long, E. B. Campbell and R. MacKinnon, *Science*, 2005, **309**, 897–902.
70. X. Tao, A. Lee, W. Limapichat, D. A. Dougherty and R. MacKinnon, *Science*, 2010, **328**, 67–73.
71. C. S. Schwaiger, P. Bjelkmar, B. Hess and E. Lindahl, *Biophys. J.*, 2011, **100**, 1446–1454.
72. E. Vargas, F. Bezanilla and B. Roux, *Neuron*, 2011, **72**, 713–720.
73. J. Payandeh, T. Scheuer, N. Zheng and W. A. Catterall, *Nature*, 2011, **475**, 353–358.
74. J. Payandeh, T. M. Gamal El-Din, T. Scheuer, N. Zheng and W. A. Catterall, *Nature*, 2012, **486**, 135–139.
75. X. Zhang, W. Ren, P. DeCaen, C. Yan, X. Tao, L. Tang, J. Wang, K. Hasegawa, T. Kumasaka, J. He, D. E. Clapham and N. Yan, *Nature*, 2012, **486**, 130–134.
76. (a) E. C. McCusker, N. D'Avanzo, C. G. Nichols and B. A. Wallace, *J. Biol. Chem.*, 2011, **286**, 16386–16391; (b) E. C. McCusker, C. Bagnéris, C. E. Naylor, A. R. Cole, N. D'Avanzo, C. G. Nichols and B. A. Wallace, *Nat. Commun.*, 2012, **3**, 1102.
77. R. Dingledine, K. Borges, D. Bowie and S. F. Traynelis, *Pharmacol. Rev.*, 1999, **51**, 7–61.
78. S. F. Traynelis, L. P. Wollmuth, C. J. McBain, F. S. Menniti, K. M. Vance, K. K. Ogden, K. B. Hansen, H. Yuan, S. J. Myers and R. Dingledine, *Pharmacol. Rev.*, 2010, **62**, 405–496.
79. G. Q. Chen, C. Cui, M. L. Mayer and E. Gouaux, *Nature*, 1999, **402**, 817–821.
80. N. Armstrong, Y. Sun, G. Q. Chen and E. Gouaux, *Nature*, 1998, **395**, 913–917.
81. R. Jin, S. K. Singh, S. Gu, H. Furukawa, A. I. Sobolevsky, J. Zhou, Y. Jin and E. Gouaux, *Embo J.*, 2009, **28**, 1812–1823.
82. A. Clayton, C. Siebold, R. J. Gilbert, G. C. Sutton, K. Harlos, R. A. McIlhinney, E. Y. Jones and A. R. Aricescu, *J. Mol. Biol.*, 2009, **392**, 1125–1132.
83. J. Kumar and M. L. Mayer, *J. Mol. Biol.*, 2010, **404**, 680–696.
84. A. I. Sobolevsky, M. P. Rosconi and E. Gouaux, *Nature*, 2009, **462**, 745–756.
85. Y. Sun, R. Olson, M. Horning, N. Armstrong, M. L. Mayer and E. Gouaux, *Nature*, 2002, **417**, 245–253.
86. P. S. Miller and T. G. Smart, *Trends Pharmacol. Sci.*, 2010, **31**, 161–174.
87. J. P. Changeux, M. Kasai, M. Huchet and J. C. Meunier, *C. R. Seances Acad. Sci., Ser. D*, 1970, **270**, 2864–2867.

88. C. Toyoshima and N. Unwin, *Nature*, 1988, **336**, 247–250.
89. N. Unwin, *J. Mol. Biol.*, 1993, **229**, 1101–1124.
90. A. Miyazawa, Y. Fujiyoshi and N. Unwin, *Nature*, 2003, **423**, 949–955.
91. N. Unwin, *J. Mol. Biol.*, 2005, **346**, 967–989.
92. K. Brejc, W. J. van Dijk, R. V. Klaassen, M. Schuurmans, J. van der Oost, A. B. Smit and T. K. Sixma, *Nature*, 2001, **411**, 269–276.
93. A. B. Smit, N. I. Syed, D. Schaap, J. van Minnen, J. Klumperman, K. S. Kits, H. Lodder, R. C. van der Schors, R. van Elk, B. Sorgedrager, K. Brejc, T. K. Sixma and W. P. Geraerts, *Nature*, 2001, **411**, 261–268.
94. R. J. Hilf and R. Dutzler, *Nature*, 2008, **452**, 375–379.
95. R. J. Hilf and R. Dutzler, *Nature*, 2009, **457**, 115–118.
96. N. Bocquet, H. Nury, M. Baaden, C. Le Poupon, J. P. Changeux, M. Delarue and P. J. Corringer, *Nature*, 2009, **457**, 111–114.
97. J. Pan, Q. Chen, D. Willenbring, K. Yoshida, T. Tillman, O. B. Kashlan, A. Cohen, X. P. Kong, Y. Xu and P. Tang, *Nat. Commun.*, 2012, **3**, 714.
98. I. Zimmermann and R. Dutzler, *PLoS Biology*, 2011, **9**, e1001101.
99. A. J. Thompson, M. Alqazzaz, C. Ulens and S. C. Lummis, *Neuro-pharmacology*, 2012, **63**, 761–767.
100. H. Nury, C. Van Renterghem, Y. Weng, A. Tran, M. Baaden, V. Dufresne, J. P. Changeux, J. M. Sonner, M. Delarue and P. J. Corringer, *Nature*, 2011, **469**, 428–431.
101. R. J. Howard, S. Murail, K. E. Ondricek, P. J. Corringer, E. Lindahl, J. R. Trudell and R. A. Harris, *Proc. Natl. Acad. Sci. U. S. A.*, 2011, **108**, 12149–12154.
102. M. S. Prevost, L. Sauguet, H. Nury, C. Van Renterghem, C. Huon, F. Poitevin, M. Baaden, M. Delarue and P. J. Corringer, *Nat. Struct. Mol. Biol.*, 2012, **19**, 642–649.
103. R. E. Hibbs and E. Gouaux, *Nature*, 2011, **474**, 54–60.
104. M. Alqazzaz, A. J. Thompson, K. L. Price, H. G. Breitinger and S. C. Lummis, *Biophys. J.*, 2011, **101**, 2912–2918.
105. R. J. Hilf, C. Bertozzi, I. Zimmermann, A. Reiter, D. Trauner and R. Dutzler, *Nat. Struct. Mol. Biol.*, 2010, **17**, 1330–1336.
106. S. Amiri, K. Tai, O. Beckstein, P. C. Biggin and M. S. P. Sansom, *Mol. Memb. Biol.*, 2005, **22**, 151–162.
107. R. A. North, *Ciba Found. Symp.*, 1996, **198**, 91–105; discussion 105–109.
108. T. Kawate, J. C. Michel, W. T. Birdsong and E. Gouaux, *Nature*, 2009, **460**, 592–598.
109. S. J. Ennion and R. J. Evans, *Mol. Pharmacol.*, 2002, **61**, 303–311.
110. M. Hattori and E. Gouaux, *Nature*, 2012, **485**, 207–212.
111. T. Kawate, J. L. Robertson, M. Li, S. D. Silberberg and K. J. Swartz, *J. Gen. Physiol.*, 2011, **137**, 579–590.
112. J. Jasti, H. Furukawa, E. B. Gonzales and E. Gouaux, *Nature*, 2007, **449**, 316–323.
113. E. B. Gonzales, T. Kawate and E. Gouaux, *Nature*, 2009, **460**, 599–604.
114. O. B. Kashlan and T. R. Kleyman, *Am. J. Physiol. Renal. Physiol.*, 2011, **301**, F684–696.

115. M. M. White and C. Miller, *J. Biol. Chem.*, 1979, **254**, 10161–10166.
116. T. J. Jentsch, K. Steinmeyer and G. Schwarz, *Nature*, 1990, **348**, 510–514.
117. C. Miller, *Nature*, 2006, **440**, 484–489.
118. J. Lisal and M. Maduke, *Nat. Struct. Mol. Biol.*, 2008, **15**, 805–810.
119. R. Dutzler, E. B. Campbell, M. Cadene, B. T. Chait and R. MacKinnon, *Nature*, 2002, **415**, 287–294.
120. A. Accardi and C. Miller, *Nature*, 2004, **427**, 803–807.
121. L. Feng, E. B. Campbell, Y. Hsiung and R. MacKinnon, *Science*, 2010, **330**, 635–641.
122. R. Dutzler, E. B. Campbell and R. MacKinnon, *Science*, 2003, **300**, 108–112.
123. H. A. Lewis, S. G. Buchanan, S. K. Burley, K. Conners, M. Dickey, M. Dorwart, R. Fowler, X. Gao, W. B. Guggino, W. A. Hendrickson, J. F. Hunt, M. C. Kearins, D. Lorimer, P. C. Maloney, K. W. Post, K. R. Rajashankar, M. E. Rutter, J. M. Sauder, S. Shriver, P. H. Thibodeau, P. J. Thomas, M. Zhang, X. Zhao and S. Emtage, *EMBO J.*, 2004, **23**, 282–293.
124. H. A. Lewis, X. Zhao, C. Wang, J. M. Sauder, I. Rooney, B. W. Noland, D. Lorimer, M. C. Kearins, K. Conners, B. Condon, P. C. Maloney, W. B. Guggino, J. F. Hunt and S. Emtage, *J. Biol. Chem.*, 2005, **280**, 1346–1353.
125. H. A. Lewis, C. Wang, X. Zhao, Y. Hamuro, K. Conners, M. C. Kearins, F. Lu, J. M. Sauder, K. S. Molnar, S. J. Coales, P. C. Maloney, W. B. Guggino, D. R. Wetmore, P. C. Weber and J. F. Hunt, *J. Mol. Biol.*, 2010, **396**, 406–430.
126. J. L. Mendoza, A. Schmidt, Q. Li, E. Nuvaga, T. Barrett, R. J. Bridges, A. P. Feranchak, C. A. Brautigam and P. J. Thomas, *Cell*, 2012, **148**, 164–174.
127. P. H. Thibodeau, C. A. Brautigam, M. Machius and P. J. Thomas, *N. Nat. Struct. Mol. Biol.*, 2005, **12**, 10–16.
128. M. F. Rosenberg, L. P. O'Ryan, G. Hughes, Z. Zhao, L. A. Aleksandrov, J. R. Riordan and R. C. Ford, *J. Biol. Chem.*, 2011, **286**, 42647–42654.
129. R. J. Dawson and K. P. Locher, *Nature*, 2006, **443**, 180–185.
130. C. Alexander, A. Ivetac, X. Liu, Y. Norimatsu, J. R. Serrano, A. Landstrom, M. Sansom and D. C. Dawson, *Biochemistry*, 2009, **48**, 10078–10088.
131. Y. Norimatsu, A. Ivetac, C. Alexander, J. Kirkham, N. O'Donnell, D. C. Dawson and M. S. Sansom, *Biochemistry*, 2012, **51**, 2199–2212.
132. J. P. t. Clement, K. Kunjilwar, G. Gonzalez, M. Schwanstecher, U. Panten, L. Aguilar-Bryan and J. Bryan, *Neuron*, 1997, **18**, 827–838.
133. S. Shyng and C. G. Nichols, *J. Gen. Physiol.*, 1997, **110**, 655–664.
134. A. L. Gloyn, E. R. Pearson, J. F. Antcliff, P. Proks, G. J. Bruining, A. S. Slingerland, N. Howard, S. Srinivasan, J. M. Silva, J. Molnes, E. L. Edghill, T. M. Frayling, I. K. Temple, D. Mackay, J. P. Shield, Z. Sumnik, A. van Rhijn, J. K. Wales, P. Clark, S. Gorman, J. Aisenberg, S. Ellard, P. R. Njolstad, F. M. Ashcroft and A. T. Hattersley, *N. Engl. J. Med.*, 2004, **350**, 1838–1849.

135. J. V. Sagen, H. Raeder, E. Hathout, N. Shehadeh, K. Gudmundsson, H. Baevre, D. Abuelo, C. Phornphutkul, J. Molnes, G. I. Bell, A. L. Gloyn, A. T. Hattersley, A. Molven, O. Sovik and P. R. Njolstad, *Diabetes*, 2004, **53**, 2713–2718.

136. S. Trapp, S. Haider, M. S. P. Sansom, F. M. Ashcroft and P. Jones, *EMBO J.*, 2003, **22**, 2903–2912.

137. M. V. Mikhailov, J. Campbell, H. de Wet, K. Shimomura, B. Zadek, R. F. Collins, M. S. P. Sansom, R. C. Ford and F. M. Ashcroft, *EMBO J.*, 2005, **24**, 4166–4175.

138. Y. H. Chen, L. Hu, M. Punta, R. Bruni, B. Hillerich, B. Kloss, B. Rost, J. Love, S. A. Siegelbaum and W. A. Hendrickson, *Nature*, 2010, **467**, 1074–1080.

139. G. Chang, R. H. Spencer, A. T. Lee, M. T. Barclay and D. C. Rees, *Science*, 1998, **282**, 2220–2226.

140. R. B. Bass, P. Strop, M. Barclay and D. C. Rees, *Science*, 2002, **298**, 1582–1587.

141. W. Wang, S. S. Black, M. D. Edwards, S. Miller, E. L. Morrison, W. Bartlett, C. Dong, J. H. Naismith and I. R. Booth, *Science*, 2008, **321**, 1179–1183.

142. Z. Liu, C. S. Gandhi and D. C. Rees, *Nature*, 2009, **461**, 120–124.

143. J. R. Schnell and J. J. Chou, *Nature*, 2008, **451**, 591–595.

144. A. L. Stouffer, R. Acharya, D. Salom, A. S. Levine, L. Di Costanzo, C. S. Soto, V. Tereshko, V. Nanda, S. Stayrook and W. F. DeGrado, *Nature*, 2008, **451**, 596–599.

145. M. Sharma, M. Yi, H. Dong, H. Qin, E. Peterson, D. D. Busath, H. X. Zhou and T. A. Cross, *Science*, 2010, **330**, 509–512.

146. H. E. Kato, F. Zhang, O. Yizhar, C. Ramakrishnan, T. Nishizawa, K. Hirata, J. Ito, Y. Aita, T. Tsukazaki, S. Hayashi, P. Hegemann, A. D. Maturana, R. Ishitani, K. Deisseroth and O. Nureki, *Nature*, 2012, **482**, 369–374.

147. P. A. Lobo and F. Van Petegem, *Structure*, 2009, **17**, 1505–1514.

148. C. C. Tung, P. A. Lobo, L. Kimlicka and F. Van Petegem, *Nature*, 2010, **468**, 585–588.

149. M. D. Seo, S. Velamakanni, N. Ishiyama, P. B. Stathopulos, A. M. Rossi, S. A. Khan, P. Dale, C. Li, J. B. Ames, M. Ikura and C. W. Taylor, *Nature*, 2012, **483**, 108–112.

150. L. G. Cuello, V. Jogini, D. M. Cortes, A. Sompornpisut, M. D. Purdy, M. C. Wiener and E. Perozo, *FEBS Lett.*, 584, 1133–1138.

151. A. Alam and Y. Jiang, *Nat. Struct. Mol. Biol.*, 2009, **16**, 35–41.

152. A. Alam and Y. Jiang, *Nat. Struct. Mol. Biol.*, 2007, **16**, 30–34.

153. J. A. Butterwick and R. MacKinnon, *J. Mol. Biol.*, 2010, **403**, 591–606.

154. G. M. Clayton, S. Altieri, L. Heginbotham, V. M. Unger and J. H. Morais-Cabral, *Proc. Natl. Acad. Sci USA*, 2008, **105**, 1511–1515.

155. V. N. Bavro, R. De Zorzi, M. R. Schmidt, J. R. Muniz, L. Zubcevic, M. S. Sansom, C. Venien-Bryan and S. J. Tucker, *Nat. Struct. Mol. Biol.*, 2012, **19**, 158–163.

156. J. L. Robertson, L. Kolmakova-Partensky and C. Miller, *Nature*, 2010, **468**, 844–847.

CHAPTER 5

Voltage-gated Sodium Channels: Structure, Function, and Molecular Pharmacology

WILLIAM A. CATTERALL

Department of Pharmacology, University of Washington, Campus Mailstop 357280, Seattle, WA 98195-7280
Email: wcatt@uw.edu

5.1 Sodium Channel Function

The primary physiological function of sodium channels is initiation and conduction of action potentials in nerve, muscle, endocrine, and other excitable cells.[1,2] In neurons, sodium channels are localized in high density in axon initial segments, where the action potential begins, and in nodes of Ranvier in myelinated nerves, where they support rapid conduction.[3,4] Lower densities are found in cell bodies, dendrites, and unmyelinated axons.[3,4] The functional properties of sodium channels have been very well defined by biophysical studies using the voltage clamp technique.[1,2] Depolarization of excitable membranes causes activation of sodium channels followed within a few milliseconds by inactivation. Sodium entry through voltage-gated sodium channels is responsible for the rising phase of the action potential in excitable cells and determines the rate of movement of the action potential along nerve and skeletal muscle fibers and across cardiac muscle. Sodium channels have two parallel inactivation processes. Fast inactivation is engaged immediately upon depolarization and is complete within a few milliseconds in most sodium channels. It is reversed within a few

RSC Drug Discovery Series No. 39
Ion Channel Drug Discovery
Edited by Brian Cox and Martin Gosling
© The Royal Society of Chemistry 2015
Published by the Royal Society of Chemistry, www.rsc.org

milliseconds upon repolarization. Slow inactivation becomes prominent in the time frame of 100 milliseconds to 1 second, and it requires seconds of repolarization for recovery. The molecular mechanisms of fast and slow inactivation are distinct as discussed below.

5.2   Sodium Channel Subunit Structure

Sodium channel proteins were first purified from mammalian brain using neurotoxins as specific probes to label the channels. These purified channels are composed of a 260 kD α-subunit in association with one or two auxiliary β-subunits of 33 to 36 kD in size (Figure 5.1).[5,6] The α subunit forms the transmembrane pore and has the primary sites for neurotoxin binding. All three subunits are intrinsic membrane glycoproteins. The purified complex of α- and β-subunits is sufficient to reconstitute voltage-dependent ion conductance in phospholipid vesicles and planar bilayers (Figure 5.1).

The primary amino acid sequence determined from cDNA cloning and sequencing predicted that the sodium channel α-subunit folds into four internally repeated domains (I-IV), each of which contains six transmembrane segments (S1-S6) (Figure 5.2A).[6,7] In each domain, the S1 through S4 segments serve as the voltage-sensing module, and the S5 and S6 segments and the reentrant P loop that folds into the transmembrane region between them serve as the pore-forming module. A large extracellular loop connects the S5 or S6 transmembrane segments to the P loop in each domain, whereas the other extracellular loops are small. Large intracellular loops link

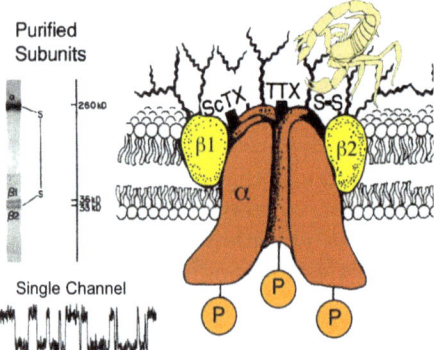

Figure 5.1   Subunit structure of voltage-gated sodium channels. Left: SDS polyacrylamide gel electrophoresis patterns illustrating the α and β subunits of the brain sodium channels. Right: Sodium channel purified from rat brain showing the α, β1, and β2 subunits and their molecular weights. As illustrated, the α and β2 subunits are linked by a disulfide bond. Tetrodotoxin and scorpion toxins bind to the α subunits of sodium channels as indicated and were used as molecular tags to identify and purify the sodium channel protein.[84–86] Inset: Single channel currents conducted by a single purified sodium channel incorporated into a planar bilayer.[87]

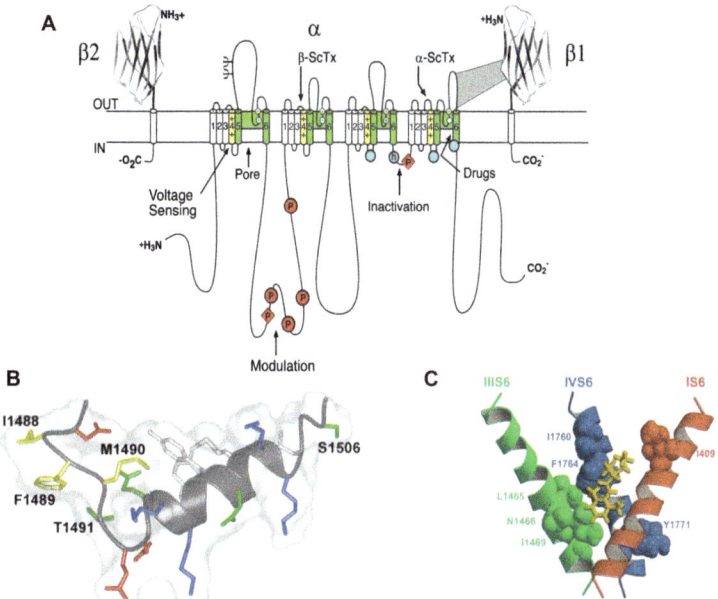

Figure 5.2   Transmembrane organization of sodium channel subunits. (A) The primary structures of the subunits of the voltage-gated ion channels are illustrated as transmembrane folding diagrams. Cylinders represent alpha helical segments. Bold lines represent the polypeptide chains of each subunit with length approximately proportional to the number of amino acid residues in the brain sodium channel subtypes. The extracellular domains of the β1 and β2 subunits are shown as immunoglobulin-like folds. Ψ, sites of probable N-linked glycosylation; P, sites of demonstrated protein phosphorylation by PKA (circles) and PKC (diamonds); shaded, pore-lining S5-P-S6 segments; white circles, the outer (EEDD) and inner (DEKA) rings of amino residues that form the ion selectivity filter and the tetrodotoxin binding site; ++, S4 voltage sensors; h in blue circle, inactivation particle in the inactivation gate loop; blue circles, sites implicated in forming the inactivation gate receptor. Sites of binding of α- and β-scorpion toxins and a site of interaction between α and β1 subunits are also shown. The colors in the transmembrane segments are correlated with the three-dimensional model in Figure 5.3. White and yellow, the voltage-sensing module composed of S1-S4; green, the pore-forming module composed of S5, P, and S6. (B) Inactivation gate structure. Three-dimensional structure of the central segment of the inactivation gate as determined by multi-dimensional NMR. Isoleucine 1488, phenylalanine 1489, and methionine 1490 (IFM) are illustrated in yellow. Threonine 1491, which is important for inactivation, and serine 1506, which is a site of phosphorylation and modulation by protein kinase C, are also indicated. (C) Model of etidocaine binding to the local anesthetic receptor site formed by transmembrane segments IS6, IIIS6, and IVS6 of the Na$_v$1.2 channel. Three-dimensional model of proposed orientation of amino acid residues within the Na$^+$ channel pore with respect to the local anesthetic etidocaine. Only transmembrane segments IS6 (red), IIIS6 (green), and IVS6 (blue) are shown. Residues important for etidocaine binding are shown in space-filling representation.

the four homologous domains, and the large N-terminal and C-terminal domains also contribute substantially to the mass of the internal face of sodium channels. This view of sodium channel transmembrane architecture, originally derived from hydrophobicity analysis and molecular modeling, has been largely confirmed by biochemical, electrophysiological, and structural experiments (see below).

5.3   Sodium Channel Genes

Sodium channels are the founding members of an ion channel superfamily that includes voltage-gated Ca^{2+} channels, TRP channels, voltage-gated, inward rectifying, and two-pore-domain K^+ channels, and cyclic nucleotide-regulated CNG and HCN channels.[8] In evolution, the four-domain sodium channel was last among the voltage-gated ion channels to appear, and it is only found in multicellular organisms. It is thought that sodium channels evolved by two rounds of gene duplication from ancestral single-domain bacterial sodium channels like NaChBac found in *Baccillus halodurans*.[9] Voltage-gated sodium channel genes encoding four-domain channels are present in a wide variety of metazoan species including fly, leech, squid, jellyfish, and all vertebrates.[10] The biophysical properties, pharmacology, gene organization, and even intron-splice sites of the invertebrate sodium channels are largely similar to the mammalian sodium channels, adding further support for the idea that the primordial sodium channel was established before the evolution of vertebrates.

The human, mouse, and rat sodium channel genes and proteins are the best characterized to date. Ten related sodium channel genes are found in vertebrates (Figure 5.3).[11,12] Genes encoding sodium channels $Na_v1.1$, $Na_v1.2$, $Na_v1.3$, and $Na_v1.7$ are localized on chromosome 2 in human and mouse, and these channels share similarities in sequence, biophysical characteristics, blockage by nanomolar concentrations of tetrodotoxin, and broad expression in neurons. A second cluster of genes encoding $Na_v1.5$, $Na_v1.8$, and $Na_v1.9$ channels is localized to human chromosome 3p21-24 and to chromosome 3 in mouse. Although they are more than 75% identical in sequence to the group of channels on chromosome 2, these sodium channels all contain amino acids substitutions that confer varying degrees of resistance to the pore blocker tetrodotoxin. In $Na_v1.5$, the principal cardiac isoform, a single amino acid change from phenylalanine to cysteine in the pore region of domain I is responsible for a 200-fold reduction in TTX-sensitivity compared to those channels on chromosome 2.[13] At the identical position in $Na_v1.8$ and $Na_v1.9$, the amino acid residue is serine, and this change results in even greater resistance to TTX.[14] These two channels are preferentially expressed in peripheral sensory neurons. In comparison to the sodium channels on chromosomes 2 and 3, $Na_v1.4$ which is expressed in skeletal muscle and $Na_v1.6$ which is highly abundant in CNS have greater than 85% sequence identity and similar functional properties, including TTX-sensitivity in the nanomolar concentration range. $Na_v1.4$, and $Na_v1.6$

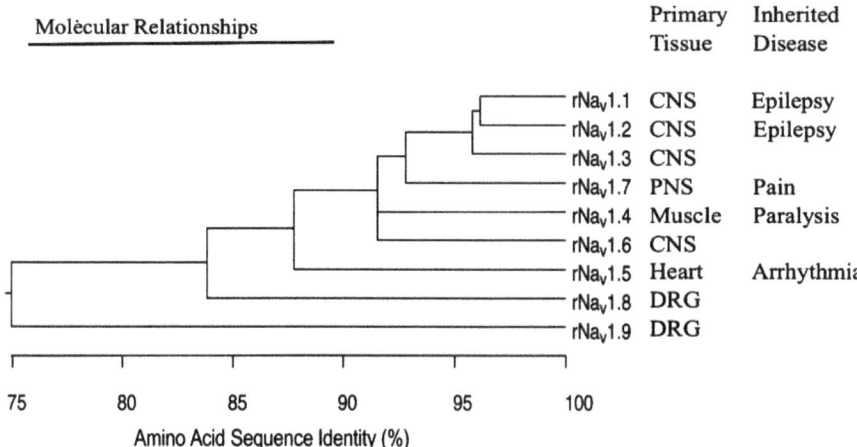

Figure 5.3   Amino acid sequence similarity and phylogenetic relationships of voltage-gated sodium channel α subunits. Phylogenetic relationships of rat sodium channel sequences $Na_v1.1$ - $Na_v1.9$ and Na_x. To perform the analysis, the amino acid sequences for all of the isoforms were aligned and identical residues identified using Clustal W. Divergent portions of the terminal regions and the cytoplasmic loops between domains I-II and II-III were excluded from the analysis. The tree was rooted by including the invertebrate sodium channel sequences.

are located on human chromosome 11 (mouse chromosome 17) and human chromosome 15 (mouse chromosome 12), respectively.

A tenth sodium channel, Na_x, whose gene is located near the sodium channels of chromosome 2, is evolutionarily more distant.[11] Key differences in functionally important regions of voltage sensor and inactivation gate and lack of functional expression of voltage-gated sodium currents in heterologous cells suggest that Na_x may not function as a voltage-dependent sodium channel. Consistent with this conclusion, targeted deletion of the Na_x gene in mice causes functional deficits in sensing plasma salt levels but not in electrical excitability.[15]

The $Na_v\beta$ subunits are encoded by four distinct genes in mammals.[16–19] The β1 and β3 subunits associate noncovalently with the α subunits, whereas β2 and β4 are covalently linked by a disulfide bond. All four β subunits have a single transmembrane segment, a large N-terminal extracellular domain and a small intracellular C-terminal domain (Figure 5.2A). The $Na_v\beta$ subunits are dual-function proteins. Co-expression of $Na_v\beta$ subunits alters the kinetics and voltage dependence of activation and inactivation of sodium channel complexes. Moreover, their extracellular domain forms an immunoglobulin-like fold, similar to many cell adhesion molecules, and the β subunits are involved in interactions of sodium channels with proteins in the extracellular matrix, with other cell adhesion molecules, and with intracellular regulatory proteins like protein kinases and phosphoprotein phosphatases.[17,20–23] The extracellular domains of these type-I single-membrane-spanning proteins are predicted to fold in a similar manner to myelin protein P0,[24] whose immunoglobin-like fold is known to be formed by a sandwich of two beta sheets held together by hydrophobic interactions (Figure 5.2A). The importance of the functions of the $Na_V\beta$ subunits is underscored by the strong effects of deletion of the genes encoding $Na_V\beta1$ and $Na_V\beta2$, which include impairments of myelination and axonal conduction, epileptic seizures, and premature death.[25,26]

5.4   Expression and Localization of Sodium Channel Subtypes

The sodium channel subtypes are differentially expressed in tissues and differentially localized within individual cells.[3,4,27–30] $Na_V1.3$ is highly expressed in fetal nervous tissues, whereas $Na_V1.1$, $Na_V1.2$, and $Na_V1.6$ are abundant in juvenile and adult CNS. Generally, $Na_V1.1$ and $Na_V1.3$ are primarily localized in neuronal cell bodies and axon initial segments where they may control the neuronal excitability through integration of synaptic impulses to set the threshold for action potential generation and propagation to the dendritic and axonal compartments. $Na_V1.2$ is primarily localized in unmyelinated axons. During development, $Na_V1.6$ replaces $Na_V1.2$ in maturing nodes of Ranvier in myelinated axons. $Na_V1.1$, $Na_V1.2$, and $Na_V1.6$ are also expressed in the peripheral nervous system. However, the three isoforms that have been cloned from sympathetic and dorsal root ganglion neurons, $Na_V1.7$, $Na_V1.8$ and $Na_V1.9$, are more abundant and are the principal sodium channels in the peripheral nervous system. The $Na_V1.7$ channel is localized in axons, where it functions in initiating and conducting the action potential. More restricted expression patterns are observed for $Na_V1.8$ and $Na_V1.9$; these channels are highly expressed in small sensory neurons of dorsal root ganglia and trigeminal ganglia where they have a key role in nociception. $Na_V1.4$ and $Na_V1.5$ are the primary muscle sodium channels that control the excitability of the skeletal and cardiac myocytes, respectively. $Na_V1.5$ is transiently expressed in developing skeletal muscle but is replaced by $Na_V1.4$ in the adult.

5.5   Molecular Basis of Sodium Channel Function

Classical work by Hodgkin and Huxley[1] defined the three key features of sodium channels: (1) voltage-dependent activation, (2) rapid inactivation, and (3) selective ion conductance. Building upon this foundation, recent structure–function studies employing molecular, biochemical, structural and electrophysiological techniques have provided clear understanding of the molecular basis of sodium channel function.[7,31–34] The narrow outer pore is formed by the reentrant P loops between transmembrane segments S5 and S6 of each domain (Figure 5.2A). Mutation of a set of four residues in analogous positions in each domain (aspartate in domain I, glutamate in domain II, lysine in domain III, and alanine in domain IV, DEKA) to glutamates confers calcium selectivity, indicating that the side chains of these amino acid residues are likely to interact with sodium ions as they are conducted through the ion selectivity filter of the pore.

The steep voltage dependence of activation of sodium channels derives from outward movement of gating charges as a consequence of depolarization of the membrane and consequent reduction of membrane electric field.[35] The S4 segments of each homologous domain serve as the primary voltage sensors for activation.[31,36] They contain repeated motifs of a positively charged amino acid residue followed by two hydrophobic residues creating a transmembrane spiral of positive charges. Upon depolarization, outward movement and rotation of S4 is thought to initiate a conformational change that opens the sodium channel pore. This 'sliding helix' model is supported by strong evidence (reviewed in[37]), as described in more detail below.

Fast inactivation of the sodium channel is a critical process that occurs within milliseconds of channel opening.[1,2] The generally accepted model of this process involves a conserved inactivation gate formed by the intracellular loop connecting domains III and IV, which serves as a hinged-lid that binds to the intracellular end of the pore and blocks it (Figure 5.2A,B and see below).[6,31–33] Intracellular perfusion of proteases or intracellular application of antibodies directed to this loop prevents fast inactivation. The latch of this fast inactivation gate is formed by three key hydrophobic residues, IFM, and an adjacent threonine (T) (Figure 5.2B). Mutations of these amino acid residues destabilize the inactivated state, and peptides harboring the IFM motif can restore inactivation to sodium channels having mutated inactivation gates. The closed inactivation gate is thought to make multiple interactions with hydrophobic amino acid residues near the intracellular mouth of the pore that may constitute the inactivation gate receptor. Scanning mutagenesis experiments implicate hydrophobic residues in intracellular S4-S5 loops of domains III and IV, as well as the intracellular end of the S6 transmembrane segment in domain IV in forming the inactivation gate receptor.

5.6   Sodium Channel Pharmacology

Sodium channels are the molecular targets for a large number of distinct families of neurotoxins, which paralyze prey by preventing conduction of action potentials in nerve and/or muscle cells.[38,39] These toxins act at six or more distinct receptor sites on sodium channels, formed by different regions of the channel protein (Table 5.1). Functional effects of sodium channel toxins include pore block, enhanced activation, and slowed or blocked inactivation. Pore blockers like tetrodotoxin and saxitoxin prevent action potential generation and conduction directly, whereas gating modifier toxins that enhance activation or inhibit inactivation prevent normal action potential generation and conduction by depolarization block of nerve and muscle fibers.

Table 5.1   Receptor sites on sodium channels.

Receptor site	Toxin or drug	Location (Domain/Segment)
Neurotoxin receptor site 1	Tetrodotoxin	I/P, II/P, III/P, IV/P
	Saxitoxin	I/P, II/P, III/P, IV/P
	μ-Conotoxin	
Neurotoxin receptor site 2	Veratridine	I/S6, IV/S6
	Batrachotoxin	
	Grayanotoxin	
Neurotoxin receptor site 3	α-Scorpion toxins	I/S5–IS6, IV/S1–S2, IV/S3–S4
	Sea anemone toxins	IV/S3–S4
Neurotoxin receptor site 4	β-Scorpion toxins	II/S1–S2, II/S3–S4
Neurotoxin receptor site 5	Brevetoxins	I/S6, IV/S5
	Ciguatoxins	
Neurotoxin receptor site 6	δ-Conotoxins	IV/S3–S4
Local anesthetic receptor site	Local anesthetic drugs	I/S6, III/S6, IV/S6
	Antiarrhythmic drugs	
	Antiepileptic drugs	

Sodium channels are also the molecular targets for drugs used in prevention of acute pain in dentistry and surgery and in treatment of cardiac arrhythmias, epilepsy, and bipolar disorder,[40] and sodium channel-blocking drugs are in development for treatment of chronic pain.[41] Local anesthetics bind to a specific receptor within the pore of sodium channels, formed by the S6 segments in domains I, III, and IV (Figure 5.2(C)).[42–44] Their binding blocks ion movement through the pore and stabilizes the inactivated state of sodium channels. Antiarrhythmic drugs and anti-epileptic drugs share similar, overlapping receptor sites.[45] Complete blocking of sodium channels would be lethal. However, these drugs selectively block sodium channels in depolarized and/or rapidly firing cells, such as axons carrying high-intensity pain information and rapidly firing nerve and cardiac muscle cells that drive epileptic seizures or cardiac arrhythmias. This selective block arises because the drugs can reach their binding site in the pore of the sodium channel more rapidly when the pore is repetitively opened and they bind with high affinity to inactivated sodium channels that are generated in rapidly firing or depolarized cells.[46] This use-dependent action of the sodium channel blocking drugs is essential for their therapeutic efficacy.

5.7   High-resolution Structure of Sodium Channels

Both α and β subunits of Nav channels are posttranslationally processed, including extensive glycosylation, palmitoylation, and phosphorylation. It has not been possible to express these proteins in high yield or to produce homogeneous preparations that would be appropriate for crystallization, despite much effort. The pathway forward to resolve this problem was provided by discovery of the NaChBac family of bacterial Na channels by Clapham and colleagues.[9] These bacterial channels have structural and functional characteristics of both Nav and Cav channels and likely are the molecular ancestor of both.[8] They have typical voltage-activated Na-selective conductance, and they are inhibited by local anesthetics in a state-dependent manner that resembles mammalian cardiac Nav channels.[9,47] They are small proteins of less than 300 amino acid residues, which contain six transmembrane segments and are structurally analogous to one domain of a mammalian Nav channel.[9] They form functional Nav channels as homotetramers. They are the ideal target for Na channel structural biology.

5.7.1 Structure of NavAb in a Membrane-like Environment

The NavAb protein from *Arcobacter butzleri* was identified as a putative member of the bacterial Nav channel family and electrophysiological characterization confirmed that NavAb indeed functions as a voltage-gated sodium-selective ion channel.[48] It is activated rapidly by depolarizing pulses and inactivates rapidly as well. Its kinetics and voltage dependence of activation and inactivation are similar to vertebrate Nav channels. Purified NavAb was solubilized and purified in detergent and crystallized in a novel lipid-based bicelle system.[48] Phosphoplipids are required to maintain the functional integrity of purified vertebrate Nav channels and it was necessary to include phospholipids in the crystallization of NavAb. The orientation of NavAb in crystals suggests lipid bilayers throughout the crystal lattice and ~28 lipid molecules are bound to each channel tetramer. The NavAb voltage-sensing domains (VSDs) interact noncovalently with the pore module of a neighboring subunit (Figure 5.4A),[48] as also seen in lipid-based crystal structures of the $K_V1.2/2.1$ channel chimera.[49] Two NavAb molecules interact with each other through their extracellular domains in these crystals (Figure 5.4(C)), but there are few points of crystal contact between the two protein monomers. The central pore is surrounded by the four pore-forming modules (Figure 5.4(D), blue), whereas the four voltage-sensing modules are associated with the periphery of the pore domain (Figure 5.4(D), warmer colors). The crystallographic temperature factors that are color-coded highlight the dynamic nature of the voltage-sensing modules relative to the pore-forming modules (Figure 5.4(C) and (D)), despite their close interactions with the pore module.

5.7.2 The NavAb Voltage Sensor is Activated

In NavAb crystals, the four voltage-sensing modules are in an activated state with a conformation similar to previous K_V channel structures (Figure 5.4(B)).[48] The four conserved gating charges are all arginine residues (R1-R4), and they are arrayed across the membrane in the NavAb structure (Figure 5.4(B)).[48] A hydrophobic constriction site (HCS) forms a tight barrier to ion flow through the voltage sensor (Figure 5.4(B)).[48] Gating charges R1-R3 are outside of the HCS, interacting with the negatively charged side

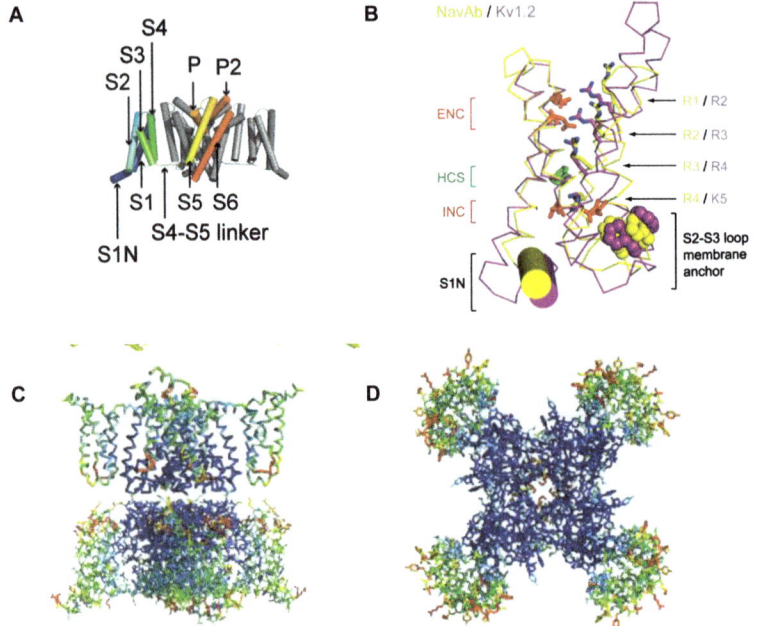

Figure 5.4 NavAb crystallized in a membrane-like environment. (A) Organization
of the transmembrane segments. (B) Voltage sensor structure of NavAb
(yellow) compared to $K_V1.2$ (purple) in an activated state. ENC, red,
extracellular negative cluster; HCS, green, hydrophobic restriction site;
INC, red, intracellular negative cluster. Gating charges R1-R4 of NavAb,
yellow; gating charges R2-R4 and K5 of $K_V1.2$, purple. (C) Side view of
NavAb colored according to crystallographic temperature factors of the
main-chain (blue < 50 Å2 to red > 150 Å2). (D) Top view of NavAb
colored as in panel (C) The selectivity filter region of the PD is the most
rigid portion of the structure when compared to the peripherally
located VSDs.

chains of the extracellular negative cluster (ENC), whereas the R1 gating
charge is on the intracellular side of the HCS interacting with the negatively
charged side chains of the intracellular negative cluster (INC,
Figure 5.4(B)).[48] This structure is well suited to move the gating charges
through the membrane from interactions with the INC in the resting state
toward interactions with the ENC in the activated state (see below). Additional interactions of the gating charges with hydrophilic side chains and
backbone carbonyls are observed in this high-resolution structure of the
voltage sensor structure.

5.7.3   Architecture of the Pore of NavAb

The NavAb structure elucidates the basis for the selectivity and high conductance rates of Nav channels. NavAb forms an ion conduction pore along
the center of its four transmembrane subunits (Figure 5.4(D)).[48] In agreement with classical depictions of Nav channels, the NavAb pore module

consists of an outer funnel-like vestibule, a narrow selectivity filter, a central cavity, and an intracellular activation gate (Figure 5.5(A)).[48] The pore-lining S6 helices form an exceptionally large central cavity (Figure 5.5(A)). In NavAb, the pore (P)-helices are positioned to potentially stabilize cations within the central cavity through helical dipole interactions (Figure 5.5A), as suggested for Kv-channels.[50] A second pore-helix (P2-helix) not seen in Kv-channels forms an extracellular funnel at the entryway to the selectivity filter in NavAb (Figure 5.5(A)).[48] The selectivity filter forms the narrowest constriction of the pore near the extracellular side of the membrane (Figure 5.5(A)), while the activation gate formed by the intracellular ends of the S6 segments controls access to the pore from the intracellular side.

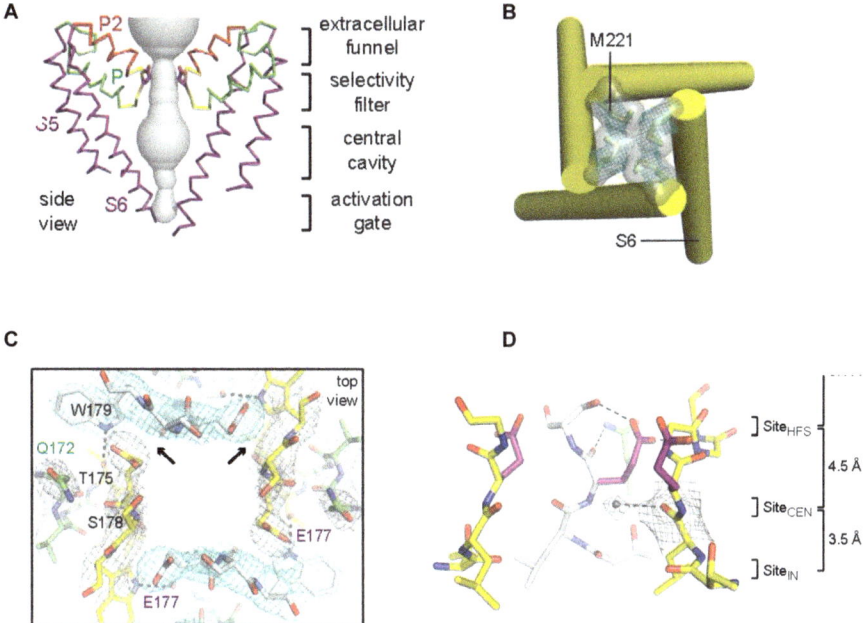

Figure 5.5   Structural elements of the NavAb pore. (A) Architecture of the pore. S5 and S6 helices (purple), pore turret loop and P-helix (green), selectivity filter signature sequence (yellow), P2-helix (red), pore and central cavity volume in gray. Two subunits are omitted for clarity. (B) The closed activation gate. Intracellular ends of the S6 segments, yellow; space-filling representation of four Met221 residues, showing that their side chains fill the pore. (C) Top view of the selectivity filter. Symmetry-related molecules are colored blue and pink; P-helix residues are colored yellow. The key hydrogen-bond between Thr175 and Trp179 is indicated by a dashed line (pink). Electron-density from an Fo-Fc omit map is shown contoured at 4.5 σ. (D) Side view of the selectivity filter. Glu177 (purple) interactions with Gln172, Ser178 and the backbone of Ser180 are shown in the far subunit. F_o-F_c omit map, 4.75 σ (blue); putative cations or water molecules (red spheres, Ion$_{EX}$). Electron-density around Leu176 (grey; F_o-F_c omit map at 1.75 σ) and a putative water molecule is shown (grey sphere). Na$^+$-coordination sites: Site$_{HFS}$, Site$_{CEN}$ and Site$_{IN}$.[48]

5.7.4 The NavAb Activation Gate is Closed

The central pore of NavAb is in a closed conformation (Figure 5.5(B)), providing the first view of a closed pore in a voltage-gated ion channel. Met221 at the intracellular end of the pore-lining S6 helices completely occludes the ion conduction pathway (Figure 5.5(B)). The positions of the S6 helices in NavAb are clearly different from the open-pore structure of $K_V1.2$.[49,50] This comparison indicates that a relatively subtle iris-like dilation of the activation gate is sufficient to open the pore in NavAb.[48] It was initially surprising to have a closed pore in a Nav channel with activated voltage-sensors at 0 mV potential. However, a pre-open state with four activated voltage-sensors and a closed pore is an expected intermediate in Na channel gating that would be present at 0 mV, because the voltage sensors in all four domains or subunits of the channel must activate to generate the pre-open state and then the pore opens in a single concerted conformational change.[51,52]

5.7.5 Ion Conductance and Selectivity in NavAb

Classic permeation studies suggested dimensions of $\sim 3.1 \times 5.1$ Å for the narrowest portion of the selectivity filter in Nav channels, and $\sim 5.5 \times 5.5$ Å in Cav channels.[53-56] Extensive mutagenesis studies have also implicated glutamate side-chains as key determinants of ion selectivity in these vertebrate channels.[34] In concert with these findings, the four Glu177 side-chains that form the outer end of the selectivity filter of NavAb define a $\sim 6.5 \times 6.5$ Å rectangle measured at the centers of the oxygen atoms and a $\sim 4.6 \times 4.6$ Å orifice measured from their van der Waals surfaces (Figure 5.5(C)).[48] The NavAb selectivity filter remains wide throughout its length (~ 8 Å) normal to the plane of the membrane (Figure 5.5(D)).[48] Remarkably, Glu177 aligns exactly with the glutamate residues that determine ion selectivity in Nav and Cav channels.[48] The crucial Glu177 side-chains of NavAb are supported by an elaborate architecture of hydrogen bonds and appear rigid in this structure. The P-helix ends with the conserved Thr175 side-chain engaging in a hydrogen bonding interaction (3.0 Å) with the conserved side-chain of Trp179 from a neighboring subunit (Figure 5.5(C)).[48] This landmark interaction staples together adjacent subunits at the selectivity filter. The intervening residues line the filter in an extended conformation, forming a tight turn between Thr175 and Trp179, which leaves the backbone carbonyls of Thr175 and Leu176 exposed to conduction pathway and available for ion binding (Figure 5.5(D)).[48]

Analysis of the NavAb pore radius indicates that a partially hydrated Na ion can be accommodated at the high field-strength site formed by the Glu177 side-chains (Site$_{HFS}$; Figure 5.5(D)).[48] Structural superposition further demonstrates the spacious nature of the NavAb selectivity filter, because a Kv-channel selectivity filter can be fit inside it. A fully-hydrated Na ion could interact with the backbone carbonyls of either Leu176 (Site$_{CEN}$) or Thr175 (Site$_{IN}$).[48] The Na ion would be surrounded by a square array of four water molecules in the plane of the carbonyls at either of these sites (Figure 5.5(D)).

Hille developed a single ion-pore model for Nav channels with three ion-binding sites separated by four energy barriers.[55] A high field-strength anionic site was proposed to partially dehydrate the permeating Na ion. The NavAb structure fits closely with expectations of Hille's model. Site$_{HFS}$ would be the first interaction point for a permeating ion, whose waters of hydration would be partially shed to allow interaction with Glu177 (Figure 5.5(C) and (D)). The Na ion could fit in-plane between the Glu177 side-chains, with one side-chain coordinating the Na ion directly, and neighboring Glu177 side-chains acting as hydrogen bond acceptors for one or two in-plane (Na bound) water molecules. When the permeating ion escapes Site$_{HFS}$, full rehydration would occur as the ion interacts with the water-lined sites formed by the carbonyls of Leu176 (Site$_{CEN}$) and the carbonyls of Thr175 (Site$_{IN}$; Figure 5.5D). Free diffusion would then allow the hydrated Na ion to enter the central cavity and move past the open activation gate into the cytoplasm.

5.7.6 Fenestrations Provide Hydrophobic Access to the Pore of NavAb

The NavAb structure presents a completely unexpected type of lipid interaction, which has profound implications for many ion channel proteins. Inspection of the NavAb central cavity reveals four lateral openings leading from the membrane to the lumen of the closed pore (Figure 5.6). These fenestrations measure ~8×10 Å (Figure 5.6). Fatty acyl chains were observed

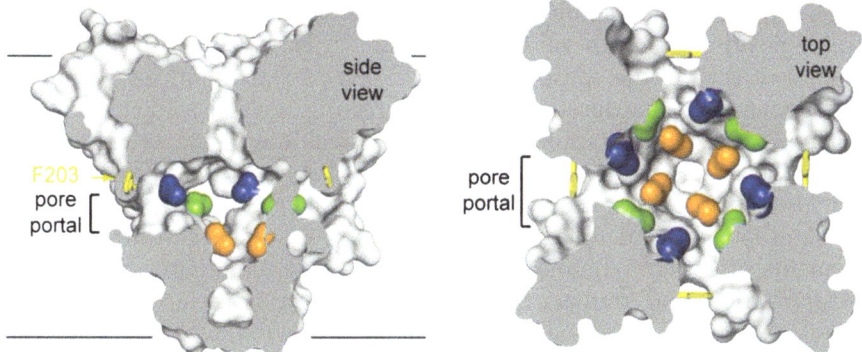

Figure 5.6 Drug binding site in the central cavity in NavAb. Left: Side-view through the pore module illustrating fenestrations (portals) and hydrophobic access to central cavity. Phe203 side-chains, yellow sticks. Surface representations of NavAb residues aligning with those implicated in drug binding and block, Thr206, blue; Met209, green; Val213, orange. Membrane boundaries, grey lines. Electron-density from an F$_o$-F$_c$ omit map is contoured at 2.0 σ. Right: Top-view sectioned below the selectivity filter, colored as in left-hand panel.

penetrating these fenestrations in the NavAb crystals, and other small, hydrophobic molecules could reach the lumen of the pore *via* this route. These fenestrations provide a pathway for potential lipid regulators of the sodium channel to reach binding sites within the pore itself.

5.8   Structural Basis for Voltage-dependent Gating

5.8.1   The Rosetta Sliding Helix Model of Voltage Sensing

The S4 transmembrane segments of sodium channels contain four to seven repeated motifs of a positively charged amino acid residue (usually arginine) followed by two hydrophobic residues, and four of those positive charges are highly conserved in voltage sensors of all sodium channels. The structure of NavAb defines the positions of these charged amino acid residues in an activated state of the voltages sensor, but the position of the gating charges in the resting state of the voltage sensor has not yet been determined from high-resolution structures. These positively charged residues in the S4 segments were proposed to carry the gating charges of sodium channels in the *sliding helix* or *helical screw* model of voltage sensing.[36,57–59] In these models, the S4 segments are proposed to be in a transmembrane position in both resting and activated states, the gating charges are stabilized in their transmembrane position by forming ion pairs with neighboring negatively charged residues, and their outward movement is catalyzed by exchange of these ion pair partners.[36,57–59] The transmembrane position of the S4 segment in sodium channels has been confirmed by mapping the receptor sites for scorpion toxins in detail and showing that these toxins bind to the outer end of the S3-S4 loop of the voltage sensors in both resting and activated states, thereby establishing that the S4 segment remains in a transmembrane position in both of these states.[60–65] High-resolution molecular models of the voltage sensors with rigid α- and β-scorpion toxins bound provide templates for the structures of the resting and activated states of the voltage sensor.[64,65] Mutation of the arginine residues in the S4 segment of sodium channels reduces the steepness of voltage-dependent gating, consistent with the idea that these residues serve as gating charges.[31,66] Covalent labeling and voltage clamp fluorescence studies show that the S4 segments of sodium channels move outward and rotate upon membrane depolarization and transport the gating charges from an inner water-accessible vestibule to an outer water-accessible vestibule.[67–69] Based on the X-ray crystal structures of the $K_V1.2$ and NavAb channels with activated voltage sensors, *ab initio* molecular modeling using the Rosetta algorithm has provided a detailed structural model of the resting states of the voltage sensor and defined the sequence of conformational changes and the sequence of gating charge interactions with negative charges and hydrophilic groups in the voltage sensor during activation of the channel (Figure 5.7).[58,70] These structural models show that the S4 segment and its gating charges move through a narrow gating pore that focuses the

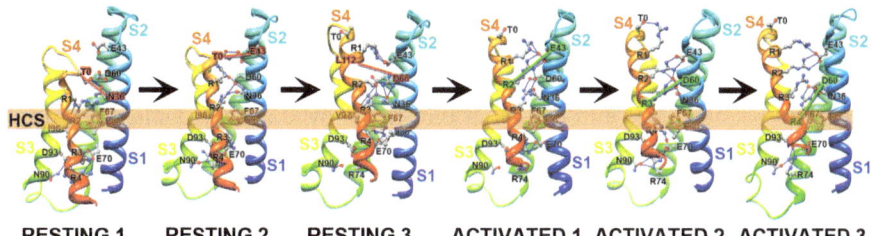

RESTING 1   RESTING 2   RESTING 3   ACTIVATED 1 ACTIVATED 2 ACTIVATED 3

Figure 5.7   Model of conformational changes in the voltage sensor during gating. Transmembrane view of the ribbon representation of Rosetta models of three resting and three activated states of the VSD of NaChBac. Segments S1 through S4 colored individually and labeled. Side chain atoms of the gating charge carrying arginines in S4 (colored dark blue), negatively charged residues in S1, S2, and S3 segments (colored red), polar residues in S1, S3, and S4 (colored purple), and key hydrophobic residues in S1, S2, and S3 (colored gray) are shown in sphere representation and labeled. The HCS is highlighted by the orange bar.

transmembrane electric field to a distance of approximately 5 Å normal to the membrane and allows the gating charges to move from an intracellular aqueous vestibule to an extracellular aqueous vestibule with a short transit through the channel protein. This mechanism of outward movement of the gating charges during the activation process has been confirmed by extensive disulfide-locking studies of the ion-pair interactions in sodium channels predicted from the *sliding helix* model of gating.[70–73] This model is also consistent with metal ion and sulfhydryl crosslinking studies of *Shaker* potassium channels.[74–76] Together, these studies define the detailed mechanism of voltage-sensing and voltage-dependent activation of the voltage sensor of sodium channels through a series of resting and activated states at the atomic level (Figure 5.7).[70]

5.8.2   Slow Inactivation

Voltage-gated sodium channels in eukaryotes have two distinct inactivation mechanisms. Fast inactivation takes place on the millisecond time scale and reverses with similar speed upon repolarization.[1,2] In contrast, slow inactivation requires depolarization for hundreds of milliseconds and reverses in seconds.[77] Slow inactivation controls the frequency and duration of trains of action potentials, as it builds up from action potential to action potential during high-frequency firing and eventually extinguishes the train.[78] Bacterial sodium channels do not have the structural equivalent of the fast inactivation gate because they are composed of four homologous domains without inter-domain linkers. However, they do have a multi-phase inactivation process, which is similar to slow inactivation in that it is modified by mutations in the pore module.[79] Recent studies have revealed the structural basis for slow inactivation of two distinct bacterial Nav channels.[80,81] For NavAb, the crystals contained pairs of channels in two closely related conformations in a single unit cell.[80] Both conformations (designated AB and CD) have a characteristic difference from the previous structure of pre-open NavAb channels—one pair of opposing S6 segments has collapsed toward the central axis and the other pair has moved away to yield a striking dimer-of-dimers organization (Figure 5.8).[80] The selectivity filter is distorted to more rectangular shape (Figure 5.8(A)), the central cavity takes on the shape of a parallelogram (Figure 5.8(B)), and the closed activation gate is rectangular rather than square (Figure 5.8(C)).[80] The different contours of the S6 segments in the AB and CD inactivated conformations are compared to the pre-open state in Figure 5.8(D), which illustrates the movement along the entire length of the S6 segments.

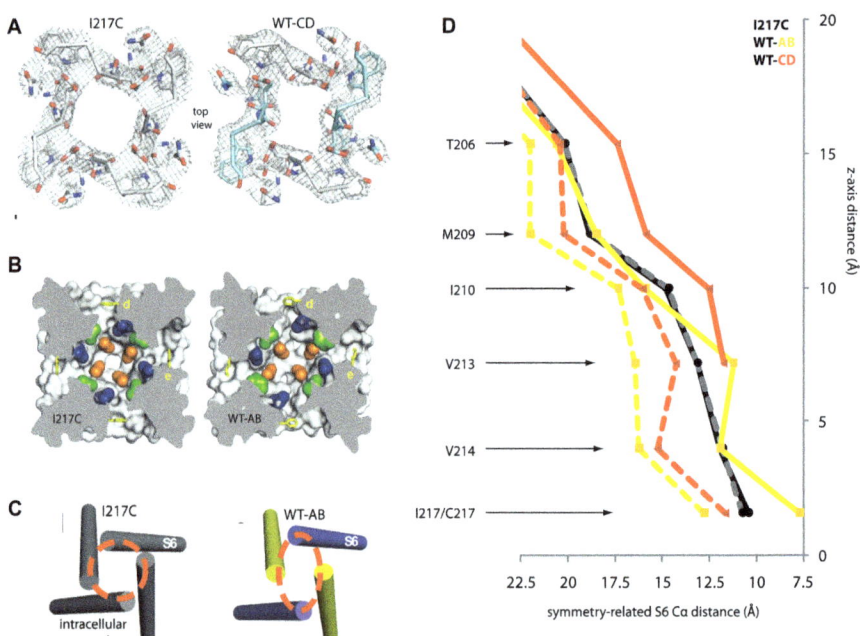

Figure 5.8   Collapse of the structural elements of the pore in the slow-inactivated state. (A) The selectivity filter viewed from the extracellular side. (B) The cavity viewed from the extracellular side. (C) The activation gate viewed from the intracellular side. (D) C_α distances along the S6 segments in NavAb. Comparison of the distances between equivalent C_α positions for pore-lining residues along the S6 segments in NavAb-Ile217Cys, NavAb-AB, and NavAb-CD models highlights the asymmetry within the pore of the NavAb-WT channels. These measurements were made between symmetry related chains; for example, the distance (in Å) between Met209 C_α in chain A and Met209 C_α in chain A'', where A'' is the chain generated through crystallographic symmetry operators. NavAb-I217C chain A is solid black, NavAb-I217C chain B is dashed grey, WT-AB chain A is dashed yellow, WT-AB chain B is solid yellow, WT-CD chain C is solid red, and WT-CD chain D is dashed red.

5.8.3 The Fast Inactivation Gate

The only high-resolution structure of vertebrate sodium channels is the solution structure of the inactivation gate in the intracellular loop connecting domains III and IV, as determined by NMR analysis of this intracellular loop expressed in bacteria (Figure 5.2(B)).[82] This analysis reveals a rigid alpha helical structure, preceded by two turns that position the key hydrophobic sequence motif IFMT such that it can interact with and block the inner mouth of the pore. The IFMT motif serves as the latch that keeps the fast inactivation gate closed until repolarization of the membrane returns the voltage sensors to their resting state and releases the fast inactivation gate from its binding position.

5.9 Structural Basis for Sodium Channel Pharmacology

5.9.1 Receptor Sites for Pore Blockers

NavAb provides a structural framework to interpret mechanisms of pharmacological action. On the extracellular side, the Glu177 side-chains represent the location of proton block in Nav channels as well as the binding site for the guanidinium moiety of the high-affinity blockers tetrodotoxin and saxitoxin (Figure 5.5(A) and (C)). From the intracellular side, Nav channels are blocked by drugs including local anesthetics, antiarrhythmics, and antiepileptics. These drugs enter through the open intracellular mouth of the pore and bind to an overlapping receptor site formed by the S6 segments, as assessed by site-directed mutagenesis studies (Figure 5.2(C)). Alignment of NavAb with vertebrate Nav and Cav channels reveals a high degree of sequence similarity along their S6 segments, and drug molecules could easily fit into the large central cavity of NavAb (Figure 5.5(A)). Use-dependent drug block is enhanced by repetitive opening of the pore to provide drug access.[46] The tight seal observed at the intracellular activation gate in NavAb illustrates why pore opening is required for the access of large or hydrophilic drugs to the local anesthetic receptor site formed by the S6 segments (Figure 5.5(B)). The three-dimensional arrangement of the amino acid residues in NavAb that are analogous to those that bind local anesthetics and related drugs in mammalian channels is illustrated by colored balls in Figure 5.6. These residues surround the ion permeation pathway in the cavity of NavAb in ideal position to bind pore-blocking drugs.

5.9.2 Drug Access in the Resting State

In addition to reaching the local anesthetic receptor site through the intracellular activation gate when it is open, small hydrophobic drugs could gain access to the local anesthetic receptor site from the membrane *via* the fenestrations. These side portals lead directly to the amino acid residues that are analogous to those that form the local anesthetic receptor site in mammalian Nav channels (Figure 5.6, colored balls). Drug entry through these pathways would give resting state block. The side chain of Phe203 is in a position to control passage through the fenestrations (Figure 5.6, yellow), and this residue is adjacent to the position of Ile1760 in cardiac Nav1.2 channels (and the homologous Ile1758 in Nav1.5 channels), which we previously showed controls the access of local anesthetics and antiarrhythmic drugs to their receptor site in the resting state of the channel.[42,83] The Navab fenestrations are compatible with the passage of small neutral or hydrophobic drugs such as phenytoin, lidocaine, and benzocaine.[46] These small hydrophobic drugs are known to gain access to their receptor site within a closed channel.[46] The pore fenestrations may be directly involved in the process of voltage-dependent drug binding and block according to the Modulated Receptor Hypothesis.[46] The unexpected finding of fenestrations in the side of the pore of NavAb highlights the potential for local anesthetic, antiarrhythmic, and antiepileptic drugs and other hydrophobic molecules to influence the conduction behavior of Nav channels from the lipid phase of the membrane.

5.9.3 Conformational Change in the Local Anesthetic Receptor Site during Slow Inactivation

The dramatic conformational change in the pore that occurs during slow inactivation reorganizes the molecular relationships within the central cavity (Figure 5.8). This structural change has major implications for drug binding because the receptor site in the central cavity is changed in shape dramatically and the size and shape of the fenestrations is also altered (Figure 5.8B).[80] Local anesthetics and related drugs bind with highest affinity to inactivated sodium channels.[46] The asymmetric conformation of the local anesthetic receptor site in the inactivated state may allow it to interact with bound drugs with higher affinity. Consistent with this idea, amino acid residues that bind local anesthetics and related drugs with high affinity are found in the S6 segments of domains I, III, and IV, but not in domain II.[43,44] It is likely that these changes in the structure of the drug binding site contribute in a major way to the state-dependence of binding and action of local anesthetic, antiarrhythmic and anti-epileptic drugs that act on sodium channels.

5.10   Looking Ahead

Emerging information on the different physiological roles of sodium channel subtypes points to many opportunities for discovery of drugs with greater efficacy and safety than current sodium channel blocking drugs. Targeting $Na_V1.7$, $Na_V1.8$, and $Na_V1.9$ channels in sensory neurons may be valuable in treatment of chronic and neuropathic pain. More highly state-dependent sodium channel blockers may be improved antiarrhythmic drugs. Drugs that target specific sodium channel subtypes expressed in the brain may be more effective anti-epileptic drugs with fewer side effects. The availability of high-resolution structural models for bacterial sodium channels, and the possibility of obtaining such information from mammalian Na$_V$ channels is an exciting prospect that should drive future drug discovery efforts.

References

1. A. L. Hodgkin and A. F. Huxley, *J. Physiol.*, 1952, **117**, 500–544.
2. B. Hille, *Ionic Channels of Excitable Membranes*, 3rd Ed, Sinauer Associates Inc., Sunderland, MA, 2001.
3. M. N. Rasband and P. Shrager, *J. Physiol*, 2000, **525 Pt 1**, 63–73.
4. H. Vacher, D. P. Mohapatra and J. S. Trimmer, *Physiol. Rev.*, 2008, **88**, 1407–1447.
5. W. A. Catterall, *Science*, 1984, **223**, 653–661.
6. W. A. Catterall, *Neuron*, 2000, **26**, 13–25.
7. W. Stühmer, *Annu. Rev. Biophys. Biophys. Chem.*, 1991, **20**, 65–78.
8. F. H. Yu and W. A. Catterall, *Sci. STKE*, 2004, **2004**, re15.
9. D. Ren, B. Navarro, H. Xu, L. Yue, Q. Shi and D. E. Clapham, *Science*, 2001, **294**, 2372–2375.
10. H. H. Zakon, *Proc. Natl. Acad. Sci. U. S. A.*, 2012, **109**(Suppl 1), 10619–10625.
11. A. L. Goldin, R. L. Barchi, J. H. Caldwell, F. Hofmann, J. R. Howe, J. C. Hunter, R. G. Kallen, G. Mandel, M. H. Meisler, Y. Berwald Netter, M. Noda, M. M. Tamkun, S. G. Waxman, J. N. Wood and W. A. Catterall, *Neuron*, 2000, **28**, 365–368.
12. W. A. Catterall, A. L. Goldin and S. G. Waxman, *Pharmacol. Rev.*, 2005, **57**, 397–409.
13. J. Satin, J. W. Kyle, M. Chen, P. Bell, L. L. Cribbs, H. A. Fozzard and R. B. Rogart, *Science*, 1992, **256**, 1202–1205.
14. L. Sivilotti, K. Okuse, A. N. Akopian, S. Moss and J. N. Wood, *FEBS Lett.*, 1997, **409**, 49–52.
15. E. Watanabe, A. Fujikawa, H. Matsunaga, Y. Yasoshima, N. Sako, T. Yamamoto, C. Saegusa and M. Noda, *J. Neurosci.*, 2000, **20**, 7743–7751.
16. L. L. Isom, K. S. De Jongh, D. E. Patton, B. F. X. Reber, J. Offord, H. Charbonneau, K. Walsh, A. L. Goldin and W. A. Catterall, *Science*, 1992, **256**, 839–842.
17. L. L. Isom, D. S. Ragsdale, K. S. De Jongh, R. E. Westenbroek, B. F. X. Reber, T. Scheuer and W. A. Catterall, *Cell*, 1995, **83**, 433–442.
18. K. Morgan, E. B. Stevens, B. Shah, P. J. Cox, A. K. Dixon, K. Lee, R. D. Pinnock, J. Hughes, P. J. Richardson, K. Mizuguchi and A. P. Jackson, *Proc. Natl. Acad. Sci. U. S. A.*, 2000, **97**, 2308–2313.
19. F. H. Yu, R. E. Westenbroek, I. Silos-Santiago, K. A. McCormick, D. Lawson, P. Ge, H. Ferriera, J. Lilly, P. S. DiStefano, W. A. Catterall, T. Scheuer and R. Curtis, *J. Neurosci.*, 2003, **23**, 7577–7585.

20. J. Srinivasan, M. Schachner and W. A. Catterall, *Proc. Natl. Acad. Sci. U. S. A.*, 1998, **95**, 15753–15757.
21. C. F. Ratcliffe, Y. Qu, K. A. McCormick, V. C. Tibbs, J. E. Dixon, T. Scheuer and W. A. Catterall, *Nat. Neurosci.*, 2000, **3**, 437–444.
22. C. F. Ratcliffe, R. E. Westenbroek, R. Curtis and W. A. Catterall, *J. Cell Biol.*, 2001, **154**, 427–434.
23. W. J. Brackenbury and L. L. Isom, *Front. Pharmacol.*, 2011, **2**, 53.
24. L. Shapiro, J. P. Doyle, P. Hensley, D. R. Colman and W. A. Hendrickson, *Neuron*, 1996, **17**, 435–449.
25. C. Chen, V. Bharucha, Y. Chen, R. E. Westenbroek, A. Brown, J. D. Malhotra, D. Jones, C. Avery, P. J. Gillespie, 3rd, K. A. Kazen-Gillespie, K. Kazarinova-Noyes, P. Shrager, T. L. Saunders, R. L. Macdonald, B. R. Ransom, T. Scheuer, W. A. Catterall and L. L. Isom, *Proc. Natl. Acad. Sci. U. S. A.*, 2002, **99**, 17072–17077.
26. C. Chen, R. E. Westenbroek, X. Xu, C. A. Edwards, D. R. Sorenson, Y. Chen, D. P. McEwen, H. A. O'Malley, V. Bharucha, L. S. Meadows, G. A. Knudsen, A. Vilaythong, J. L. Noebels, T. L. Saunders, T. Scheuer, P. Shrager, W. A. Catterall and L. L. Isom, *J. Neurosci.*, 2004, **24**, 4030–4042.
27. W. A. Catterall, E. Perez-Reyes, T. P. Snutch and J. Striessnig, *Pharmacol. Rev.*, 2005, **57**, 411–425.
28. R. E. Westenbroek, D. K. Merrick and W. A. Catterall, *Neuron*, 1989, **3**, 695–704.
29. R. E. Westenbroek, J. L. Noebels and W. A. Catterall, *J. Neurosci.*, 1992, **12**, 2259–2267.
30. S. Beckh, M. Noda, H. Lübbert and S. Numa, *EMBO J.*, 1989, **8**, 3611–3616.
31. W. Stuhmer, F. Conti, H. Suzuki, X. Wang, M. Noda, N. Yahadi, H. Kubo and S. Numa, *Nature*, 1989, **339**, 597–603.
32. P. M. Vassilev, T. Scheuer and W. A. Catterall, *Science*, 1988, **241**, 1658–1661.
33. J. W. West, D. E. Patton, T. Scheuer, Y. Wang, A. L. Goldin and W. A. Catterall, *Proc. Natl. Acad. Sci. U. S. A.*, 1992, **89**, 10910–10914.
34. S. H. Heinemann, H. Terlau, W. Stühmer, K. Imoto and S. Numa, *Nature*, 1992, **356**, 441–443.
35. C. M. Armstrong, *Physiol. Rev.*, 1981, **61**, 644–682.
36. W. A. Catterall, *Annu. Rev. Biochem.*, 1986, **55**, 953–985.
37. W. A. Catterall, *Neuron*, 2010, **67**, 915–928.
38. W. A. Catterall, *Annu. Rev. Pharmacol. Toxicol.*, 1980, **20**, 15–43.
39. W. A. Catterall, S. Cestele, V. Yarov-Yarovoy, F. H. Yu, K. Konoki and T. Scheuer, *Toxicon*, 2007, **49**, 124–141.
40. W. A. Catterall, *Trends Pharmacol. Sci.*, 1987, **8**, 57–65.
41. B. T. Priest and G. J. Kaczorowski, *Expert Opin. Ther. Targets*, 2007, **11**, 291–306.
42. D. S. Ragsdale, J. C. McPhee, T. Scheuer and W. A. Catterall, *Science*, 1994, **265**, 1724–1728.

43. V. Yarov-Yarovoy, J. Brown, E. Sharp, J. J. Clare, T. Scheuer and W. A. Catterall, *J. Biol. Chem.*, 2001, **276**, 20–27.
44. V. Yarov-Yarovoy, J. C. McPhee, D. Idsvoog, C. Pate, T. Scheuer and W. A. Catterall, *J. Biol. Chem.*, 2002, **277**, 35393–35401.
45. D. R. Ragsdale, J. C. McPhee, T. Scheuer and W. A. Catterall, *Proc. Natl. Acad. Sci. U. S. A.*, 1996, **93**, 9270–9275.
46. B. Hille, *J. Gen. Physiol.*, 1977, **69**, 497–515.
47. Y. Zhao, T. Scheuer and W. A. Catterall, *Proc. Natl. Acad. Sci. U. S. A.*, 2004, **101**, 17873–17878.
48. J. Payandeh, T. Scheuer, N. Zheng and W. A. Catterall, *Nature*, 2011, **475**, 353–358.
49. S. B. Long, X. Tao, E. B. Campbell and R. MacKinnon, *Nature*, 2007, **450**, 376–382.
50. S. B. Long, E. B. Campbell and R. MacKinnon, *Science*, 2005, **309**, 897–903.
51. W. N. Zagotta, T. Hoshi and R. W. Aldrich, *J. Gen. Physiol.*, 1994, **103**, 321–362.
52. Y. Zhao, V. Yarov-Yarovoy, T. Scheuer and W. A. Catterall, *Neuron*, 2004, **41**, 859–865.
53. B. Hille, *J. Gen. Physiol.*, 1971, **59**, 599–619.
54. B. Hille, *J. Gen. Physiol.*, 1972, **59**, 637–658.
55. B. Hille, *J. Gen. Physiol.*, 1975, **66**, 535–560.
56. E. W. McCleskey and W. Almers, *Proc. Natl. Acad. Sci. U. S. A.*, 1985, **82**, 7149–7153.
57. W. A. Catterall, *Trends Neurosci.*, 1986, **9**, 7–10.
58. V. Yarov-Yarovoy, D. Baker and W. A. Catterall, *Proc. Natl. Acad. Sci. U. S. A.*, 2006, **103**, 7292–7297.
59. H. R. Guy and P. Seetharamulu, *Proc. Natl. Acad. Sci. U. S. A.*, 1986, **508**, 508–512.
60. W. A. Catterall, *J. Gen. Physiol.*, 1979, **74**, 375–391.
61. J. C. Rogers, Y. Qu, T. N. Tanada, T. Scheuer and W. A. Catterall, *J. Biol. Chem.*, 1996, **271**, 15950–15962.
62. S. Cestèle, Y. Qu, J. C. Rogers, H. Rochat, T. Scheuer and W. A. Catterall, *Neuron*, 1998, **21**, 919–931.
63. S. Cestele, V. Yarov-Yarovoy, Y. Qu, F. Sampieri, T. Scheuer and W. A. Catterall, *J. Biol. Chem.*, 2006, **281**, 21332–21344.
64. J. Z. Zhang, V. Yarov-Yarovoy, T. Scheuer, I. Karbat, L. Cohen, D. Gordon, M. Gurevitz and W. A. Catterall, *J. Biol. Chem.*, 2011, **286**, 33641–33651.
65. J. Wang, V. Yarov-Yarovoy, R. Kahn, D. Gordon, M. Gurevitz, T. Scheuer and W. A. Catterall, *Proc. Natl. Acad. Sci. U. S. A.*, 2011, **108**, 15426–15431.
66. K. J. Kontis, A. Rounaghi and A. L. Goldin, *J. Gen. Physiol.*, 1997, **110**, 391–401.
67. N. Yang and R. Horn, *Neuron*, 1995, **15**, 213–218.
68. N. Yang, A. L. George, Jr. and R. Horn, *Neuron*, 1996, **16**, 113–122.
69. B. Chanda and F. Bezanilla, *J. Gen. Physiol.*, 2002, **120**, 629–645.

70. V. Yarov-Yarovoy, P. G. Decaen, R. E. Westenbroek, C. Y. Pan, T. Scheuer, D. Baker and W. A. Catterall, *Proc. Natl. Acad. Sci. U. S. A.*, 2012, **109**, E93–E102.

71. P. G. DeCaen, V. Yarov-Yarovoy, Y. Zhao, T. Scheuer and W. A. Catterall, *Proc. Natl. Acad. Sci. U. S. A.*, 2008, **105**, 15142–15147.

72. P. G. DeCaen, V. Yarov-Yarovoy, E. M. Sharp, T. Scheuer and W. A. Catterall, *Proc. Natl. Acad. Sci. U. S. A.*, 2009, **106**, 22498–22503.

73. P. G. DeCaen, V. Yarov-Yarovoy, T. Scheuer and W. A. Catterall, *Proc. Natl. Acad. Sci. U. S. A.*, 2011, **108**, 18825–18830.

74. F. V. Campos, B. Chanda, B. Roux and F. Bezanilla, *Proc. Natl. Acad. Sci. U. S. A.*, 2007, **104**, 7904–7909.

75. A. Broomand and F. Elinder, *Neuron*, 2008, **59**, 770–777.

76. M. C. Lin, J. Y. Hsieh, A. F. Mock and D. M. Papazian, *J. Gen. Physiol.*, 2011, **138**, 155–163.

77. B. Rudy, *J. Physiol. (Lond.)*, 1978, **283**, 1–21.

78. D. B. Carr, M. Day, A. R. Cantrell, J. Held, T. Scheuer, W. A. Catterall and D. J. Surmeier, *Neuron*, 2003, **39**, 793–806.

79. E. Pavlov, C. Bladen, R. Winkfein, C. Diao, P. Dhaliwal and R. J. French, *Biophys. J.*, 2005, **89**, 232–242.

80. J. Payandeh, T. M. Gamal El-Din, T. Scheuer, N. Zheng and W. A. Catterall, *Nature*, 2012, **486**(7401), 135–139.

81. X. Zhang, W. Ren, P. DeCaen, C. Yan, X. Tao, L. Tang, J. Wang, K. Hasegawa, T. Kumasaka, J. He, J. Wang, D. E. Clapham and N. Y. Yan, *Nature*, 2012, **486**(7401), 130–134.

82. C. A. Rohl, F. A. Boeckman, C. Baker, T. Scheuer, W. A. Catterall and R. E. Klevit, *Biochemistry*, 1999, **38**, 855–861.

83. Y. Qu, J. Rogers, T. Tanada, T. Scheuer and W. A. Catterall, *Proc. Natl. Acad. Sci. U. S. A.*, 1995, **270**, 25696–25701.

84. D. A. Beneski and W. A. Catterall, *Proc. Natl. Acad. Sci. U. S. A.*, 1980, **77**, 639–643.

85. R. P. Hartshorne, D. J. Messner, J. C. Coppersmith and W. A. Catterall, *J. Biol. Chem.*, 1982, **257**, 13888–13891.

86. R. P. Hartshorne and W. A. Catterall, *J. Biol. Chem.*, 1984, **259**, 1667–1675.

87. R. P. Hartshorne, B. U. Keller, J. A. Talvenheimo, W. A. Catterall and M. Montal, *Proc. Natl. Acad. Sci. U. S. A.*, 1985, **82**, 240–244.

AMPA Receptor Positive Allosteric Modulators – a Case History

SIMON E WARD

Professor of Medicinal Chemistry, Translational Drug Discovery Group, School of Life Sciences, University of Sussex, Brighton, United Kingdom, BN1 9RB
Email: simon.ward@sussex.ac.uk

6.1 Introduction

6.1.1 Ionotropic Glutamate Receptors

Significant advances have been made over recent years in our understanding of the tremendous complexity underlying the function of the human brain, in particular gaining insight into the mechanisms of synaptic plasticity which are key to developmental, adaptive and learning processes. Glutamate, the major excitatory neurotransmitter in the central nervous system (CNS), is of critical importance to these processes, acting at chemical synapses on two major classes of receptors—the metabotropic family of G-protein coupled receptors (mGluRs 1–8) and the ionotropic family of ion channel forming receptors (iGluRs).[1,2] The latter comprises the α-amino-3-hydroxyl-5-methyl-4-isoxazole-propionic acid (AMPA), N-methyl-D-aspartate (NMDA) and kainate receptors which are made up of distinct subunits which have been systematically renamed as GluA1-4 (AMPA), GluN1, GluN2A-D and GluN3A-B (NMDA) and GluK1-5 (kainate) respectively.[3] Despite similarities

RSC Drug Discovery Series No. 39
Ion Channel Drug Discovery
Edited by Brian Cox and Martin Gosling
© The Royal Society of Chemistry 2015
Published by the Royal Society of Chemistry, www.rsc.org

shared within this ion channel family, there exist clear structural and pharmacological differences which underlie their individual modes of action. At the gross receptor level, all glutamatergic ion channels are tetrameric although, whilst functional AMPA and kainate receptors can be formed from homo- or hetero-tetramers, the NMDA receptors are obligate hetero-tetramers. Furthermore, all these ionotropic glutamate receptor channels contain distinct topological regions comprising amino-terminal, ligand binding, trans-membrane and C-terminal domains.[4] The amino-terminal domain (ATD) contains binding sites for molecules and proteins that are implicated in receptor assembly, localisation and trafficking and recent crystallographic studies propose mechanisms by which the ATD guides subfamily-specific receptor assembly.[5,6] The ligand binding domain (LBD) is made up of two regions termed S1 (located between trans-membrane domain M1 and ATD) and S2 (extracellular loop between trans-membrane domains M3 and M4) as illustrated in Figure 6.1. Together these form the classic glutamate binding clamshell region and include the binding sites for a number of ionotropic glutamate agonists and antagonists. The trans-membrane domain consists of three membrane spanning regions (M1, M3 and M4) and a re-entrant loop (M2), which corresponds to the analogous P-loop region in the voltage-gated potassium channel family. Finally the

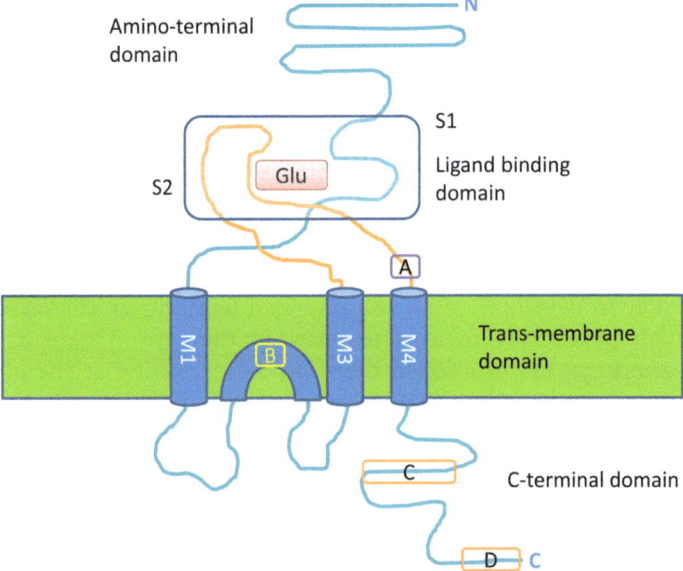

Figure 6.1   Schematic of single subunit of AMPAR key domains, LBD, with S1 and S2 polypeptides coloured differently. Key residue regions highlighted: A: flip/flop splice variants; B: post transcriptional modification sites regulating Ca^{2+} permeability; C: NSF binding domain; D: PDZ binding domain.
(Figure reproduced in part with permission from Ward *et al.*[7] Copyright 2010 American Chemical Society).

intracellular C-terminal domain is involved in interactions with accessory proteins governing receptor localisation and regulation.

Despite these common structural motifs, the iGluR family receptors display profoundly different pharmacological properties. Both AMPA and kainate receptors open rapidly upon binding glutamate and also display subsequently fast kinetics of deactivation and desensitisation. In contrast, NMDA receptors activate, deactivate and desensitise more slowly. Furthermore, in addition to the binding of glutamate, opening of NMDA receptor channels also requires membrane depolarisation to relieve the magnesium ion channel block as well as the binding of glycine as a co-agonist.[8]

6.1.2   AMPA Receptors (AMPARs)

Whilst, as stated, AMPARs are functionally active as homomers, which provides valuable opportunities for establishing recombinant cell lines for screening, native receptors are almost exclusively heteromeric,[9–11] which leads to a diversity of receptor subunit composition in the human brain. This diversity influences both the biophysical and functional properties of the AMPAR *via* effects on kinetics,[12] channel open time,[13] internalisation and trafficking.[14,15] In addition to this subunit complexity, there are also significant additional post-transcriptional modifications and splice variants. Specifically, in the mature brain, RNA editing of GluA2 replaces the genomically encoded glutamine (Q) in the M2 re-entrant loop with a positively charged arginine (R), thereby causing a reduction in both the block by intracellular polyamines and the passage of divalent cations.[16] Consequently, in adults, the majority of GluA2 subunits contain the arginine residue which limits the flux of calcium ions through the AMPAR channel, allowing just the passage of sodium and potassium ions instead, which is an important determinant of synaptic plasticity.[17] Furthermore, all AMPA subunits can exist as two alternative forms arising from differential splicing of the extracellular S2 domain. These forms, termed flip and flop, vary in their expression through development and, in tandem with the properties of the individual subunits also affect a number of biophysical and kinetic parameters of the AMPARs.[2]

An additional important characteristic of the regulation of all AMPAR variants is the presence of accessory proteins which interact with motifs present on the C-terminal domain. These motifs are able to phosphorylate the trans-membrane AMPA regulatory proteins (TARPs) as well as interact with the more recently identified accessory proteins such as cornichon proteins CNIH-2 and CNIH-3, all of which serve to regulate AMPARs by increasing their expression and altering receptor kinetics.[18,19]

Understanding of the mechanism and structure of the AMPA receptor (AMPAR) has been advanced by extensive electrophysiological and crystallographic studies, driven by many research groups. Firstly, a number of key papers have been published showing that desensitisation of the AMPAR is driven by agonist concentrations that are lower than those required for activation and that the process of restoring the receptor from the desensitised state to one from which it can be reactivated proceeds approximately ten times slower than the rate of recovery from a deactivated state.[20,21] This correlates with an increase in agonist affinity, which is a classic feature of the desensitised state, which for AMPARs is approximately 20 fold above resting state affinity.[22] A series of site directed mutagenesis experiments further elucidated the mechanisms behind these processes and in particular demonstrated the importance of the dimer interface through the key mutation of a leucine to a tyrosine (L483Y). This single modification at the interface of the two subunits essentially blocked desensitisation of the receptor and formed a stable GluA2 dimer with a dissociation constant of 30 nM.[23] Later reports also identified that the substitution of a conserved positively charged arginine (R) with a negatively charged glutamate in the linker between the pore-forming M3 segment and the S2 lobe, a region outside the ligand-binding domain, also blocks desensitisation in homomeric AMPA receptors composed of GluA2(flip) subunits.[24]

Subsequent to the initial pharmacology studies, the rat orthologue of the GluA2 S1S2 LBD construct has been extensively studied, specifically through crystallisation and electrophysiology, and has enabled the formulation of hypotheses for receptor dynamics. This construct contains a peptide bridge between the sequences that link to trans-membrane domains M1 and M3 and it forms similar dimers in both the resting and activated state. The receptors are activated by the binding of glutamate to the clamshell-like LBDs, which causes a closure of the LBD within each subunit, and this increases the distance between the portions of the LBDs that are linked to the ion channel. During this process of activation, the dimer interface remains intact and the strain caused by the closure of the agonist binding domain is transferred to the gate resulting in the opening of the channel. Receptor desensitisation occurs through reorganisation of the LBD dimer interface allowing the domain closure to be decoupled from the channel gate and hence allowing the ion channel to close. The energy barrier to this process of desensitisation is higher than that required for activation, hence the observation that desensitisation proceeds at a slower rate than activation. However, the resulting desensitised state is also more stable than the originating activated state, and so on prolonged activation, the majority of receptors will be driven across to the desensitised state.[6]

Most recently, the outstanding efforts from the group of Eric Gouaux have culminated in the publication of the X-ray structure of the full length GluA2 homomeric receptor at 3.6 Å in complex with a competitive antagonist.[24] This structure showed the receptor to have an overall axis of two-fold symmetry, with an additional four-fold symmetry present in the ion channel region. Furthermore, as well as the predicted organisation of the extracellular domains into pairs of dimers, a crossover of these domains was also observed between subunits. Importantly, the widely characterised GluA2 LBD hybrid structure was very similar in structure to the LBD region of the full-length protein, validating its utility as a model system for studying ligand interactions.

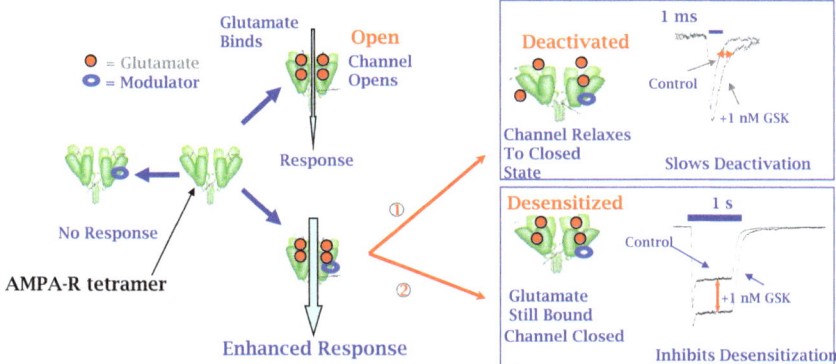

Figure 6.2   Schematic of the AMPAR with the arrow representing the entry of cations from the extracellular to the intracellular matrix. No response is observed upon binding the modulator alone, and the glutamate response is amplified in the presence of the modulator. Sample electrophysiology traces are shown for deactivation and desensitisation modes.

6.1.3   AMPAR Positive Allosteric Modulators

Many molecules have been described as AMPAR positive modulators which are able to potentiate ion flux through the AMPA channel.[25,26] Importantly, AMPAR positive modulators are not able to activate the channel alone, and are only able to do so in the presence of a bound agonist—thereby, for therapeutic use, providing a means to control both the spatial and temporal control of AMPAR activation (Figure 6.2). AMPAR positive modulators have been shown to potentiate AMPAR-mediated currents by slowing the rate at which the receptor either desensitises in the continued presence of glutamate, and/or deactivates after removal of glutamate. The resulting prolongation of AMPAR-mediated synaptic transmission relieves the voltage-dependent Mg^{2+} block of NMDA receptors so that they too can contribute to fast synaptic transmission and, by enabling Ca^{2+} influx into the postsynaptic neurone, modulate synaptic plasticity mechanisms such as long term potentiation (LTP) and long term depression (LTD) thought to underlie mnemonic processing.[8,27]

Crystal structures have shown positive allosteric modulators binding at the LBD dimer interface between two subunits, leading to the thinking that they slow deactivation by stabilising the clamshell dimer in its closed cleft glutamate bound conformation and to slow desensitisation by stabilising the dimer interface. Given that the inter-domain hinges are located at the dimer interface, it is not surprising that desensitisation and deactivation are interconnected, and many AMPAR modulators slow both processes.[28]

6.2   Clinical Landscape

Growing preclinical and clinical data support the utility of positive modulators of the AMPAR as potential broad spectrum therapies for many psychiatric and neurological diseases. The progenitor of all AMPAR positive modulators was the nootropic agent aniracetam, which had been shown to be effective in improving learning and memory in the rat.[29] Since then, the first of the modern generation of AMPAR positive modulators, CX516, was reported to improve cognitive performance in both healthy young and aged volunteers[30] with inconsistent cognitive benefits being observed in schizophrenics.[31,32] Moreover, several AMPAR positive modulators have exhibited efficacy in a range of rodent cognition tasks probing different domains and neuroanatomical substrates including: fear conditioning (associative learning, amygdala based), novel object recognition (episodic visual recognition memory, perirhinal cortex, hippocampus based) and attentional set-shifting (cognitive flexibility/executive function, prefrontal cortex based).[33]

More recently, compelling data have been reported from a small (n = 23–28) crossover study comparing two doses of CX717 (200 or 800 mg twice daily, for 3 weeks) with placebo in adults with moderate to severe ADHD symptoms. At the highest dose tested significant improvements in hyperactivity and inattention indices (assessed using the ADHD-Rating Scale) were observed with limited side-effects although no neurocognitive data were disclosed from this study.[34] Importantly, evidence from [^{18}F]-FDG-PET imaging studies shows that CX717 strengthens transmission in frontostriatal circuitry which is believed to be glutamatergic in nature and underactivated in disease states such as ADHD[35] and is central in the control of locomotion, affect, impulsivity, attention, cognition and emotion.[36] Thus, the improvement of working memory (delayed match to sample task) observed in Rhesus monkeys following administration of CX717 is paired with a robust activation of the dorsolateral prefrontal cortex (and also the medial temporal lobe).[37]

In addition to these, a number of other molecules from the Cortex laboratories have been progressed into clinical trials. Cortex had licensed faramaptor (Org24448; CX691; SCH900460) to Organon (subsequently Merck) for the potential treatment of depression. However, the only apparently registered trial showed, in 2008, that it had been suspended by the National Institute of Mental Health pending results of cardiac safety from a previous clinical study and Merck have subsequently discontinued all licensing agreements with Cortex for progression of AMPAR positive modulators in depression. More recently, CX1739 was reported to be in a Phase II study for Alzheimer's disease and respiratory depression, with reported intent to initiate a Phase II ADHD trial, although no recent updates have been made available. CX717 remains in clinical trials and is reported to now be within Valeant's portfolio to treat drug-induced respiratory depression. A recent press release from Cortex and Servier disclosed a new molecule CX1632 (S47445) which was about to enter Phase I for Alzheimer's disease.

Few other trials listed for AMPAR positive modulators appear to have completed successfully. Two registered Phase II trials for ORG26576 in major depressive disorder (studying 2 doses opened in October 2007) and adult ADHD (doses ranging from 100 to 300 mg twice daily opened in

January 2008) have been marked as complete in 2010, although no data are available on whether the trial ran or was discontinued. The Lilly Phase II trial with LY451395 remains open, targeting assessment of the aggression and agitation symptoms in Alzheimer's disease in 180 patients, initially using 3 mg twice-daily for 12 weeks. The completion data for this trial has been postponed from 2010 to October 2011, and follows the earlier reported negative trial for assessment of the cognitive deficits in Alzheimer's disease. GlaxoSmithKline recently disclosed the structure of their first clinical candidate, *N*-[(2*S*)-5-(6-fluoro-3-pyridinyl)-2,3-dihydro-1*H*-inden-2-yl]-2-propane-sulfonamide, to enter Phase I for the cognitive deficits in schizophrenia. This molecule demonstrated potent activity in behavioural models of cognition in the rat as well as a very clean developability profile. Initial pharmacokinetic data were also reported from the Phase I study which showed the development molecule to be very well absorbed, but also to have a significantly longer half-life than observed in pre-clinical species.

Unfortunately, many AMPAR modulators have failed to deliver significant efficacy in the clinic due to specific properties of each molecule such as too short half-life, insufficient potency, lack of tolerability or inability to dose high enough to achieve sufficient exposure.[38,39]

6.3 Discovery Landscape and Choice of Screening Methodology

From this broad assessment of the clinical development landscape, it is clear that there still exists an unmet clinical need for novel AMPAR positive modulators (Figure 6.3) which, based on preclinical evidence,[40] should, when tested in man, exhibit a good pharmacokinetic profile and efficacy as well as being safe and well tolerated and possess the potential to redress the deficiencies in the glutamatergic system which appear to underlie a plethora of disease states.

To this end, many research groups have been active in bringing forward new molecules for various indications. The main chemical classes investigated to date cover the benzamides **1–3**, benzothiadiazines **4,5**, sulfonamides **6–8** and pyrazoles, with a few less-well exemplified classes described, and whilst the reader is referred to recent reviews[25,26,41] for a detailed perspective, the series are discussed here to inform our choice of screening platform and cascade.

6.3.1 Benzamides

The oldest class, the benzamides, originating from aniracetam **1** and subsequently the early Cortex (University of California) clinical compounds, such as **2** and **3** described earlier, was principally investigated by studying AMPA-induced Ca^{2+} uptake in neuronal cultures and potentiation of steady state currents using patch clamp electrophysiology. These molecules have

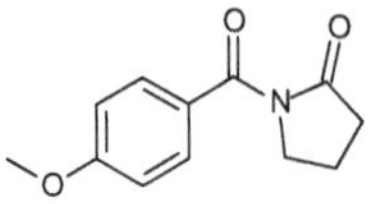

1, aniracetam
Roche

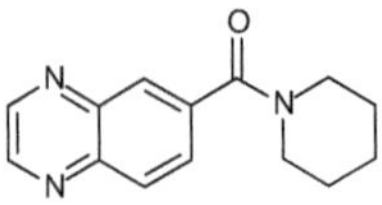

2, CX516
Cortex

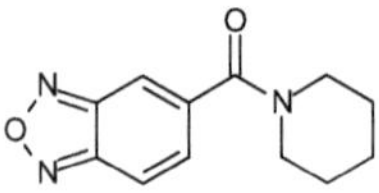

3, CX691, ORG2448
Cortex - Schering Plough

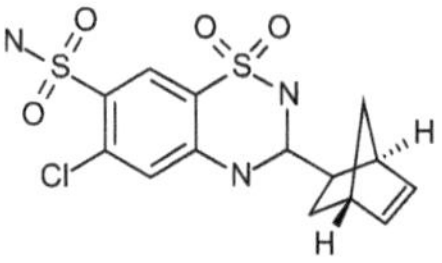

4, cyclothiazide

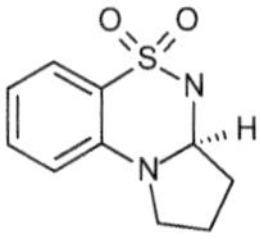

5, S-18986
Servier

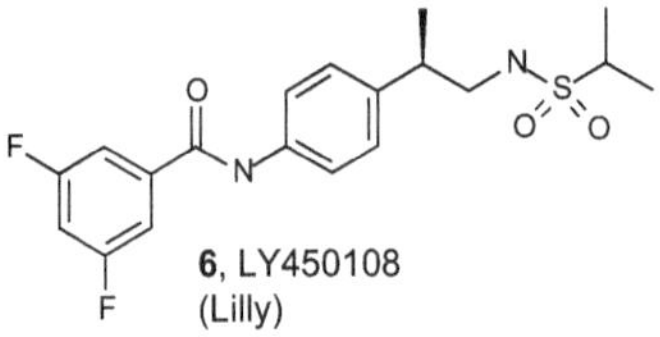

6, LY450108
(Lilly)

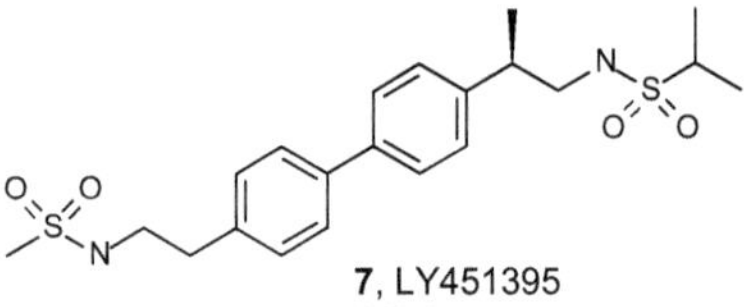

7, LY451395
(Lilly)

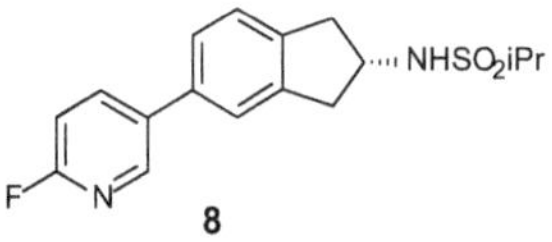

8
(GlaxoSmithKline)

Figure 6.3 Chemical structures for AMPAR positive modulators which have entered clinical development.

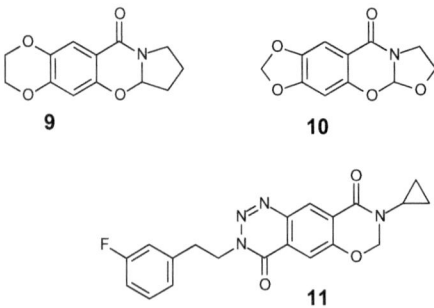

9

10

11

Figure 6.4 Structures of more recent benzamide derivatives.

spawned several generations of analogues, principally from Cortex and Neurosearch, which, in general, have increased both in terms of chemical complexity and potency through molecules **9** to **11** (Figure 6.4). During these investigations, the electrophysiology methodology used elucidated deactivation kinetics, *via* fast glutamate application, and was often followed by *ex vivo* slice recordings of excitatory postsynaptic currents (EPSPs).

Conformational constraint of the earlier reported systems led to **9** and **10** which were reported to lead to greater potentiation of EPSPs and slow de-activation kinetics.[42] Most recently, Cortex and Servier have filed patent applications around the fused triazinone chemical class describing whole cell patch clamp, rat hippocampal slice and *in vivo* electrophysiology characterisation. Related patents also include data for inhibition of D-amphetamine-stimulated locomotion linked with potentiation of the field EPSP in the rat dentate gyrus, showing significant improvement on CX516 used as a standard control.

6.3.2 Benzothiadiazines

This second major class has its origins in the diuretic agent cyclothiazide **4** which, although not a CNS penetrant molecule, has been very widely characterised as an AMPAR positive modulator and used in many mechanistic and structural studies. The related analogue **5** from Servier is claimed to be a CNS penetrant, and has been evaluated clinically[47] and has demonstrated activity in the object recognition model in the rat, from 0.3 mg/kg when dosed orally. More recent analogues have been thoroughly reviewed by Pirotte.[41]

6.3.3 Phenethyl Sulfonamides

This class originated from a high throughput screen conducted by Lilly using human GluA4 flip receptors expressed in HEK-293 cells assessing their ability to potentiate responses mediated by 100 μM L-glutamate *via* changes in intracellular calcium ion concentration. The molecules which have progressed to clinical evaluation, **6** and **7** above, are typical of the series, being among the most potent potentiators described to date. The nitrile analogue **12**, whilst not progressed into clinical development, has nonetheless been extensively characterised and has provided enabling characterisation for the potential of AMPA receptor positive modulators in a number of disease indications. In particular, **12** has been described to be a potent potentiator in the aforementioned assay (pEC50 6.5) and displayed similar potency against GluA2 with lower potency against GluA1 and GluA3.[43,44] Whole-cell patch clamp studies in native tissue preparations from rat cortical neurons indicated that **12** potentiates the AMPAR by blocking desensitisation of the channel, and studies in Purkinje neurons confirmed the superior potency of this chemical class to its predecessors. Importantly, the studies were progressed to bridge between the *in vitro* data and the behavioural characterisation by developing an *in vivo* electrophysiology model, which allowed **12** to demonstrate a dose-dependent increase in rat hippocampal neuronal firing (ED_{50} 12 μg/kg). This allowed confident association of the recorded *in vitro* data with the data obtained from behavioural models or working memory and also with an increase of BDNF levels in rat.

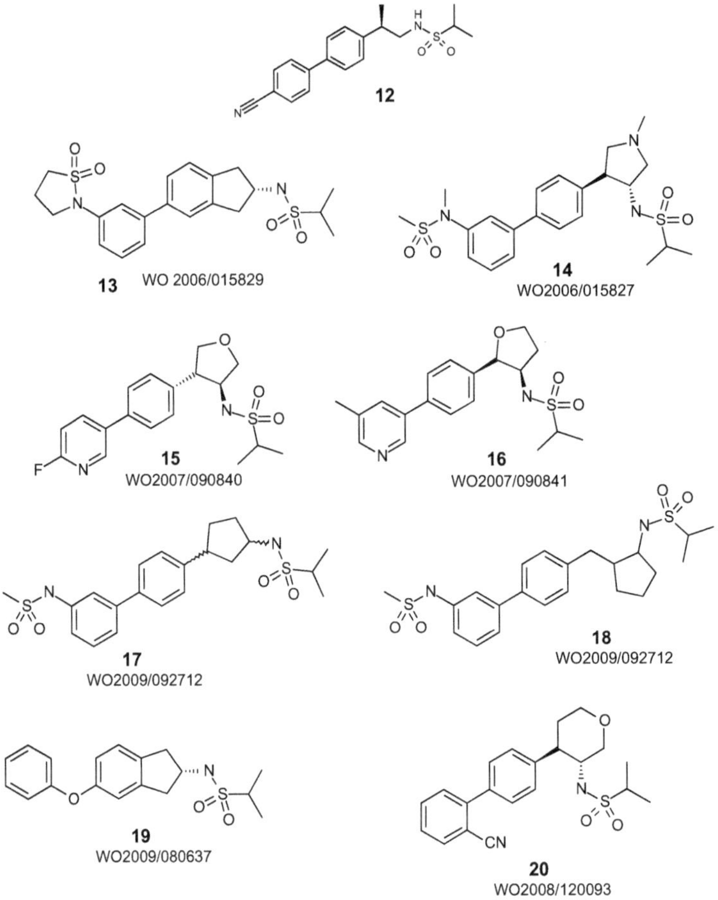

Figure 6.5 Structures of new phenethyl sulfonamide derivatives.

Further elaboration within this sulfonamide series has delivered a number of related chemotypes outlined in Figure 6.5, principally from GlaxoSmithKline, but also more recently from Pfizer. These chemotypes have considerably broadened the physicochemical properties of the core and have explored a range of conformational constraints and substitutions of the phenyl ring of the phenethylsulfonamide core, but, very significantly, from the published articles and associated patents, it is clear that the secondary sulfonamide unit with a small alkyl group is a key for activity (see further discussion around X-ray structures below).

6.4   Selection of Screening Platforms and Cascades

During the ongoing research programme we reasoned that a component of the lack of success in progressing AMPAR positive modulators through to Phase III was the lack of diversity in the available chemotypes as described above. Indeed, given the number of publications and the longevity of interest in the field of AMPAR modulator discovery, it is surprising that only the above series discovered by Lilly represented the first identification of a significant, new chemotype of AMPAR positive modulator. This was of particular significance as, prior to that, the two main chemotype series, benzamides and benzothiadiazines, both originated from the discovery that existing clinical molecules aniracetam and cyclothiazide possessed activity at the AMPAR.

However, this being said, given that the majority of work in this field was driven *via* patch clamp electrophysiology systems which were very low throughput, it was not until Lilly developed its high throughput screen that a very diverse range of structures was able to be screened. Consequently, when we started research efforts in this area, we too wanted to run a diversity screen to maximise our chances of uncovering a novel chemotype. This need was driven in part by the desire to uncover new chemical space which could allow for identification of unique, highly potent molecules with strong developability and drug-like characteristics. In particular, the benzothiadiazine and sulfonamide chemical classes described above contain inherently polar functionality which make it difficult to introduce hydrogen-bonding functionality elsewhere, without reducing the molecules' ability to permeate through the blood brain barrier. The exacting nature of the structure–activity relationship (SAR) developed around the secondary sulfonamide group has been well characterised and is clearly a key binding motif.[45]

We were interested by the choice of screening platform developed by Lilly and chose to establish a similar system. However, contrary to Lilly's approach to screen at homomeric recombinant GluA4, we instead chose to screen against homomeric GluA2 (flip splice variant) from analysis of the distribution patterns of the two subunits. Specifically, GluA2 is more highly expressed in cortical and subcortical areas of the brain than GluA4, which is predominant in the cerebellum.[46,47] The choice of GluA2 as the screening system is significant because, as was discussed above, the GluA2 subunit plays a critical role in the permeability of heteromeric AMPARs to calcium ions. Given that GluA2 receptors containing the arginine residue are impermeable to calcium we needed to configure our screen using the unedited GluA2 (Q) form to allow calcium permeability. The assay we developed was configured for the fluorescent imaging plate reader (FLIPR) which allows for high throughput detection of calcium ion concentrations by monitoring the fluorescence that occurs upon binding calcium to a membrane permeable fluorescent dye as shown in Figure 6.6. A dual addition protocol was used, whereby the compound solution was added 5 minutes before the addition of 100 µM of glutamate, to assay both the potential for the molecules to activate the channel *per se* and their ability to potentiate the glutamate response.

Intracellular calcium change can be detected by calcium-sensitive dyes using the fluorescent imaging plate reader (FLIPR). HEK293 cells expressing hGluA2 receptor are loaded with Fluo4-AM. The AM-ester is cleaved by

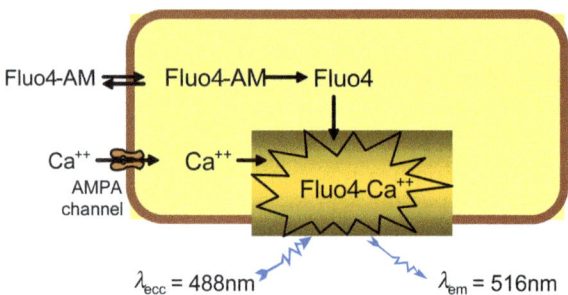

Figure 6.6 Schematic representation of FLIPR assay to detect new AMPA positive modulators.

intracellular esterases. Fluo4 binds to intracellular Ca^{2+} and increases fluorescence.

Although this assay provided us with the methodology to screen GlaxoSmithKline's compound collection, we were concerned that the data produced would not necessarily be meaningful in the context of the native system. In particular, as outlined in the introduction, the native receptors are likely to be heteromeric, can exist in multiple forms due to splice variants and sites for post-transcriptional modification, and will also be associated with multiple regulatory accessory proteins. To this end, it was imperative that we established a means of assaying potentiation of the AMPA receptor in native tissue to gain confidence in our ability to run a high throughput screen which could be reliably interpreted. Thus we developed a rat hippocampal electrophysiology assay using neuronal cultures from the dissected hippocampi of embryonic rat brains, in which the membrane potential of an individual neuron was held at -70 mV throughout the recording period. Control AMPA receptor-mediated currents were established by rapidly moving from normal extracellular solution to extracellular solution containing 30 µM AMPA, (corresponding to the previously determined approximate EC_{50}) for 2 seconds then returning to normal extracellular solution for 30 seconds. This cycle of solution changes was repeated continuously throughout the baseline recording period. For the assay itself, the lowest concentration of test compound was added to both perfusion tubes and the solution change cycle repeated until a stable current was measured. The inward currents induced by AMPA application were complex in that their magnitude and duration depended on three different mechanisms: (1) opening of the AMPA receptor channel measured in terms of peak amplitude, (2) desensitisation of AMPA receptors measured as the degree of relaxation of the inward current from its initial peak amplitude to a reduced steady state plateau level in the continued presence of AMPA and (3) deactivation of AMPA receptors measured as the rate at which the inward current decayed back to baseline following termination of the AMPA application.

As shown in Figure 6.7, the peak current amplitude was measured as the difference between baseline current and the maximum inward current

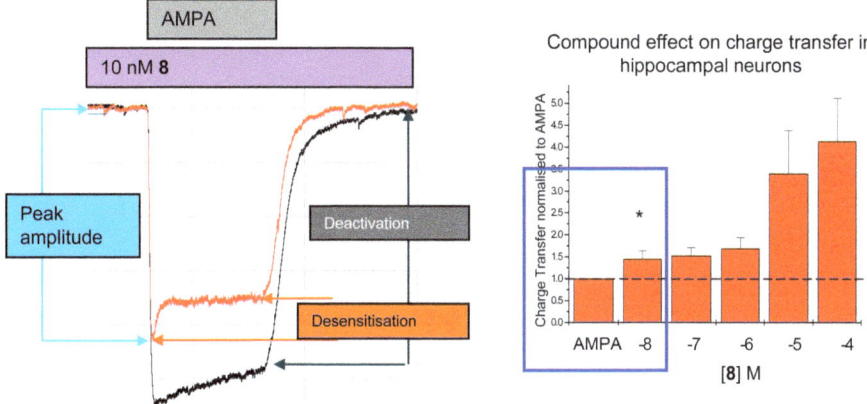

Figure 6.7 Sample data from electrophysiological characterisation in rat hippo-campal neurons. Right hand panel contains data for the charge transfer measured as the area under the curve for the left hand panel (excluding the deactivation phase) at log concentrations, *i.e.*, 10 nM [**8**] through to 10 μM [**8**].

detected following fast perfusion application of either AMPA alone or AMPA + test molecule. This peak current amplitude, measured at each concentration in the presence of 30 μM AMPA, was normalised to the peak current amplitude after application of AMPA alone. The desensitisation of the AMPA receptors was measured by the reduction from peak inward current amplitude to a steady-state current influx in the continued presence of AMPA. The deactivation was measured as the decay constant, tau, of a mono-exponential curve fitted to the relaxing inward current after AMPA application has stopped. The tau value, measured after each concentration in the presence of 30 μM AMPA, was normalised to the value of tau after application of AMPA alone. From these data we were able to generate a detailed biophysical profile of our lead molecules, and in particular, to understand the SAR against the various components of potentiation detailed above as well as the crude sum effect observed in the FLIPR experiment.

In addition to the rat hippocampal neuron electrophysiology assay, which allows comparisons to be made between AMPAR modulator potency *in vitro* and measured concentrations of AMPAR modulator in rat behavioural models, an intermediate bridging assay was developed using electro-physiology characterisation on the stable, recombinant human GluA2 homomeric cell line. This allowed us to develop increased confidence in the data gleaned from two assays, FLIPR and electrophysiology, using the same human recombinant cell line, specifically that the biophysical profile would be replicated in man.

With these assays in place, we constructed a preliminary screening cascade based on their respective throughputs and also to take account of the key parameters against which we wanted to optimise. In particular, we made

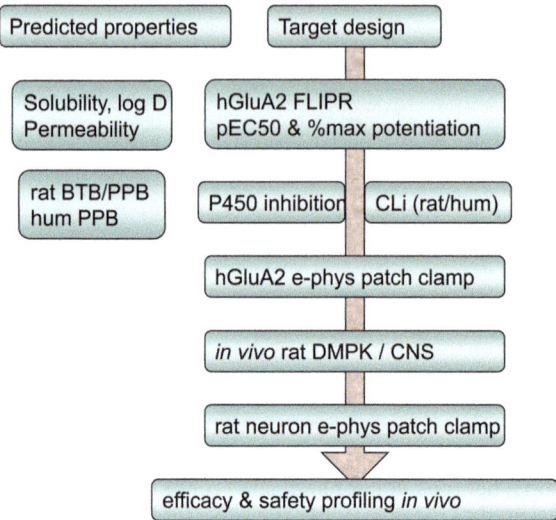

Figure 6.8 Early screening cascade used for initial stages of discovery programme.

extensive use of predicted physicochemical and DMPK properties to allow us to hone our target design, and refined these models as we generated data within a given chemical class. Given the complexities inherent in the pharmacology of this system, we wanted to ensure that we controlled as many aspects of the profile of the molecules as possible—in particular to endow them with the requisite properties to afford high free brain concentrations at target to drive pharmacodynamic effect (Figure 6.8).

6.5 Integration of X-ray Crystallography

In addition to the use of the biological characterisation, we were very interested in developing the reports from Eric Gouaux's laboratory on the use of the ligand binding domain construct. Clearly, in addition to providing significant additional information on the structure of the LBD and the mechanism of both activation by agonists and potentiation by allosteric modulators, these papers also provided us with the opportunity to consider a structure-based drug design approach, which in itself, is an unusual opportunity within an ion channel discovery programme. To this end, we followed the literature precedent and established our own high resolution crystal structure data for the rat GluA2 S1S2 LBD (specifically the pET15b-ratGluA2flopS1-GlyThr-S2 N754S construct) transformed into *E. coli* BL21(DE3) for expression into inclusion bodies or alternatively transformed into Origami B (DE3) where the protein was expressed in soluble form with an intact disulfide.[48] The protein was then expressed, isolated and purified as described, affording, according to LC-MS and *N*-terminal analyses, the correct species with a molecular weight of 29154 Da. For use in our lead optimisation programme, molecules were crystallised by screening against a grid of twelve conditions in a 96-well MRC sitting drop plate. The crystals could then be harvested, and from single frozen crystals for which data could be collected and processed with programmes from the CCP4 suite (CCP4, 1994). The structures were subsequently refined with Refmac,[49] initially with rigid body refinement, starting from related structures in the same $P2_12_12$ cell (*e.g.* pdb code 1ftj[50]) using the Coot programme was used for the model building.[51]

In the course of these studies, we also determined the crystal structure of the human GluA2 S1S2 LBD. As predicted from site directed mutagenesis studies and sequence alignments, the residues forming the binding site for positive modulators are the same in rat and human proteins, and moreover, the binding modes for compounds appear to be identical. Due to the finding that the rat protein was easier to prepare in high quality and the derived crystals were easier to grow, we routinely screened against the rat protein rather than the human. Prior to, and to a greater degree following, our work to generate in house crystal structure data, a number of structures have been published and their coordinates deposited which allows useful cross-series comparisons to be made (reviewed in Stawski *et al.,* 2010).[52]

6.6 Ion Channel Lead Optimisation Case History

From the high throughput screen that we ran, we identified a number of potential starting points, and initiated the early evaluation of a number of these in parallel. Given our prior experience in this field which led to the identification of (*N*-[(2*S*)-5-(6-fluoro-3-pyridinyl)-2,3-dihydro-1*H*-inden-2-yl]-2-propanesulfonamide) we recognised the importance of selecting the highest quality start point. Furthermore, given that the potencies of the AMPAR positive modulators reported vary widely and also, in our experience, do not necessarily correlate with downstream pharmacodynamic or behavioural model activity, we chose to place greater emphasis on the developability and drug-like characteristics of the lead above the inherent potency. Whilst this theme has received considerably more attention in recent years, for us it was nonetheless a bold step to favour our early optimisation efforts on one of the lower potency hits from the screen to maximise the opportunity to achieve an attractive development candidate.[53] Clearly, part of this thinking was driven by the potential of the molecule to be optimised further, and in particular with the ability to generate ligand-bound crystal structures, we reasoned that we would be able to accomplish this aspect of the development candidate discovery process in a more informed manner. Indeed, the generation of a high resolution crystal structure of our initial hit bound to the LBD construct was key to the successful initiation of this hit to lead optimisation process. To this end, we prepared crystals of **21** bound to the protein and analysed the structure obtained with reference to a number of standard AMPAR positive modulators we had crystallised during the early phase of the programme. Our initial structure, at 1.55 Å resolution, showed that there was a single molecule bound at the dimer interface at the position which is common to all of the reported structures for AMPAR positive modulators. This site is an inverted U-shape which is formed by two hydrophobic pockets which are linked by a hydrophobic 'saddle' region. Due to the symmetry of the LBD protein dimer construct discussed above, these hydrophobic pockets are identical, and indeed the difference in density for our molecules obtained from these crystallography experiments was an average of the ligand in both alternative orientations. In the absence of modulator ligand, the hydrophobic pockets are occupied by a network of water molecules, and across the classes of ligands that have been studied, structures have been determined for molecules binding to both, one or neither of the hydrophobic pockets. Aniracetam **1** binds only to the central part of the U, and doesn't occupy either of the hydrophobic pockets, which are instead each occupied by the network of four water molecules found in the parent protein structure.[28] Conversely, the structure obtained with the highly characterised allosteric modulator cyclothiazide **4** has two molecules bound to the protein dimer rather than the single molecule in the case of aniracetam.[54] Furthermore, the molecules bind into the hydrophobic pocket, in which the norbornenyl group displaces the network of water molecules and the polar functionality present in cyclothiazide extends out towards the solvent region making potential hydrogen bonds with side chains Ser497 and Ser754 and main chains Pro494 and Ser497. Consistent with the structure analogues in which the norbornenyl group are replaced by smaller hydrophobic groups are less potent.[55,56] A third mode of binding is exhibited by the phenethyl sulfonamide class described earlier. Data are shown for our earlier clinical development candidate **8**, which, in common with our in house and also the reported structures for the Lilly unconstrained analogues, binds to both the hydrophobic pocket and occupies the central part of the inverted U. For these molecules, the isopropyl sulfonamide group is present in the hydrophobic pocket, and the proximal phenyl ring occupies the 'saddle' region. The main-chain carbonyl oxygen of Pro494 is observed to make a hydrogen-bond with the amide nitrogen of the sulfonamide consistent with the exacting nature of the SAR observed for this sulfonamide group.

From the structure of our initial hit **21**, we observed a similar interaction of the phenyl ring on the twofold axis (*i.e.* at the centre of the inverted U) making a hydrophobic interaction with the Pro494 residues that creates this saddle-like structure (Figure 6.9(c)). The hydrophobic pocket, which was occupied by the sulfonamide group in the phenethylsulfonamide class or the norbornenyl moiety in cyclothiazide, is occupied by the trifluoromethyl indazole group. The carbonyl group pendant to the phenyl points down towards the other pocket of the inverted U and appears to make a single hydrogen bond to one of the four waters in the pocket (similar to the hydrogen bonding interaction described for aniracetam). Whilst there are a number of serine residues surrounding the binding pocket, there is no evidence for any obvious hydrogen bonds between **21** and the protein. However, there is a 2.9 Å contact from the main-chain NH function of Gly 731

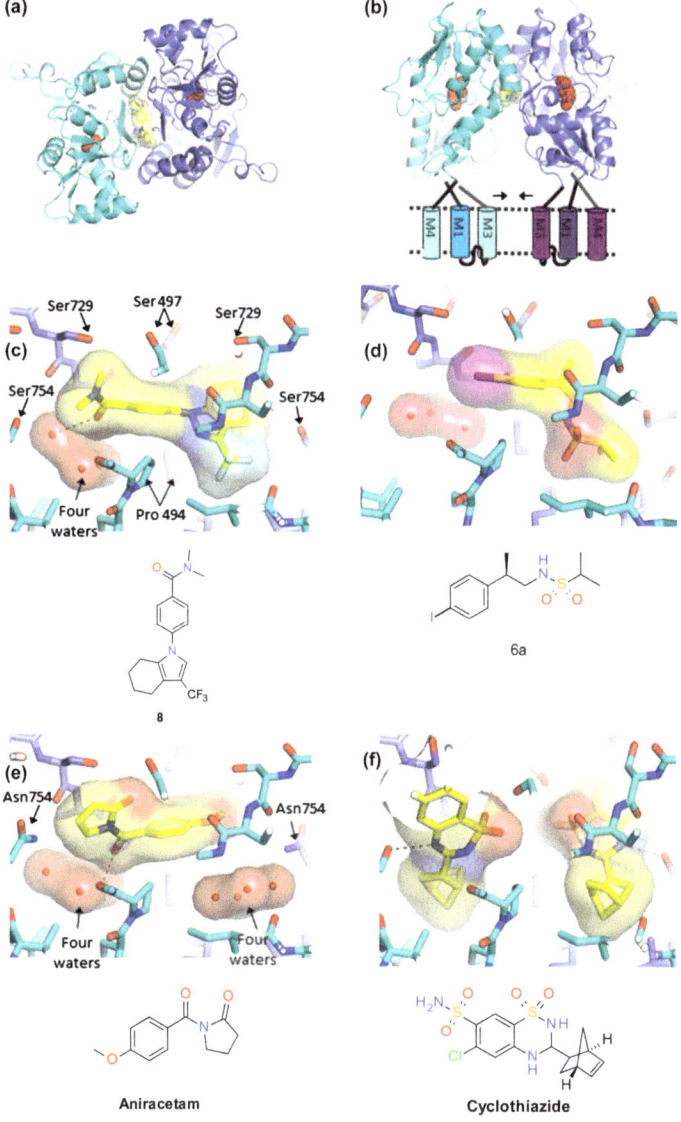

Figure 6.9 Orthogonal views of the 1.55 Å structure of **21** (**21** is in yellow; glutamate in red); (**b**) for orientation, schematic of the ion channel shown in an open state. The directions of the arrows indicate directions of movement for closure; (**c–f**) comparison of the binding mode of **21** (panel **c**) with other allosteric modulators; (**d**) **8** in complex with the human GluA2 S1S2 LBD;[28] (**e**) Aniracetam (pdb code: 2al5); (**f**) cyclothiazide (pdb code: 1lbc). The binding pocket resembles an inverted U. In the Aniracetam structure (**e**) both ends of the pocket are occupied by a cluster of four water molecules (small red spheres). H-bonds to ligands indicated by dotted lines.
(Figure reproduced in part with permission from Ward *et al.*[7] Copyright 2010 American Chemical Society).

to one of the fluorine atoms of **21**, which indicates that this fluorine could potentially be acting as an unusual hydrogen bond acceptor.[57] Interestingly, in later work published by Jamieson, they observe, consistent with the cyclothiazide crystal structure, that their trifluoromethyl interaction for a structurally related series is purely hydrophobic in nature.[58] In this way, we observe this trifluoromethyl group of **21**, having displaced the cluster of four waters, also makes van der Waals interactions with residues Leu 751, Ile481 and the carbonyl backbone of Lys 493. The tetrahydroindazole ring is also largely buried in the pocket, and although the upper part (oriented as in Figure 6.9) makes contact with solvent, this suggested to us that limited substitutions would be tolerated at this position. Conversely, the dimethyl amide group which is only pointing towards the other hydrophobic pocket above the expected waters, should have greater potential for modification and optimisation, potentially to pick up further hydrogen bonding interactions through the waters, Ser729 or Ser754.

In addition to our understanding of the binding mode of **21**, we also characterised this molecule extensively before initiation of hit to lead activities. Characterisation of the molecule in our FLIPR assay described above against the human, recombinant GluR2 cell line showed it to be weakly potent with measured pEC50 4.6 and 94% maximum potentiation (which was the fitted maximum response relative to 100% set by the presence of cyclothiazide standard included on each screening plate). Importantly, and in line with the earlier comments, we established that this molecule was able to potentiate AMPA-mediated currents in electrophysiology on both the human recombinant cell line and the rat native tissue neuronal cells. In addition to this characterisation at the AMPA receptor, we performed a wide selectivity screen and found that **21** exhibited a broadly clean profile and also appeared to be equipotent at the other homomeric AMPAR cell lines, GluA1, GluA3 and GluA4. Physiochemical data confirmed that the molecule had encouraging properties for initiation of lead optimisation, with moderate molecular weight (337 Da), particularly given the presence of the trifluoromethyl substituent, and lipophilicity (daylight clogP 3.0; measured logD 3.0) and low polar surface area (PSA 38 Å^2). The latter was particularly attractive in light of the observations that exploration in both the cyclothiazide and phenethylsulfonamide classes struggle to maintain sufficiently low PSA to confer high passive permeability across the blood–brain barrier due to the inherently polar nature of primary and secondary sulfonamide groups. Passive permeability was good, although solubility was moderate (approximately 140 µg/mL measured kinetic solubility from DMSO stock solution in pH 7.4 phosphate buffered saline). Furthermore, **21** had an acceptable profile against a number of early liability screens, such as hERG and P450 inhibition, although it did show rapid turnover in both rat and human microsomal incubations (12.7 mL/min/g and 3.5 mL/min/g, respectively) which mirrored the data obtained from a subsequent *in vivo* rat pharmacokinetic study in which measured concentrations were limited by a relatively high rate of clearance (estimated CLb 61 mL/min/kg, approximately 72% of liver blood flow in the rat). Encouragingly, however, good CNS penetration was measured (AUC $_{0-t}$ brain:blood ratio of 1.3) in line with the low PSA value, with no measurable PGP interaction or inhibition. Pleasingly, **21** also had a moderate free fraction in both rat plasma and rat brain (95.2% and 98.2% bound, respectively).

Given these data and the understanding from the X-ray crystal structure obtained, the initial focus to this optimisation centred on improving the metabolic stability which was presumed to be due to the presence of the terminal dimethylamide substituent. We prepared a number of cyclic amide derivatives which were able to stabilise the chemical template, but which, for the six-membered ring amides, also significantly reduced AMPAR activity. As shown in Table 6.1, morpholine **22** and pyrrolidine **23** had reduced turnover in both rat and human microsomes (data not generated for piperidine **24**). From the crystal structure determinations we had made, we were also able to construct models of the binding site, into which we could attempt to dock our new structures using various software packages. Attempts to dock **22** and **24** were hampered by the rigidity of the system and the resulting difficulty in accommodating the six-membered ring in the region pointing towards the hydrophobic pocket, due, in the main, to the limited scope for positioning the ligands afforded by the proline saddle region. Further modelling analysis suggested that introducing conformational flexibility would enhance

Table 6.1 Biological and *in vitro* DMPK data for early hit to lead analogues.[a]

Cpd	n	Q	pEC50	Asym. max	MWt	logD[b]	Sol[c]	CLi (mL/min/g)[d] Rat	Human
21	0	NMe$_2$	4.6	94%	337	3.0	138	12.7	3.5
22	0	Morpholine	<4.7	81%	379	3.1	123	<0.5	1.3
23	0	Pyrrolidine	5.0	123%	363	3.0	133	2.4	<0.5
24	0	Piperidine	<4	100%	377	1.6	16		
25	0	Me	4.5	94%	308	3.0	13		
26	1	NMe$_2$	5.7	107%	351	2.3	118	4.3	8.8
27	1	Pyrrolidine	5.5	82%	377	3.3	77	14.6	24.9

[a]FLIPR generated pEC$_{50}$ against hGluA2 flip isoform. All values are ± 0.2 and $n \geq 3$. Asym. max is the fitted maximum response, relative to 100% defined as the maximal response of cyclothiazide standard.
[b]Measured logD values.
[c]Kinetic solubility from DMSO stock solution in pH 7.4 phosphate buffered saline.
[d]Intrinsic clearance in rat and human liver microsomes.

activity, allowing the phenyl substituent remote from the tetrahydroindazole unit to be able to more effectively reach into the other arm of the hydrophobic pocket. Consequently, we prepared analogues such as **26** and **27** with an inserted methylene spacer unit, which showed approximately a ten-fold increase in potency. However, as shown in Table 6.1, although the switch from dimethyl amide (**26**) to pyrrolidine (**27**) amide was able to confer metabolic stability *in vitro* in the benzamide structures, preparing the analogous molecules in the phenyl acetamide (**21** and **23**) series did not afford the same gain. Further SAR was developed around the requirements for the various atoms in the amide, and interesting the carbonyl was the key motif, with ketone **25** being equipotent with the amide **21**. This was also true for the acetophenone derivative, although associated *in vitro* PK data were considerably worse.

However, and as was frequently taken advantage of during the course of these investigations, we were able to generate further ligand-bound X-ray crystal structures to enable us to both further refine our computational docking models and also to increase the fine definition of the binding pocket and potential for further modifications to the ligands. These would enable us, from analysis of the high resolution data such as shown in Figure 6.10 to be able to build hypotheses maps around the molecule such as the one shown in Figure 6.11. We would then prepare specific analogues to challenge these hypotheses and assist in their continued refinement,

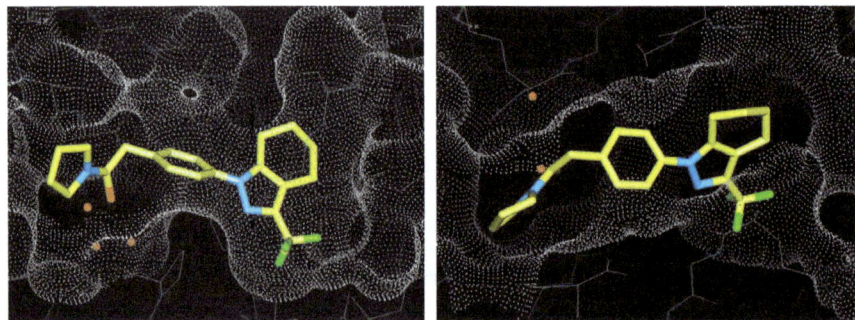

Figure 6.10   Crystal structure of **37** bound to rat GluA2S1S2 LBD.

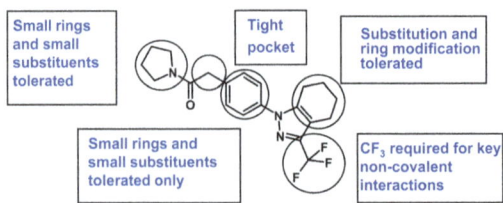

Figure 6.11   Hypotheses generated from crystal structure and SAR determination.

and were then able to use these models prospectively to filter proposed target molecules and to design new ligands. Indeed, the combined use of virtual docking and also of predictive models of DMPK and developability parameters using in house software greatly reduced the number of molecules we actually prepared. We predicted properties for clearance, absorption and CNS penetration through to solubility, permeability and developability risks, with the main physicochemical determinants being the molecular weight, lipophilicity, aromatic ring count and polar surface area.

Consequently, when we needed to prepare targets to address a specific issue, such as the metabolic instability, we were able to pre-filter our targets for likely activity and good developability and so prosecute the lead optimisation activities with the need for fewer molecules to be prepared. For example, our strategy to reduce the metabolic clearance of this series focussed on the changes to the molecule that had occurred in moving from **22** to **27**. Specifically, we designed molecules to block the benzylic methylene (**28** to **31**) which is commonly a site for oxidative metabolism and to reduce the electron density on both the phenyl ring (**32**) and the pyrrolidine ring (**33**). However, all the molecules shown in Figure 6.12, whilst active at the AMPAR, unfortunately possessed high intrinsic clearance in both rat and human microsomes.

Indeed, the only improvement in the metabolic stability that could be determined within this series was from the reversed amide analogues prepared to maintain the conformational flexibility of the homologated analogues above. **34** and **35** shown in Table 6.2 had significantly lower intrinsic clearance, which could be reduced even further by replacing the amide with a sulfonamide as in **36** and **37**.

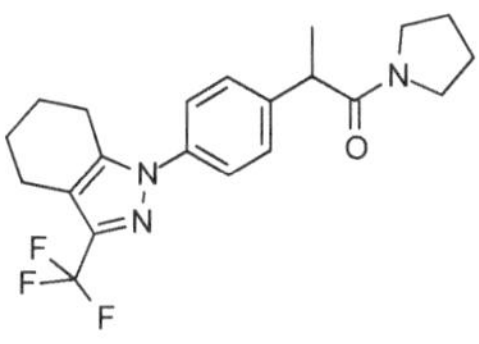

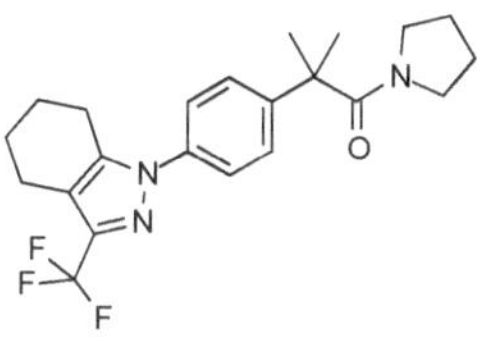

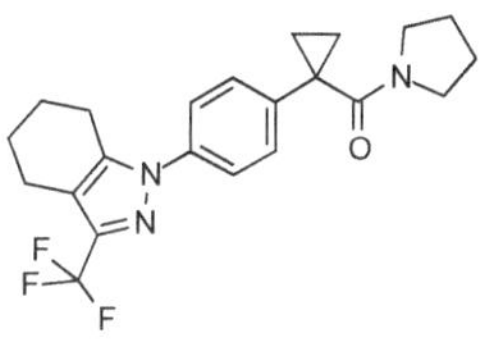

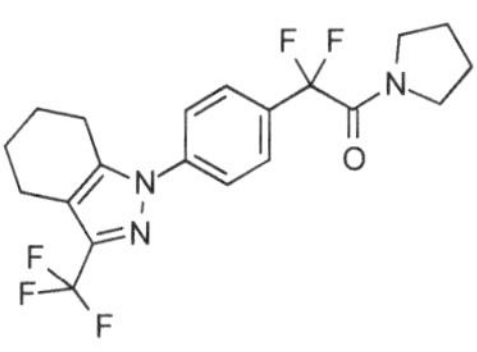

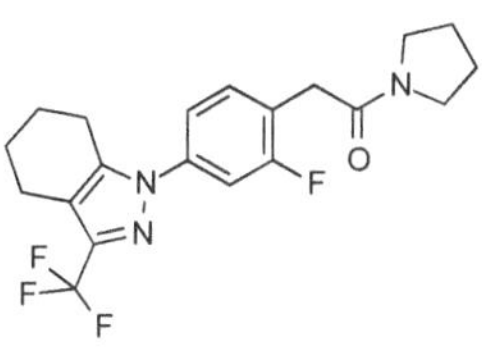

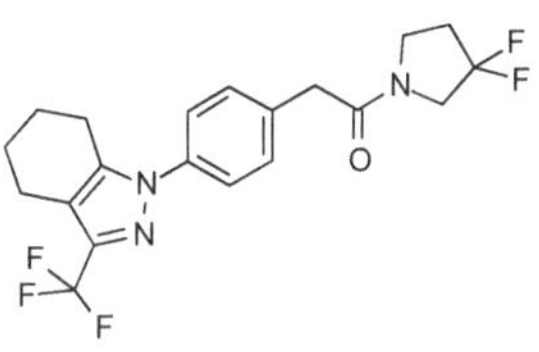

Figure 6.12   Analogues designed to reduce metabolic stability.

Table 6.2   Biological and *in vitro* DMPK data for early hit to lead analogues (see Table 6.1 legend for assay details).

							CLi (mL/min/g)	
Cpd	X	NRY	pEC50	Asym. max.	MWt	logD	Sol	Rat	Human
34	H	Pyrrolidinone	6.0	132%	363	2.4	110	2.4	9.1
35	F	Pyrrolidinone	6.0	112%	381	3.3	88	2	2.9
36	H	NHSO$_2$Me	5.1	112%	373	2.5	103	1.6	<0.7
37	H	NHSO$_2^i$Pr	5.0	79%	401	2.8	11	2.5	0.8

To more fully understand the various parameters driving the metabolic instability, we incubated a number of our molecules with rat and human microsomes and analysed the metabolites by mass spectrometry. Interestingly, whilst we did observe metabolism occurring in and around the amide groups, we saw a major pathway of oxidative metabolism occurring on the aliphatic distal carbons of the tetrahydroindazole ring, and for some molecules such as **35**, we observed only the tetrahydroindazole oxidation. Clearly, from these data and our understanding of the protein environment around the tetrahydroindazole part of the molecules, we designed molecules to address this clearance pathway, by replacing one of the vulnerable methylene units with a heteroatom. Our earlier attempts to modify the trifluoromethyl pyrazole ring had been unsuccessful, and indeed for all the molecules that we prepared and for which we subsequently generated ligand-bound crystal structures, we found that the key binding motif was through the occupancy of the trifluoromethyl pyrazole in the hydrophobic pocket. So, focussing design on the aliphatic ring analogues, we prepared a number of molecules including **38** to **40** and found that these had generally improved metabolic stability both *in vitro* and *in vivo*. Furthermore, as would be expected from the changes in physicochemical properties, the molecules had improved solubility and reduced non-specific protein binding, although surprisingly, this significant difference in electrostatic character was well tolerated for binding to the protein (Table 6.3).

These modifications, whilst clearly representing only a subset of the molecules prepared, allowed us to identify a sufficient number of analogues (Figure 6.13) to be able to progress with confidence into the late lead optimisation phase, during which molecules were subjected to further characterisation to assess their suitability for progression as clinical drug candidates.

Table 6.3 Characterisation of indazole derivatives containing heteroatoms (see Table 6.1 legend for assay details).

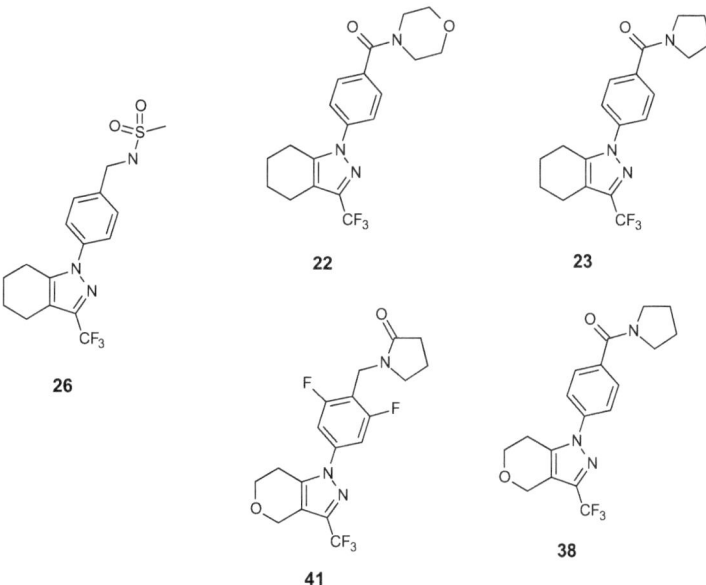

Cpd	X	Y	pEC50	Asym max	logD	Bl Cmax[a] ng/mL	CLb[b] mL/ min/ kg	rBTB %	Perm[c] nm/s	Sol	CLi (mL/min/g) Rat	CLi (mL/min/g) Human
38	CH$_2$	O	5.0	138%	2.4	808	e41	87.9	470	215	0.8	0.5
39	O	CH$_2$	5.3	116%	2.6	331	ND	94.9	440	123	<0.5	1.7
40	NMe	CH$_2$	4.4	102%	2	131	e22	63.0	430	93	<0.5	<0.5

Key to assay details in Table 6.1 legend. Additionally, [a]from 3 mg/kg oral dose; [b]e = Estimated clearance value, based on *in vivo* hepatic extraction determined following a 3 mg/kg oral dose and blood sampling *via* the hepatic portal vein and heart and using 85 mL/min/kg as liver blood flow in rat; [c]Artificial membrane permeability assay.

Figure 6.13 Molecules selected for late lead optimisation phase.

6.7   Challenges in Lead Optimisation and Selection of Clinical Discovery Candidate

Our goal in integrating X-ray crystallography into this lead optimisation programme was to be able to rationalise the observed affinities and design a superior next generation of analogues. To this end, from our ligand bound crystal structures in this series we were able to rationalise the potency differences on the basis of either additional protein–ligand interactions, or more commonly, by assessing the number of water molecules that were displaced from the hydrophobic pockets. As an illustration of this, homologating benzamide **21** to phenyl acetamide derivative **23** displaced two additional buried water molecules which can potentially be the explanation for the increase in potency from pEC50 4.5 to 5.4. The supplementary potency increase observed in moving to reversed amides **34** and **35** was associated with the observation that in addition to the displacement of the two buried water molecules, these molecules also displaced two further solvent exposed waters. This being said, these crystal structures and their derived computational docking models could not be used with any great success for the *de novo* design of new ligands, and no accurate predictive model for binding affinity could be established. However, their value lay principally in defining the regions of the molecules open to modification and in gross classification of ligands into inactive, low and high potency. Clearly, a component of this lies in the gap that exists between the binding affinity of a molecule for a LBD protein construct and the functional response required to demonstrate potentiation at the AMPA receptor. This same gap was evident in the progression of molecules down the screening cascade through initially electrophysiological characterisation on human recombinant GluA2 cell lines and rat hippocampal neurones and subsequently *in vivo* pharmacodynamic and behavioural models. During the lead optimisation and characterisation process, it was important to remember that the FLIPR data, whilst providing a surrogate functional readout for potentiation of the AMPA, are nonetheless generated for a recombinant homomeric cell line, and does not reflect the subunit heterogeneity, splice variants or post-transcriptional modifications present in the native state. Furthermore, and given the increasingly important roles being ascribed to the accessory proteins, the regulation of signalling through the AMPA receptor based on the channel alone is clearly an over-simplification. However, despite these concerns, a screening cascade could be constructed to address these concerns and in essence build confidence in the validity of the early screening data. As discussed, the *in vitro* electrophysiology was key to this, and building on these data, we also developed an *in vivo* electrophysiology paradigm. In this model, a concentric, bipolar stimulating electrode implanted into the medial perforant pathway was used to generate stable field excitatory postsynaptic potentials which could be recorded by a second tungsten recording electrode implanted into the ipsilateral dentate gyrus. The population spike amplitudes that were recorded were primarily a measure of synchronous monosynaptic AMPAR activation and deactivation and gave us confidence that we were able to join up the steps in our screening cascade and associate the subsequent behavioural rodent model data with AMPAR-mediated activity.

We chose to screen for effects on learning and memory to two different models—passive avoidance and novel object recognition (NOR). For the former, the task requires the rats to remember an aversive stimulus, and this recall is impaired by post-training treatment with the muscarinic antagonist scopolamine. The rats were pre-treated with the test compounds over a dose range at a pre-treatment time which would allow the best estimate of achieving maximal brain concentrations from a given dose. The scopolamine was administered 6 hours post training and the task recall occurred a further 18 hours later. This model was used, along with the *in vivo* electrophysiology, to select the most efficacious molecules for profiling in the NOR model. This model involves placing rats in a test arena with two identical objects for 3 minutes, then 24 hours later, returning the rats to the same environment for 3 minutes with one of the original, familiar objects and a novel, unfamiliar object. The time spent exploring the objects during the two test phases was recorded, and, given that rats display an innate preference to explore novel objects over familiar ones, an assessment could be made of the rats' ability to remember the object from the first day. The impact on this recall by our test compounds was assessed by administration of compounds across a suitable dose range and with appropriate pre-treatment time prior to both test arena phases.

In our hands, we found this model to be extremely reliable and reproducible, and as such provided us with the key behavioural data from which we derived efficacious concentrations and established a PK/PD profile to allow us to predict the dosing range for use in the clinic. However, we did encounter an additional complication in this regard, in that when we ran this model following a sub-chronic dosing regime, in which we dosed our compounds once or twice daily for 7 days and then prior to the test arena phases on days 8 and 9, we saw a 3–10 fold reduction in doses (and measured concentrations) required to drive improved performance. Initially we had committed to this repeat dosing paradigm to discharge the concern that we might lose sensitivity to the AMPAR potentiation over time. However, on the contrary, repeat exposure to the AMPAR positive allosteric modulators appears to facilitate the improved learning and memory that allows for enhanced performance in the novel object recognition task.

In moving our molecule set described in Figure 6.13 into the phase to select the candidate to progress to pre-clinical development, we had intentionally selected those which passed all our early criteria in terms of potency, selectivity, early developability and DMPK. Moving further down the screening path, we generated additional information against wider selectivity, DMPK in non-rodent species, bio-activation and wider P450 interaction profiles, *in vitro* toxicology screens covering genotoxicity and mutagenicity, hepatotoxicity and cardiotoxicity and *in vivo* tolerability profiling as well as the pharmacodynamic and behavioural models described above. From this, two observations were key to being able to successfully identify a high quality clinical candidate to move into development. Firstly, given the earlier comments regarding the difficulty in fully establishing the biophysical profile of the molecules across the various assay systems and, as should be emphasised, limited understanding of exactly the biophysical profile that would be most desirable, the only pragmatic approach was to profile a number of molecules in the animal models to select the most effective. From retrospective analysis of the data, this selection could not be made on purely DMPK and *in vitro* potency considerations. Secondly, we aimed to cover as much physiochemical drug-like space as possible with our molecule set, and from those identified, whilst they clearly fall within a single chemical class, they nonetheless possessed varied lipophilicity, electrostatic surface field potential and polar surface area. This potentially contributed to the varied profile we saw through the late stage screening activities, in which a range of tolerability profiles and liabilities were uncovered (details reported in Ward *et al.*).[7]

Ultimately, these analyses allowed us to progress **23** into the full development candidate activities required for initiation of clinical development. This molecule represented the best overall profile identified, being effective in the NOR model acutely from 0.3 mg/kg (equivalent to 25 ng/mL blood; 37 ng/g brain exposure) and in the sub-chronic paradigm from 0.1 mg/kg (equivalent to 7 ng/mL blood; 9 ng/g brain exposure) and possessing an overall strong developability profile.

6.8   Future Perspectives

Whilst we have been successful in identifying clinical development candidates which are AMPAR positive allosteric modulators, the clarity around the exact mechanism of action is not fully defined. Indeed this is not unexpected given that, in recent years, the literature covering AMPARs has rapidly expanded to include far greater appreciation of the roles of accessory systems, regulatory proteins and mechanisms and subunit composition and localisation. Although from a drug discovery programme perspective we are able to build confidence in our assay systems by relating data from early screens to those of increasing complexity and relevance to the native state, it is noteworthy that a fuller understanding of the regulation of trafficking and function may lead us to either identify new therapeutic targets or to be able to refine our activities to identify superior AMPAR modulators. In particular, it is now clear that attached to the core of the AMPAR are many auxiliary subunits comprising the TARPS (stargazin/g-2, g-3, g-4, g-5, g-7, g-8), cornichon 2&3 and CKAMP44 (Shisa9) as well as a whole range of transmembrane auxiliary subunits.[59] Furthermore, regulation of these interactions by proteins such as PDS-95 like membrane-associated guanylate kinases is developing[60,61] and from these understandings, new models are being proposed to describe synaptic recruitment of AMPARs.[62]

In addition to these new developments, attention has also been focussed on the potential of targeting subunit specific interactions. Through our profiling, we generally saw broadly similar activity against the various GluA1-4 homomeric subunits, although as understanding increases around the differential properties of the subunits[63] and their differential expression[64] this is clearly an area for potential future refinement of both disease area and tolerability.

It is noteworthy that positive allosteric modulation of the AMPAR has been cited as an attractive therapeutic approach to a wide range of neurological and psychiatric disease states, and it is to be anticipated that molecules which are able to either directly modulate the biophysical properties of the channel or indirectly modulate its regulation, expression or trafficking could all be viable approaches. Considerable work remains to be done to both progress an AMPAR modulator into Phase III trials and beyond and also to further delineate the various accessory and subunit complexities discussed, in particular to identify new protein targets. Furthermore, more recent attempts to allosterically modulate both the NMDA and kainate channels will hopefully be able to learn from the body of experience with the AMPA channel and potentially uncover new approaches of generic applicability.

Finally, it is imperative, given the complexities inherent in the ion channel structure and pharmacology, that any discovery programmes maintain flexibility in their approach—both in their willingness to incorporate additional information sources such as can be provided by structural determination and also to challenge the relevance of the assay systems, to ensure that the biological data being optimised against is of significance.

References

1. S. Ozawa, H. Kamiya and K. Tsuzuki, *Prog. Neurobiol.*, 1998, **54**, 581–618.
2. R. Dingledine, K. Borges, D. Bowie and S. F. Traynelis, *Pharmacol. Rev.*, 1999, **51**, 7–61.
3. G. L. Collingridge, R. W. Olsen, J. Peters and M. Spedding, *Neuropharmacology*, 2009, **56**, 2–5.
4. P. E. Chen and D. J. Wyllie, *Br. J. Pharmacol.*, 2006, **147**, 839–853.
5. Y. Sun, R. Olson, M. Horning, N. Armstrong, M. Mayer and E. Gouaux, *Nature*, 2002, **417**, 245–253.
6. E. Gouaux, *J. Physiol.*, 2004, **554**, 249–253.
7. S. E. Ward, M. Harries, L. Aldegheri, N. E. Austin, S. Ballantine, E. Ballini, D. M. Bradley, B. D. Bax, B. P. Clarke, A. J. Harris, S. A. Harrison, R. A. Melarange, C. Mookherjee, J. Mosley, G. Dal Negro, B. Oliosi, K. J. Smith, K. M. Thewlis, P. M. Woollard and S. P. Yusaf, *J. Med. Chem.*, 2010, **53**(15), 5801–5812.
8. J. N. Kew and J. A. Kemp, *Psychopharmacology (Berl)*, 2005, **179**, 4–29.

9. J. K. Christensen, A. V. Paternain, S. Selak, P. K. Ahring and J. Lerma, *J. Neurosci.*, 2004, **24**, 8986–8993.

10. W. Lu, Y. Shi, A. C. Jackson, K. Bjorgan, M. J. During, R. Sprengel, P. H. Seeburg and R. A Nicoll, *Neuron*, 2009, **62**, 254–268.

11. C. Mulle, A. Sailer, G. T. Swanson, C. Brana, S. O'Gorman, B. Bettler and S. F. Heinemann, *Neuron*, 2000, **28**, 475–484.

12. J. D. Leever, S. Clark, A. M. Weeks and K. M. Partin, *Mol. Pharmacol.*, 2003, **64**, 5–10.

13. R. M. Klein and J. R. Howe, *J. Neurosci.*, 2004, **24**, 4941–4951.

14. I. H. Greger, E. B. Ziff and A. C. Penn, *Trends Neurosci.*, 2007, **30**, 407–416.

15. G. L. Collingridge, J. T. Isaac and Y. T. Wang, *Nat. Rev. Neurosci.*, 2004, **5**, 952–962.

16. B. Sommer, M. Kohler, R. Sprengel and P. H. Seeburg, *Cell*, 1991, **67**, 11–19.

17. P. H. Seeburg, F. Single, T. Kuner, M. Higuchi and R. Sprengel, *Brain Res.*, 2001, **907**, 233–243.

18. C. Tigaret and D. Choquet, *Science*, 2009, **323**, 1295–1296.

19. J. Schwenk, N. Harmel, G. Zolles, W. Bildl, A. Kulik, B. Heimrich, O. Chisaka, P. Jonas, U. Schulte, B. Fakler and N. Klocker, *Science*, 2009, **323**, 1313–1319.

20. B. Y. D. Colquhoun, P. Jonas and B. Sakmann, *J. Physiol.*, 1992, **458**, 261–287.

21. L. O. Trussell, L. L. Thio, C. F. Zorumski and G. D. Fischbach, *Proc. Natl. Acad. Sci. U. S. A.*, 1988, **85**, 4562–4566.

22. R. A. Hall, M. Kessler, A. Quan, J. Ambros-Ingerson and G. Lynch, *Brain Res.*, 1993, **628**, 345–348.

23. Y. Stern-Bach, S. Russo, M. Neuman and C. Rosenmund, *Neuron*, 1998, **21**, 907–918.

24. A. I. Sobolevsky, M. P. Rosconi and E. Gouaux, *Nature*, 2009, **462**, 745–756.

25. S. E. Ward and M. Harries, *Curr. Med. Chem.*, 2010, **17**, 3503–3513.

26. S. J. A. Grove, C. Jamieson, J. K. F. Maclean, J. A. Morrow and Z. Rankovic, *J. Med. Chem.*, 2010, **53**, 7271–7279.

27. S. E. Ward, B. D. Bax and M. Harries, *Br. J. Pharmacol.*, 2010, **160**, 181–190.

28. R. Jin, S. Clark, A. M. Weeks, J. T. Dudman, E. Gouaux and K. M. Partin, *J. Neurosci.*, 2005, **25**, 9027–9036.

29. R. Cumin, E. F. Bandle, E. Gamzu and W. E. Haefely, *Psychopharmacology (Berl)*, 1982, **78**, 104–111.

30. M. Ingvar, J. Ambros-Ingerson, M. Davis, R. Granger, M. Kessler, G. A. Rogers, R. S. Schehr and G. Lynch, *Exp. Neurol.*, 1997, **146**, 553–559.

31. D. C. Goff and J. T. Coyle, *Am. J. Psychiatry.*, 2001, **158**, 1367–1377.

32. D. C. Goff, J. S. Lamberti, A. C. Leon, M. F. Green, A. L. Miller, J. Patel, T. Manschreck, O. Freudenreich and S. A. Johnson, *Neuropsychopharmacology*, 2008, **33**, 465–472.

33. M. L. Woolley, K. A. Waters, J. E. Gartlon, L. P. Lacroix, C. Jennings, F. Shaughnessy, A. Ong, D. J. Pemberton, M. H. Harries, E. Southam, D. N. Jones and L. A. Dawson, *Psychopharmacology*, 2009, **202**, 343–354.

34. R. H. Weisler, *Expert Opin. Emerging Drugs*, 2007, **12**, 423–434.
35. S. G. Dickstein, K. Bannon, F. X. Castellanos and M. P. J. Milham, *Child Psychol. Psychiatry*, 2006, **47**, 1051–1062.
36. M. Carlsson and A. Carlsson, *Trends Neurosci.*, 1990, **13**, 272–276.
37. R. E. Hampson, R. A. España, G. A. Rogers, L. J. Porrino and S. A. Deadwyler, *Psychopharmacology*, 2009, **202**, 355–369.
38. A. S. Chappell, C. Gonzales, J. Williams, M. M. Witte, R. C. Mohs and R. Sperling, *Neurology*, 2007, **68**, 1008–1012.
39. P. M. Doraiswamy and G. L. Xiong, *Expert Opin. Pharmacother.*, 2006, 7, 1–10.
40. S. E. Ward, in *Drug Discovery for Psychiatric Disorders*, ed. Z. Rankovic, M. Bingham, E. J. Nestler and R. Hargreaves, Royal Society of Chemistry, Cambridge, 2012.
41. B. Pirotte, P. Francotte, E. Goffin, P. Fraikin, L. Danober, B. Lesur, I. Botez, D.-H. Caignard, P. Lestage and P. de Tullio, *Curr. Med. Chem.*, 2010, **17**, 3575–3582.
42. A. C. Arai, M. Kessler, G. Rogers and G. Lynch, *Mol. Pharmacol.*, 2000, **58**, 802–813.
43. J. C Quirk and E. S. Nisenbaum, *CNS Drug Rev.*, 2002, **8**, 255–282.
44. P. Miu, K. R. Jarvie, V. Radhakrishnan, M. R. Gates, A. Ogden, P. L. Ornstein, H. Zarrinmayeh, K. Ho, D. Peters, J. Grabell, A. Gupta, D. M. Zimmerman and D. Bleakman, *Neuropharmacology*, 2001, **40**, 976–983.
45. P. L. Ornstein, D. M. Zimmerman, M. B. Arnold, T. J. Bleisch, B. Cantrell, R. Simon, H. Zarrinmayeh, S. R. Baker, M. Gates, J. P. Tizzano, D. Bleakman, A. Mandelzys, K. R. Jarvie, K. Ho, M. Deverill and R. K. Kamboj, *J. Med. Chem.*, 2000, **43**, 4354–4358.
46. R. S. Petralia and R. J. Wenthold, *J. Comp. Neurol.*, 1992, **318**, 329–354.
47. M. Beneyto and J. H. Meador-Woodruff, *J. Comp. Neurol.*, 2004, **468**, 530–554.
48. P. Naur, B. Vestergaard, L. K. Skov, J. Egebjerg, M. Gajhede and J. S. Kastrup, *FEBS Lett.*, 2005, **579**, 1154–1160.
49. G. N. Murshudov, A. A. Vagin and E. J. Dodson, *Acta Crystallogr. D Biol. Crystallogr.*, 1997, **53**, 240–255.
50. N. Armstrong and E. Gouaux, *Neuron*, 2000, **28**, 165–181.
51. P. Emsley and K. Cowtan, *Acta Crystallogr. D Biol. Crystallogr.*, 2004, **60**, 2126–2132.
52. P. Stawski, H. Janovjak and D. Trauner, *Bioorg. Med. Chem.*, 2010, **18**, 7759–7772.
53. P. D. Leeson and B. Springthorpe, *Nature*, 2007, **6**, 881–890.
54. Y. Sun, R. Olson, M. Horning, N. Armstrong, M. Mayer and E. Gouaux, *Nature*, 2002, **417**, 245–253.
55. H. Hald, P. K. Ahring, D. B. Timmermann, T. Liljefors, M. Gajhede and J. S. Kastrup, *J. Mol. Biol.*, 2009, **391**, 906–917.
56. C. P. Ptak, A. H. Ahmed and R. E. Oswald, *Biochemistry*, 2009, **48**, 8594–8602.

57. J. D. Dunitz and R. Taylor, *Chem. Eur. J.*, 1997, **3**, 89–98.
58. C. Jamieson, R. A. Campbell, I. A. Cumming, K. J. Gillen, J. Gillespie, B. Kazemier, M. Kiczun, Y. Lamont, A. J. Lyons and J. K. F. Maclean, *Bioorg. Med. Chem. Lett.*, 2010, **20**, 6072–6075.
59. T. Nakagawa, *Mol. Neurobiol.*, 2010, **42**, 161–184.
60. K. S. Kim, D. Yan and S. Tomita, *J. Neurosci.*, 2010, **30**, 1064–1072.
61. A. Sumioka, D. Yan and S. Tomita, *Neuron*, 2010, **66**, 755–767.
62. P. Opazo and D. Choquet, *Mol. Cell. Neurosci.*, 2011, **46**(1), 1–8.
63. J. J. Fleming and P. M. England, *Bioorg. Med. Chem.*, 2010, **18**, 1381–1387.
64. M. R. Emond, J. M. Montgomery, M. L. Huggins, J. E. Hanson, L. Mao, R. L. Huganir and D. V. Madison, *J. Physiol.*, 2010, **588**, 1929–1946.

The Discovery of Novel Inhaled ENaC Blockers for the Treatment of Cystic Fibrosis Lung Disease

CATHERINE HOWSHAM[a] AND HENRY DANAHAY*[b]

[a] Global Discovery Chemistry, Novartis Institutes for Biomedical Research, Horsham, West Sussex, RH12 5AB, UK; [b] Respiratory Disease Area, Novartis Institutes for Biomedical Research, Horsham, West Sussex, RH12 5AB, UK
*Email: henry.danahay@novartis.com

7.1 Introduction

On a daily basis, we can inhale up to 12,000 liters of air, and with this comes a wealth of particles including environmental pollutants, allergens and microbes that impact on the surface epithelium of the lung. Despite this ever present challenge, in healthy individuals the lung maintains a sterile environment, keeping the airways clear to ensure the efficient movement of air towards the alveolus as we respire. The lung achieves this through the utilization of various systems including phagocytosis of particles by lung resident macrophages and also through mucociliary (MCC) and cough clearance (CC). Mucociliary and cough clearance serve as one of the immediate local responses to the inhalation of particulates, and are the result of the interplay of a variety of cells in the lung mucosa that regulate the production of a mucus gel, ciliary beating to propel the mucus and neuronal

RSC Drug Discovery Series No. 39
Ion Channel Drug Discovery
Edited by Brian Cox and Martin Gosling
© The Royal Society of Chemistry 2015
Published by the Royal Society of Chemistry, www.rsc.org

reflexes that enable us to cough. An increasing body of data now highlights the degree of lung mucosal hydration as being central to effective MCC and CC and the salient roles that a number of ion channels can play in the regulation of hydration.[1–3] In this chapter, we focus on one of these ion channels, the epithelial sodium channel (ENaC), and the hypothesis that blocking ENaC in the lung mucosa will result in an enhancement of mucus clearance and thereby a reduced risk of acquiring respiratory infections. To this end, we likewise review the drug discovery approaches to attenuate airway ENaC function and the status of key experimental compounds in clinical trials.

7.2 ENaC: Structure, Function and Regulation

ENaC was first cloned in 1993[4] and since then our understanding of functional regulation and how that relates to structure has greatly expanded. The channel is composed of three subunits: α, β and γ, that form a heteromultimer with a stoichiometry that was originally believed to be $2\alpha:\beta:\gamma$[5] although recent *in silico* modeling based upon the crystal structure of the related channel ASIC1 (acid sensing ion channel) supports a $1:1:1$ mix (Figure 7.1(A)).[6,7]

Each subunit consists of two transmembrane regions, a large extracellular domain and cytosolic N- and C-termini that include proline-rich regions in the C-terminus.[5]

ENaC is expressed and is functional in renal, colonic, corneal, sweat duct and respiratory epithelia, where it forms a low conductance channel (~ 4 pS) with selectivity for sodium over potassium (>10 fold). Loss and gain of function mutations in ENaC subunits induce human disease, including pseudohypoaldosteronism type 1 (PHA1),[8,9] a salt wasting disease, and Liddle's syndrome,[10,11] a disease associated with salt retention and hypertension. The recently identified δ-subunit of ENaC may also be of functional significance, although its biological role is presently unclear. ENaC is blocked by the potassium sparing diuretics amiloride (Figure 7.2(A), Ki ~ 250–500 nM) and triamterene (Figure 7.2(B), Ki ~ 10–30 μM). Amiloride blocks the channel in the pore, and mutagenesis studies have identified the key residues for binding (Figure 7.1(B)).[5]

ENaC function can be regulated by hormones such as aldosterone, vasopressin and insulin as well as PKA/cAMP, PKC, Ca^{2+} and G proteins.[12] Furthermore, ENaC regulation can be tissue-specific. For example, ENaC function in the cortical collecting duct of the kidney can be regulated by mineralocorticoids whereas ENaC in the airways appears to be refractory to hormones such as aldosterone.[13] Similarly, flow-induced shear stress in the kidney results in an increase in ENaC activity,[14] whilst shear-stress induces ATP release in the airway that can reduce ENaC function.[15,16] Gain of function Liddle's mutations in the proline rich sequence (PY-motif) domains of the ENaC subunits result in increased retention at the plasma membrane as a result of attenuated Nedd4-mediated ubiquitination and cause an

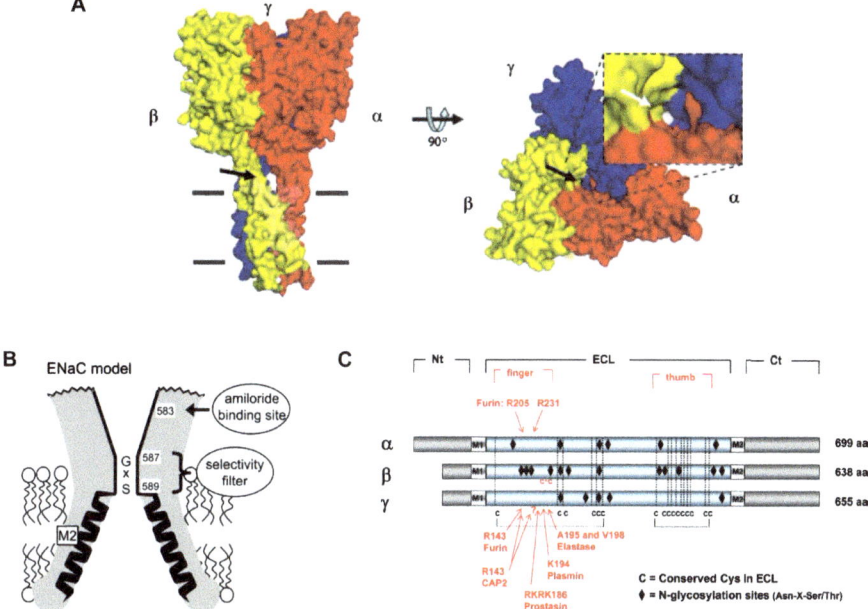

Figure 7.1   Predicted structure of oligomerized human ENaC (A) based on the structure of the related channel, ASIC1: α- (red), β- (yellow), and γ- (blue). The enlarged area shows the opening at the top of the channel to a tunnel that runs the length of the extracellular region of ENaC to the channel pore down the center. At its opening, this tunnel is 14 Å wide. At its narrowest, it is 4 Å wide (reproduced with permission from reference 7). A 20-fold change in channel affinity for amiloride caused by the mutation αS583C (B) and 1,000-fold change induced by the homologous mutations βG525 or γG537 suggests that these residues are directly involved in binding interactions with amiloride (reproduced with permission from reference 5). The interaction of a protease(s) is required to observe activation of ENaC (C). The sites for α-subunit cleavage by furin and γ-subunit cleavage by furin, prostasin (CAP1), CAP2, elastase (neutrophil and pancreatic), and plasmin are within the large extracellular loops of the sub-units (reproduced with permission from reference 23b).

increased ENaC-mediated sodium absorption from the kidney and hypertension.[17] Of note, patients with Liddle's disease do not present with a lung phenotype or evidence of enhanced ENaC function in the respiratory mucosa.[18] Together, these examples highlight the tissue-specific regulation of ENaC.

ENaC has also been reported to be regulated by the chloride channel, CFTR (cystic fibrosis transmembrane conductance regulator; see below). Biophysical data support the concept that loss of function of CFTR can potentiate the driving force for ENaC-mediated sodium transport into the cell that may contribute to some of the reported elevation of airway ENaC function in cystic fibrosis (CF) patients.[19] A more controversial model

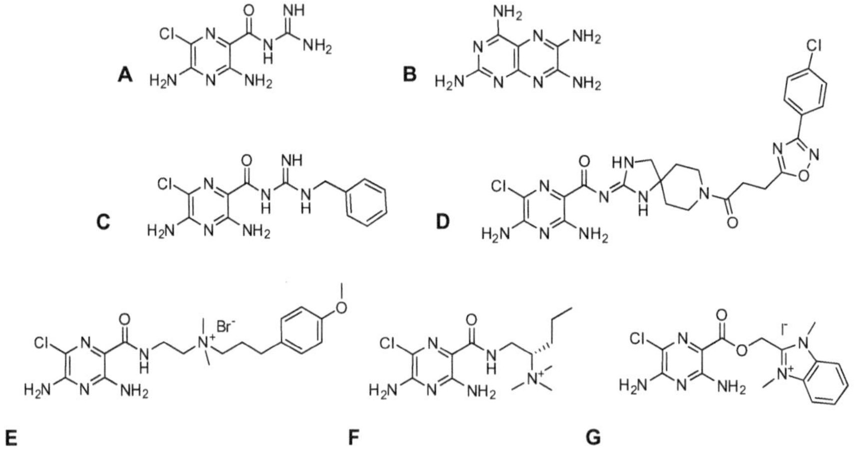

Figure 7.2   Structures of known ENaC Blockers. The prototypic pyazinoylguanidine ENaC Blocker Amiloride (A). The potassium sparing diuretic triamterene (B) inhibits ENaC with a Ki ~ 10–30 μM. Benzamil (C) has a lipophilic tail introducing ~10 fold improvements in potency over amiloride. Novartis identified alternative ENaC blockers by cyclizing the acyl guanidine moiety (D). Bioisostere approaches identified potent ENaC blockers by mimicking the protonated guanidine amiloride with a permanent positive charge (E, F, G).

suggests a direct interaction between the two channels, with CFTR exerting a negative regulation on ENaC and hence the exaggerated ENaC activity in CF.[20] Ongoing clinical studies with drug candidates that positively regulate mutated CFTR function in CF may provide important insights into this putative ENaC-CFTR interaction.[21]

Another area of recent interest has focused on the regulation of ENaC by proteases. Vallet and colleagues[22] first reported a negative regulation of xenopus ENaC by the Kunitz-type protease inhibitor aprotinin that could be reversed by an excess of trypsin. A number of proteases have since been demonstrated to activate ENaC[23a,b] including: matriptase, plasmin, neutrophil elastase and prostasin, which have been termed channel activating proteases (CAPs). Mutagenesis studies have revealed regions in the extracellular domains of the α- and γ-ENaC subunits that contain key residues required to observe CAP regulation (Figure 7.1(C)).

Whether cleavage of the extra-cellular domain is absolutely required to activate the channel or whether a steric interaction (that may culminate in cleavage) is sufficient remains to be proven. The mechanism by which a CAP-ENaC interaction/cleavage event leads to increased conduction has been proposed to be due to a loss of sodium self-inhibition and increase in open probability.[24,25] Sodium self-inhibition is a component of ENaC regulation whereby intra-cellular sodium ions will reduce channel activity and appears to be sensitive to the ENaC-CAP interaction.

7.3 ENaC: Evidence for a Role in Respiratory Function and Disease

The phenotypes of two quite different respiratory disorders perhaps provide the strongest data to support a key role for both mucosal hydration and ENaC function in the regulation of mucus clearance.[26] One of these diseases, cystic fibrosis (CF), is an autosomal recessive disease caused by loss of function mutations in CFTR resulting in a failure of both MCC and CC. CFTR is the major ion channel responsible for chloride transport across the human airway epithelium. Chloride ions are typically secreted in a serosal to mucosal direction resulting in osmotically obliged water moving in the same direction (Figure 7.3). In CF, the mucosal secretion of chloride and water are attenuated, thus limiting the available fluid for the hydration of the mucus gel (Figure 7.3(B) and (C)).[27] Indeed, clinical studies have established a relative "dehydration" of CF derived mucus (15% solids, 85% water) compared with non-CF (3% solids, 97% water). This dehydrated mucus gel displays altered rheological properties and adheres to epithelial surfaces providing a nidus for microbial colonization. Of note, a number of clinical studies have also reported evidence to suggest that ENaC activity may be enhanced in the respiratory epithelium of CF patients.[28,29] Since ENaC is responsible for the transport of sodium ions, and therefore osmotically obliged water, out of the mucosa, the activity of this channel could likewise profoundly influence the level of mucus hydration.

The reported increase of ENaC function in the respiratory epithelium of CF patients has also been proposed to correlate with disease severity.[30,31] Whether the elevated nasal potential difference in CF patients is a reliable surrogate for ENaC activity in the lung is controversial with several reports of elevated sodium transport in cultured CF-derived airway epithelia whilst others cannot replicate these findings.[19] Whether ENaC activity is elevated in the CF lung epithelium relative to non-CF is an important question, however the observation that inhaled amiloride can increase mucus clearance in CF patients is consistent with a functional significance of ENaC to the disease whether or not it is directly on the initiating pathophysiological pathway.[32,33]

Pre-clinical studies have begun to address the question as to the potential for ENaC to drive pathology in lung disease. Over expression of β-ENaC in the murine lung epithelium results in a phenotype reminiscent of CF lung disease: failed MCC, mucus accumulation, inflammation and increased infection risk that support a key role for ENaC and the importance of mucosal hydration.[34] This phenotype was attenuated by topically administered amiloride, but only when administered from birth.[35] Amiloride did not attenuate the lung phenotype when administered after the onset of disease, an observation that may be relevant for the eventual treatment of human disease, *i.e.* will we need to commence treatment early before the development of significant lung disease? An elegant transgenic pig model of CF, where CFTR has been knocked-out ($CFTR^{-/-}$) also presents with many of the key elements of CF lung disease. However, there is no evidence of elevated

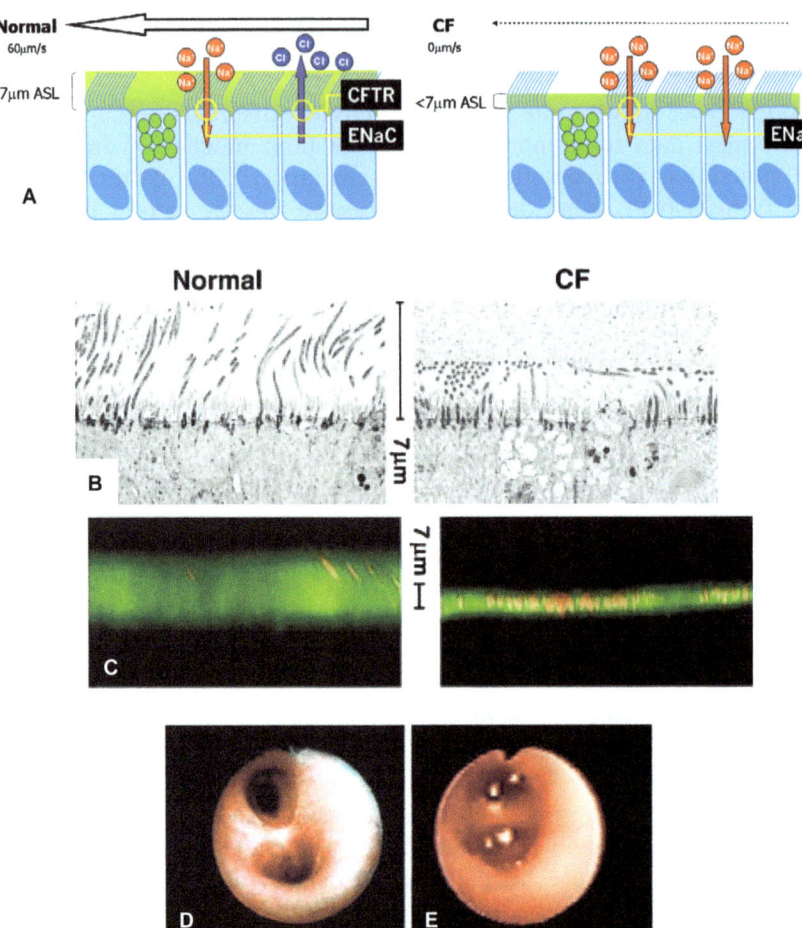

Figure 7.3   The coordinated activity of a number of ion channels and transporters
regulate airway mucosal hydration (A). Ion conduction through CFTR
and ENaC establish osmotic gradients for the movement of water to
maintain the airway surface liquid (ASL). The loss of CFTR function, as
is the case in cystic fibrosis, results in a reduced mucosal hydration.
This can be observed in cultures of primary bronchial epithelial cells
derived from CF patients (B & C). Low power EM illustrates fully
extended cilia in normals whilst cilia are bent over in the CF cultures
(B). The reduction in ASL can be seen in confocal images (C) where the
mucosal surface has been labeled with a cell impermeable fluorescent
dye (reproduced with permission from reference 3). Transbroncho-
scopic images taken of the large airways of normal (D) and a PHA
type 1 patient (E) highlights the excess of mucosal fluid in this disease
(reproduced with permission from reference 2).

ENaC activity in the new-born piglets that might challenge the significance
of ENaC as a key player in the initiation of the lung lesions.[36] It will be
fascinating to understand whether the basal level of ENaC activity in the

absence of CFTR in the CF pig ultimately contributes to the disease phenotype that manifests as they mature.

Further evidence for a prominent role of ENaC in the regulation of mucus clearance comes from a second hereditary disease, Pseudohypoaldosteronism Type 1 (PHA). PHA patients have loss of function mutations in the ENaC subunits that result in a systemic salt wasting disease.[8,9] In addition, these patients appear to have excessive fluid in the lumen of the large airways, but in the absence of any lung disease in adulthood.[2] What was surprising in these patients was that the rate of MCC was elevated three to four times higher than in non-PHA healthy controls. These data are consistent with a failure to absorb fluid from the airway mucosa (the opposite to CF) that results in an increased hydration of the mucus gel leading to an improvement in its propensity to be cleared. In support of this plasticity of the mucosal hydration-MCC axis, inhaled ENaC blockers have been described to enhance MCC in non-CF healthy individuals and also healthy sheep.[37,38]

Together, the CF and PHA data support the concept that mucosal hydration is a key node in the regulation of MCC and thus innate host defense of the lung. Genetic, clinical and pre-clinical pharmacological data are likewise consistent with ENaC function playing a key role in this process.

7.4 Inhaled ENaC Blockers for the Treatment of CF Lung Disease

As mentioned above, the proof-of-concept to highlight the relevance of ENaC block in the lung to the enhancement of MCC has been established with inhaled amiloride in CF patients. What remains to be established is whether an ENaC blocker induced enhancement of MCC can result in clinical benefit in CF patients with pre-existing lung disease. A number of large, long term (6–12 months) clinical studies have been undertaken to address this question using inhaled amiloride as the test compound.[39–42] The results have however been disappointing with a lack of compelling data to indicate any robust clinical improvement in lung disease.

The key question appears to be whether there is an issue with the compound or the hypothesis? Amiloride can be considered to be sub-optimal in terms of a suitable candidate to test this hypothesis in the lung for a number of reasons. First, amiloride is a relatively low potency blocker with an IC_{50} of approximately 400 nM on human airway epithelial cells. This means that the mass of compound to be delivered to the lung will likely be high, creating formulation and dosing problems. Second, amiloride was originally designed as an orally active potassium-sparing diuretic. The consequence of this is that drug absorbed from the lung,[43] and also the swallowed fraction of the inhaled dose, will be rapidly cleared by renal elimination resulting in a mechanism-based diuretic activity and the potentially life threatening side effect of hyperkalemia (elevated blood potassium levels).[44] The combination of these factors raised the strong possibility that there was no clinical benefit associated with amiloride in these studies because a dose suitable to provide block of target for a relevant period of the day could not be achieved. Combine this with the genetic data from CF and PHA, and a number of drug discovery companies have invested in the identification of novel ENaC blockers that are better suited to inhaled delivery and the maintenance of a therapeutic effect in the lung for a sufficient time to observe clinical efficacy.

The identification of novel classes of ENaC blockers through high throughput screening appears to have been disappointing. In Novartis, we utilized a membrane potential based assay with a murine kidney epithelial cell line (M1) that endogenously expresses all three ENaC subunits as a screening paradigm (Figure 7.4). Novel hit classes identified in this screen were limited, but included a series of 3-pyridyl-indazoles. The 3-pyridyl-indozole series was optimized to a point whereby we were able to demonstrate equivalent potency to amiloride in terms of block of ENaC function (in primary human bronchial epithelial cells). Delivery to the respiratory mucosa *in vivo* also attenuated ENaC activity[45] however this series did not display suitable drug-like properties and was not progressed. GSK have also reported on a membrane potential based HTS to identify novel ENaC regulators.[46] Using an engineered HEK293 line, multiple hit series were reported to have been discovered. As yet, we are unaware of the disclosure of these chemotypes.

The lack of additional, novel ENaC blocker scaffolds in the literature suggests a similar experience across the industry. As such, several research teams have instead focused on optimizing the existing pyrazinoylguanidine core of amiloride to design molecules suitable as inhaled drug candidates. The 3,5-diamino-6-chloro-pyrazinoyl guanidine is a privileged structure for ENaC blockade working *via* a direct exofacial block. The relative spatial arrangement of the positively charged guanidinium side chain, which interacts with a negatively charged residue inside the channel pore and the chlorine atom on the pyrazine ring, is essential to form a stable blocking complex.[47–48] Benzamil (Figure 7.2(C)), a close derivative of amiloride with a benzyl substituent on the guanidine has been shown to be approximately 10-fold more potent than amiloride, suggesting that there is an auxiliary hydrophobic binding site available in the channel pocket.[49,50]

The increased potency of benzamil led to its evaluation in two separate clinical trials. When delivered either by superfusion or aerosol delivery, benzamil showed a prolonged effect relative to amiloride in inhibiting the nasal transepithelial potential difference in CF patients.[51,52] The increased aqueous solubility of benzamil relative to amiloride enabled higher doses to be delivered in these studies that may have contributed to the improved duration of action. This led to speculation that benzamil might be useful in the long-term treatment of CF. In sheep, benzamil increased mucociliary clearance, although not significantly more than amiloride. This was rationalized to be due to the more rapid pulmonary absorption of benzamil offsetting its greater potency and questioning whether it would be any more therapeutically useful than amiloride itself.[38]

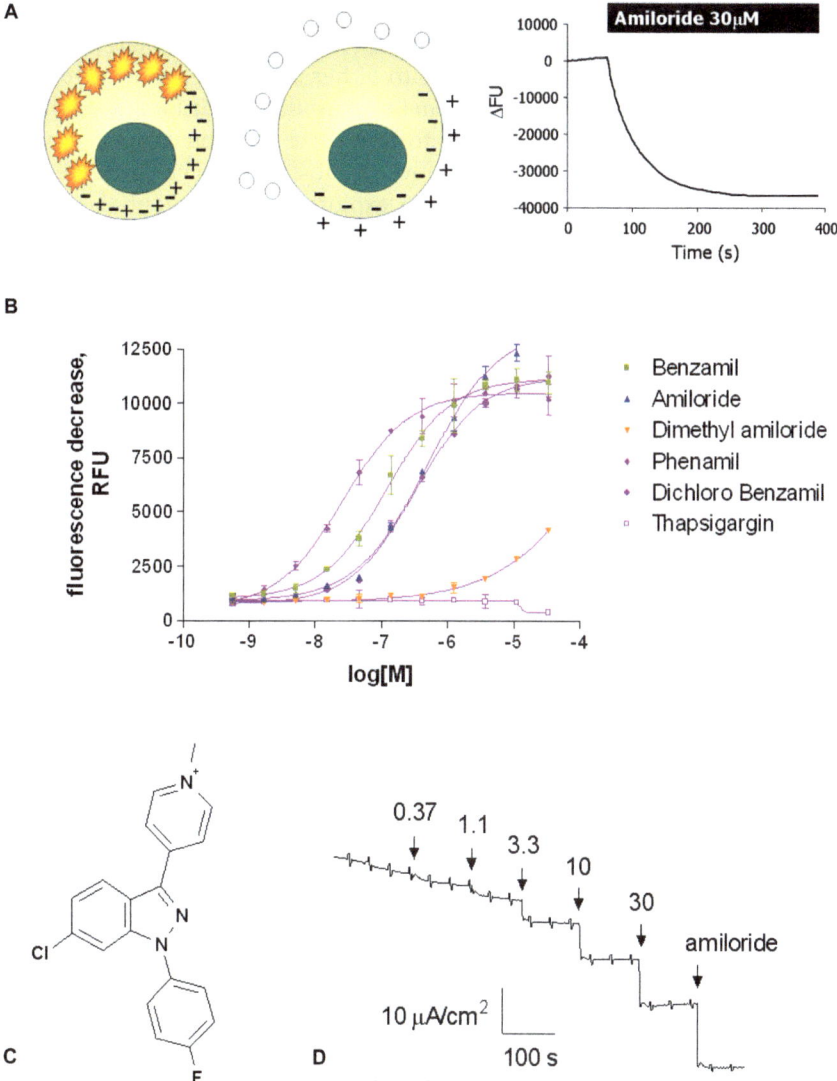

Figure 7.4   The HTS approach used by Novartis utilized the M-1 cell line that was loaded with a membrane potential sensitive dye. The addition of ENaC blockers induced an attenuation of fluorescence signal consistent with a hyperpolarization of the plasma membrane (A), with a potency order consistent with an ENaC-sensitive component to this assay (B). The 3-pyridyl indazole series was identified as a hit class (C) that validated in a short circuit current assay in primary human bronchial epithelial cells (D). Ultimately the series was optimized to provide compounds with a similar potency to amiloride so were discontinued (reproduced from reference 45).

Parion Sciences recognized that a successful ENaC blocker for the treatment of lung disease would have to show increased potency *versus* amiloride with increased duration of action due to either slower transport across the airway epithelium or decreased channel reversibility. Additionally, it should exhibit decreased renal elimination to minimize the potential of elevated blood potassium levels. A library of novel 2-substituted acylguanidine analogues of amiloride was synthesized to explore further potential auxiliary binding regions in the channel pocket.[53,54] As part of the design criteria, compounds were included which resembled endogenous metabolites in an effort to make the compounds less selective for renal clearance. Clear structure activity relationships were defined from the library, with Parion identifying a number of highly potent analogues, including compound Parion 552-02 (Figure 7.5(A)), which is 60–100 fold more potent than amiloride, 2–5 fold less reversible with a 170 fold slower k_{off} value, and significantly reduced epithelial permeability (Figure 7.5(B)). When administered as a nebulized solution to sheep, 552-02 significantly enhanced mucociliary clearance 4–6 hours after dosing while in a similar study amiloride showed no difference from control animals (Figure 7.5(C)).[54] With promising preclinical data, 552-02 was evaluated in the clinic and shown to enhance MCC in healthy volunteers.[37] The clinical data also indicated that only 2% of compound was eliminated *via* the renal route suggesting the design hypothesis had been successful. A subsequent phase IIa study

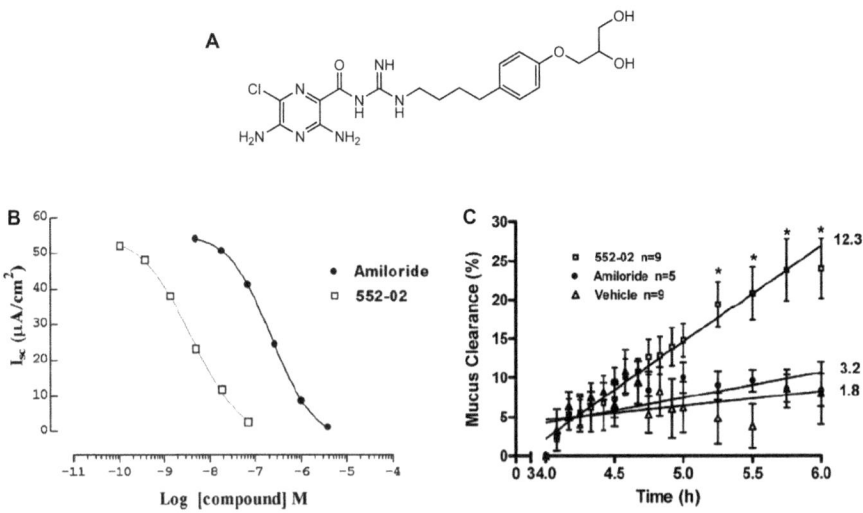

Figure 7.5 Parion's 552-02 compound (A) potently attenuates human ENaC activity *in vitro* (B). Short circuit current data using primary human bronchial epithelial cells illustrate a significantly increased potency of 552-02 compared with amiloride (B). When nebulized as a 3 mM solution to free breathing, conscious sheep, 552-02 also showed a superior profile to amiloride in terms of the rate of mucociliary clearance (C) (reproduced with permission from reference 54).

suggested a trend towards protection of loss of lung function in CF patients dosed with 552-02, although development of the compound was halted.[55] Parion and Gilead Sciences started joint development on an alternative ENaC blocker GS-9411 (undisclosed structure), which was specifically designed to have enhanced solubility in 3% hypertonic saline of $\sim$10 mg/mL, enabling the two treatments to be dosed simultaneously and take advantage of this synergistic effect. *In vitro*, GS-9411 in hypertonic saline produced a rapid initial increase in ASL volume that was sustained for longer than with either treatment alone.[56] The compound entered phase I clinical trials in 2009, however development was stopped in 2010 following observed increases in serum potassium in healthy volunteer studies.[57]

A team at Novartis used an alternative approach to optimize the pyrazinoylguanidine core for increased potency and reduced side effect potential, in this case for administration as a dry powder. Their strategy focused on minimizing the potential for hyperkalemia by making high molecular weight, low log D compounds with high protein binding to minimize passive renal clearance and favor biliary excretion.[58,59] It was thought these properties would also aid with improving lung residency. To achieve the desired profile, the team synthesized unusual dimeric compounds such as that exemplified in Figure 7.6(A), which has a high polar surface area (349) and molecular weight (731). When dosed as a solution or dry powder lactose blend directly into the airways *in vivo*, this compound attenuated the guinea pig tracheal potential difference in a dose-related fashion, without any concomitant rise in serum potassium (Figure 7.6(C) and (E)). The Parion 552-02 compound likewise attenuated airway ENaC activity in this model system with a similar potency to the dimer (Figure 7.6(B)). However, at doses required to observe efficacy in the lung, blood K+ levels were significantly elevated by 552-02 (Figure 7.6(D)). This profile would be consistent with the dimeric ENaC compounds showing a reduced potential to induce hyperkalemia and therefore represent potentially higher therapeutic index compounds for clinical evaluation. As a variation to the traditional pyrazinoylguanidine core, Novartis have also demonstrated that cyclizing the acyl guanidine portion of this structure can still afford potent ENaC blockers (Figure 7.2(D)).[60]

As a new approach to identifying inhaled ENaC blockers with a longer duration of action, Hunt *et al.*[61,62] utilized a bioisostere approach to identify a novel chemotype by rationalizing that an appropriately placed quaternary amine could mimic the positively charged acyl guanidinum functional group which is essential for blockade of ENaC. The permanent charge could also potentially reduce systemic absorption from the lung and provide the desired longer duration of action. This approach is exemplified by Figure 7.2(E) which demonstrates similar potency and efficacy to amiloride in blocking ion transport *via* ENaC in primary human bronchial epithelial cells (HBEC) (IC_{50} 200 nM), and in the guinea-pig tracheal potential difference model of ENaC blockade (ED_{50} of 44 µg/kg) *in vivo*. The same group next used the quaternary amides to exploit a stereochemical preference in the binding site of amiloride in human ENaC. A series of α-branched pyrazinoyl quaternary

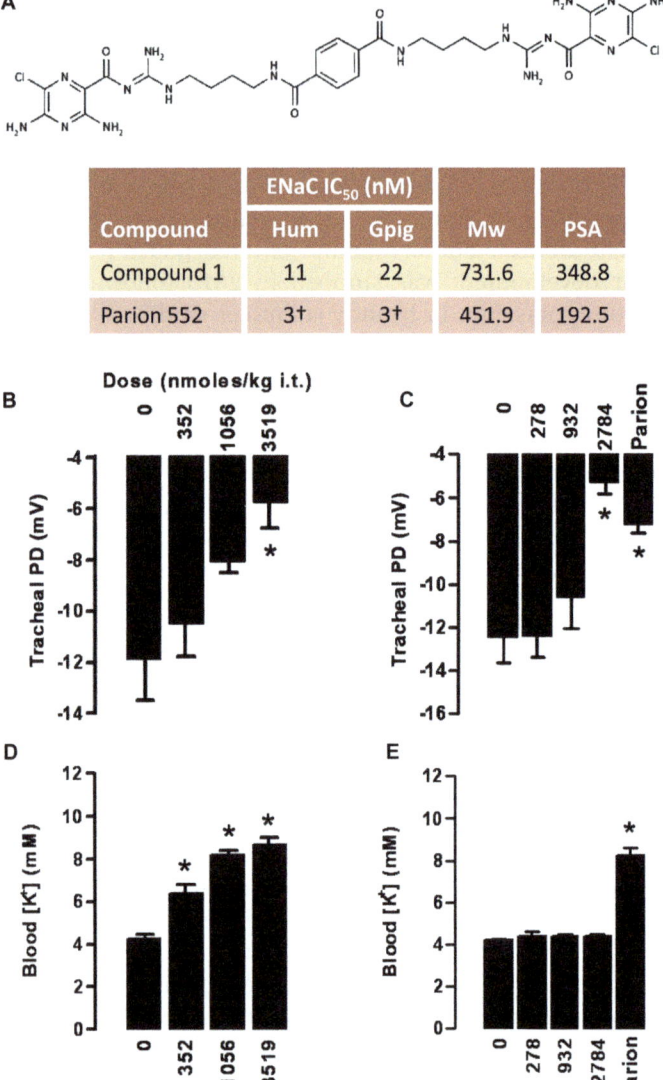

Figure 7.6 The Novartis dimeric ENaC blocker (A, patent ref) potently attenuates
both human and guinea pig ENaC function *in vitro* († taken from: Coote
et al., 2008). Compound A also attenuated ENaC function in the guinea
pig airways *in vivo*. Guinea pigs were administered lactose blends of
micronized compound by intra-tracheal instillation and the tracheal
potential difference (TPD) was measured 4h later (see Coote *et al.*, 2008
for details). Blood K+ levels were also measured at this time and
enabled an assessment of the exposure of the kidney to test compound.
Compound A attenuated TPD (C) at doses that were without effect on
blood K+ levels (E). A dry powder lactose blend of Parion's 552-02
compound likewise attenuated TPD in this model (B), although at
doses that also increased blood K+ levels (D).

amines were assessed for their ability to block ion transport *via* ENaC in HBECs, and it was found that introduction of the (S)-^{n}propyl α-substituent afforded a compound that potently inhibited sodium ion transport *via* ENaC both *in vitro* and *in vivo*, and showed ~ 10 fold improvement in potency over (R) enantiomer (Figure 7.2(F)).[62] In a similar vein, GSK have reported quaternarized benzamidazole compounds (Figure 7.2(G)) as potent ENaC blockers with $pIC_{50}s > 4.5$.[63] Whether this approach provides the desired increase in duration of action has yet to be reported.

7.5   Inhaled CAP Inhibitors for the Treatment of CF Lung Disease

The regulation of ENaC function by proteases also offers the potential to attenuate sodium absorption in the lung. With a number of serine proteases having been demonstrated to activate ENaC (Figure 7.1(C)) a key question is the identity of the lung relevant CAP or perhaps CAPs.[23a] Similarity searching to identify a human homologue of the xenopus CAP, originally described by Vallet *et al.*,[22] to activate ENaC, highlighted the trypsin-like serine protease prostasin (PRSS8) as a candidate CAP. Prostasin is expressed in the airway epithelium and the extra-cellular domain of the γ-ENaC subunit contains a predicted prostasin-dependent cleavage site.[23a] Furthermore, silencing of prostasin using an siRNA approach in a CF-derived cell line, attenuated ENaC activity.[64] Perhaps unsurprisingly, the broad-spectrum Kunitz-type protease inhibitors aprotinin and placental bikunin also inhibited prostasin.[65] Of note, both of these macromolecular inhibitors attenuated ENaC activity in cultured human airway epithelial cells from both normal and CF donors and also attenuated ENaC function in the guinea pig airways *in vivo* following topical administration to the respiratory tract.[65] In contrast, the serpin inhibitors soya bean trypsin inhibitor and $\alpha1$-anti-trypsin failed to attenuate ENaC function *in vitro* or *in vivo*. These data provide a platform of biology consistent with a Kunitz-sensitive CAP regulating ENaC function in the airway, from which to commence drug discovery efforts.

Based on the validation described above, Novartis screened for inhibitors of prostasin and identified the serine protease inhibitors camostat and nafamostat.[66] Both of these compounds were already in clinical use: camostat for the treatment of acute pancreatitis and nafamostat as an anti-coagulant. In common with the Kunitz-type inhibitors, camostat and nafamostat attenuated ENaC function in both normal and CF bronchial epithelial cells and inhibited ENaC function in the guinea pig airway *in vivo* (Figure 7.7). The observation that camostat accelerated MCC in the sheep with an improved duration of action over amiloride gated entry into clinical trials. Intra-nasal dosing with camostat was demonstrated to attenuate ENaC activity in CF patients supporting the hypothesis for a CAP-regulation of ENaC function in the respiratory epithelium.[67]

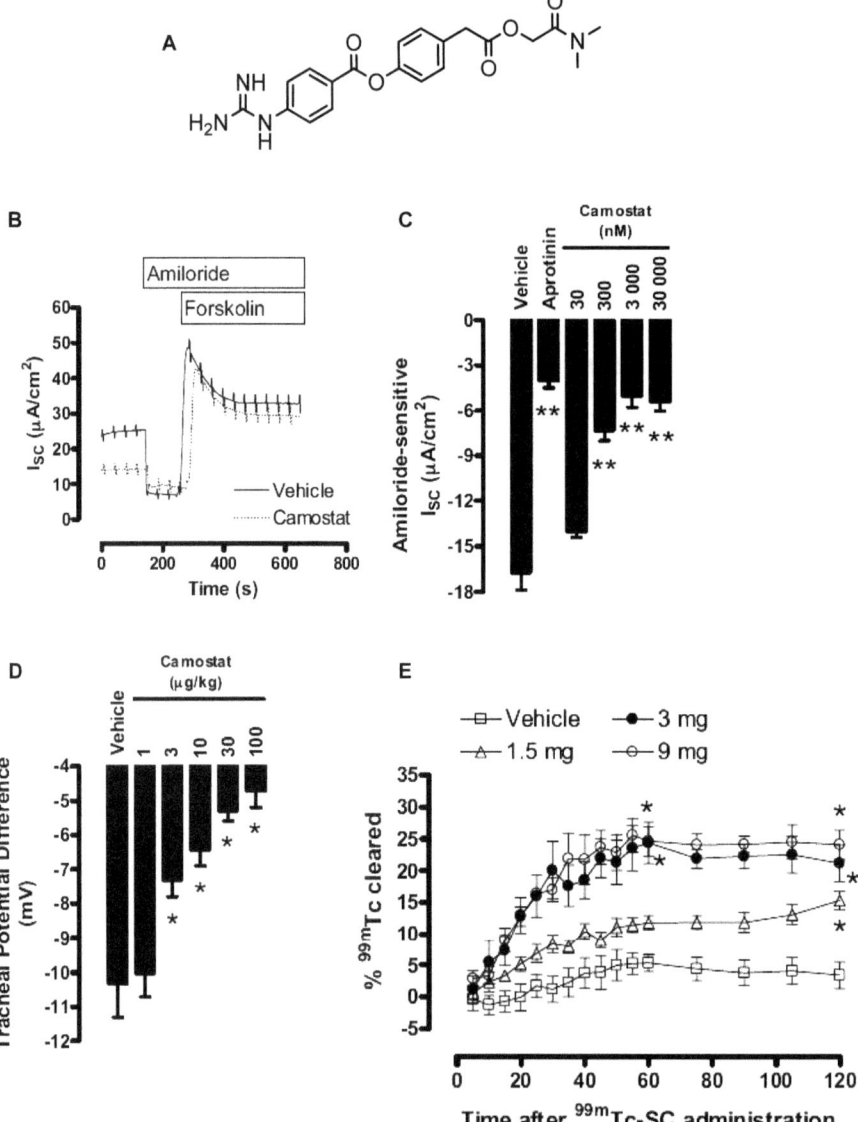

Figure 7.7 The trypsin-like protease inhibitor camostat (A) attenuates ENaC activity in primary human bronchial epithelial cells (B, C). A raw data trace illustrating the selective attenuation of the amiloride-sensitive short circuit current in human bronchial epithelial cells (B) with an IC_{50} of approximately 100 nM (C). Camostat also attenuated airway ENaC function *in vivo* in the guinea pig TPD model (D) in a dose-dependent manner. Inhaled camostat increased the rate and magnitude of clearance of ^{99m}Tc labeled particles out of the lungs of conscious free-breathing sheep (E) (reproduced with permission from reference 66).

Camostat and nafamostat are both pseudoirreversible inhibitors and therefore by definition relatively non-selective. Further efforts from the Novartis team focused on a series of peptidomimetic compounds as prostasin inhibitors that were guided by structure-based design. This approach established SAR and high potency prostasin inhibitors[68] that attenuated ENaC activity in the guinea pig airways *in vivo* and enhanced MCC in the sheep.

However, as noted above, a number of proteases are able to activate ENaC. For example, when the endogenous CAP in the human airway is inhibited, the addition of neutrophil elastase is able to activate the previously latent channel.[69] This is potentially highly relevant to the treatment of any inflammatory lung disease but especially CF where levels of airway neutrophil elastase can be very high. It is entirely plausible that a highly effective inhibitor of the endogenous airway CAP could be negated by the presence of neutrophil elastase. This protease redundancy represents a key risk associated with the CAP Inhibitor approach. If the expression of neutrophil elastase is focal, and limited to specific regions of the lung with a microbial colonization, then one might predict that an inhaled CAP inhibitor could still be efficacious in those as yet unaffected regions of the lung, maintaining MCC and reducing the risk of their eventual infection. This and other questions will have to be established in clinical studies.

7.6 Mucosal Hydration and the Future for ENaC-based Therapies

With a number of drug candidates that target ENaC presently in clinical trials, the results of key safety and efficacy studies are anticipated. Has renal elimination and the risk of hyperkalemia been sufficiently minimized with the direct ENaC blockers? Will a long lasting attenuation of ENaC function in the CF lung provide clinical benefit? Will any putative benefit be limited to patients in the early stages of disease? What clinical benefit might there be from combinations of ENaC regulators with other ion transport therapies, such as CFTR potentiators and corrector molecules? These are important, and by definition, exciting times for this class of mucokinetic therapies.

References

1. A. Caputo, E. Caci, L. Ferrera, N. Pedemonte, C. Barsanti, E. Sondo, U. Pfeffer, R. Ravazzolo, O. Zegarra-Moran and L. J. Galietta, TMEM16A, a membrane protein associated with calcium-dependent chloride channel activity, *Science*, 2008, **322**(5901), 590–594.
2. E. Kerem, T. Bistritzer, A. Hanukoglu, T. Hofmann, Z. Zhou, W. Bennett, E. MacLaughlin, P. Barker, M. Nash, L. Quittell, R. Boucher and M. R. Knowles, Pulmonary epithelial sodium-channel dysfunction and excess airway liquid in pseudohypoaldosteronism, *N. Engl. J. Med.*, 1999, **341**(3), 156–162.

3. H. Matsui, B. R. Grubb, R. Tarran, S. H. Randell, J. T. Gatzy, C. W. Davis and R. C. Boucher, Evidence for periciliary liquid layer depletion, not abnormal ion composition, in the pathogenesis of cystic fibrosis airways disease, *Cell*, 1998, **95**(7), 1005–1015.

4. C. M. Canessa, J. D. Horisberger and B. C. Rossier, Epithelial sodium channel related to proteins involved in neurodegeneration, *Nature*, 1993, **361**(6411), 467–470.

5. S. Kellenberger and L. Schild, Epithelial sodium channel/degenerin family of ion channels: a variety of functions for a shared structure, *Physiol. Rev.*, 2002, **82**(3), 735–767.

6. O. B. Kashlan and T. R. Kleyman, ENaC structure and function in the wake of a resolved structure of a family member, *Am. J. Physiol. Renal. Physiol*, 2011, **301**(4), F684–F696.

7. J. D. Stockand, A. Staruschenko, O. Pochynyuk, R. E. Booth and D. U. Silverthorn, Insight toward epithelial Na+ channel mechanism revealed by the acid-sensing ion channel 1 structure, *IUBMB Life*, 2008, **60**(9), 620–628.

8. S. S. Chang, S. Grunder, A. Hanukoglu, A. Rösler, P. M. Mathew, I. Hanukoglu, L. Schild, Y. Lu, R. A. Shimkets, C. Nelson-Williams, B. C. Rossier and R. P. Lifton, Mutations in subunits of the epithelial sodium channel cause salt wasting with hyperkalaemic acidosis, pseudohypoaldosteronism type 1, *Nat. Genet.*, 1996, **12**(3), 248–253.

9. S. S. Strautnieks, R. J. Thompson, R. M. Gardiner and E. Chung, A novel splice-site mutation in the gamma subunit of the epithelial sodium channel gene in three pseudohypoaldosteronism type 1 families, *Nat. Genet.*, 1996, **13**(2), 248–250.

10. G. W. Liddle, T. Bledsoe and W. S. Coppage Jr., A familial renal disorder simulating primary aldosteronism but with negligible aldosterone secretion, *Trans. Assoc. Am. Physicians*, 1963, **76**, 199–213.

11. M. Botero-Velez, J. J. Curtis and D. G. Warnock, Brief report: Liddle's syndrome revisited – a disorder of sodium reabsorption in the distal tubule, *N. Engl. J. Med.*, 1994, **330**(3), 178–181.

12. H. Garty and L. G. Palmer, Epithelial sodium channels: function, structure, and regulation, *Physiol Rev.*, 1997, 77(2), 359–396.

13. B. R. Grubb and R. C. Boucher, Effect of in vivo corticosteroids on Na+ transport across airway epithelia, *Am. J. Physiol.*, 1998, **275**, C303–C308.

14. M. D. Carattino, S. Sheng and T. R. Kleyman, Epithelial Na+ channels are activated by laminar shear stress, *J. Biol. Chem.*, 2004, **279**(6), 4120–4126.

15. R. Tarran, B. Button and R. C. Boucher, Regulation of normal and cystic fibrosis airway surface liquid volume by phasic shear stress, *Annu. Rev. Physiol.*, 2006, **68**, 543–561.

16. D. C. Devor and J. M. Pilewski, UTP inhibits Na+ absorption in wild-type and DeltaF508 CFTR-expressing human bronchial epithelia, *Am. J. Physiol.*, 1999, **276**(4 Pt 1), C827–C837.

17. D. Rotin and O. Staub, Role of the ubiquitin system in regulating ion transport, *Pflugers Arch.*, 2011, **461**(1), 1–21.

18. E. Baker, X. Jeunemaitre, A. J. Portal, P. Grimbert, N. Markandu, A. Persu, P. Corvol and G. MacGregor, Abnormalities of nasal potential difference measurement in Liddle's syndrome, *J. Clin. Invest.*, 1998, **102**(1), 10–14.

19. O. A. Itani, J. H. Chen, P. H. Karp, S. Ernst, S. Keshavjee, K. Parekh, J. Klesney-Tait, J. Zabner and M. J. Welsh, Human cystic fibrosis airway epithelia have reduced Cl- conductance but not increased Na+ conductance, *Proc. Natl. Acad. Sci. U. S. A.*, 2011, **108**(25), 10260–10265.

20. M. J. Stutts, C. M. Canessa, J. C. Olsen, M. Hamrick, J. A. Cohn, BC Rossier and R. C. Boucher, CFTR as a cAMP-dependent regulator of sodium channels, *Science*, 1995, **269**(5225), 847–850.

21. F Van Goor, S. Hadida, P. D. Grootenhuis, B. Burton, J. H. Stack, K. S. Straley, C. J. Decker, M. Miller, J. McCartney, E. R. Olson, J. J. Wine, R. A. Frizzell, M. Ashlock and P. A. Negulescu, Correction of the F508del-CFTR protein processing defect in vitro by the investigational drug VX-809, *Proc. Natl. Acad. Sci. U. S. A.*, 2011, **108**(46), 18843–18848.

22. V. Vallet, A. Chraib, H. P. Gaeggeler, J. D. Horisberger and B. C. Rossier, An epithelial serine protease activates the amiloride-sensitive sodium channel, *Nature*, 1997, **389**(6651), 607–610.

23. (a) B. C. Rossier and M. J. Stutts, Activation of the epithelial sodium channel (ENaC) by serine proteases, *Annu. Rev. Physiol.*, 2009, **71**, 361–379; (b) T. R. Kleyman, M. D. Carattino and R. P. Hughey, ENaC at the cutting edge: regulation of sodium channels by proteases, *J. Biol. Chem.*, 2009, **284**(31), 20447–20451.

24. A. Adebamiro, Y. Cheng, U. S. Rao, H. Danahay and R. J. Bridges, A segment of gamma ENaC mediates elastase activation of Na+ transport, *J. Gen. Physiol.*, 2007, **130**(6), 611–629.

25. A. Chraïbi and J. D. Horisberger, Na self inhibition of human epithelial Na channel: temperature dependence and effect of extracellular proteases, *J. Gen. Physiol.*, 2002, **120**(2), 133–145.

26. D. Paisley, M. Gosling and H. Danahay, Regulation of airway mucosal hydration, *Expert Rev. Clin. Pharmacol.*, 2010, **3**(3), 361–369.

27. R. C. Boucher, Evidence for airway surface dehydration as the initiating event in CF airway disease, *J. Intern. Med.*, 2007, **261**(1), 5–16.

28. M. R. Knowles, M. J. Stutts, A. Spock, N. Fischer, J. T. Gatzy and R. C. Boucher, Abnormal ion permeation through cystic fibrosis respiratory epithelium, *Science*, 1983, **221**(4615), 1067–1070.

29. P. G. Middleton, D. M. Geddes and E. W. Alton, Effect of amiloride and saline on nasal mucociliary clearance and potential difference in cystic fibrosis and normal subjects, *Thorax*, 1993, **48**(8), 812–816.

30. I. Fajac, D. Hubert, D. Guillemot, I. Honoré, T. Bienvenu, F. Volter, J. Dall'Ava-Santucci and D. J. Dusser, Nasal airway ion transport is linked to the cystic fibrosis phenotype in adult patients, *Thorax*, 2004, **59**(11), 971–976.

31. T. Leal, I. Fajac, H. L. Wallace, P. Lebecque, J. Lebacq, D. Hubert, J. Dall'Ava, D. Dusser, A. P. Ganesan, C. Knoop, J. Cumps, P. Wallemacq

and K. W. Southern, Airway ion transport impacts on disease presentation and severity in cystic fibrosis, *Clin. Biochem.*, 2008, **41**(10–11), 764–772.

32. E. M. App, M. King, R. Helfesrieder, D. Köhler and H. Matthys, Acute and long-term amiloride inhalation in cystic fibrosis lung disease. A rational approach to cystic fibrosis therapy, *Am. Rev. Respir. Dis.*, 1990, **141**(3), 605–612.

33. D. Köhler, E. App, M. Schmitz-Schumann, G. Würtemberger and H. Matthys, Inhalation of amiloride improves the mucociliary and the cough clearance in patients with cystic fibroses, *Eur. J. Respir. Dis. Suppl.*, 1986, **146**, 319–326.

34. M. Mall, B. R. Grubb, J. R. Harkema, W. K. O'Neal and R. C. Boucher, Increased airway epithelial Na+ absorption produces cystic fibrosis-like lung disease in mice, *Nat. Med.*, 2004, **10**(5), 487–493.

35. Z. Zhou, D. Treis, S. C. Schubert, M. Harm, J. Schatterny, S. Hirtz, J. Duerr, R. C. Boucher and M. A. Mall, Preventive but not late amiloride therapy reduces morbidity and mortality of lung disease in betaENaC-overexpressing mice, *Am. J. Respir. Crit. Care. Med.*, 2008, **178**(12), 1245–1256.

36. J. H. Chen, D. A. Stoltz, P. H. Karp, S. E. Ernst, A. A. Pezzulo, T. O. Moninger, M. V. Rector, L. R. Reznikov, J. L. Launspach, K. Chaloner, J. Zabner and M. J. Welsh, Loss of anion transport without increased sodium absorption characterizes newborn porcine cystic fibrosis airway epithelia, *Cell*, 2010, **143**(6), 911–923.

37. S. Donaldson, R. Smith, J. Doran, B. DiMassimo, K. Zeman, B. Bennett, H. Hurd and S. Hopkins, Safety, pharmacokinetics and effects on mucus clearance following administration of 552-02 to normal healthy volunteers, *Pediatric Pulmonology*, 2005, **40**(S28), 218 (Abstract).

38. A. J. Hirsh, J. R. Sabater, A. Zamurs, R. T. Smith, A. M. Paradiso, S. Hopkins, W. M. Abraham and R. C. Boucher, Evaluation of second generation amiloride analogs as therapy for cystic fibrosis lung disease, *J. Pharmacol. Exp. Ther.*, 2004, **311**(3), 929–938.

39. I. M. Bowler, B. Kelman, D. Worthington, J. M. Littlewood, A. Watson, S. P. Conway, S. W. Smye, S. L. James and T. A. Sheldon, Nebulised amiloride in respiratory exacerbations of cystic fibrosis: a randomised controlled trial, *Arch. Dis. Child.*, 1995, **73**(5), 427–430.

40. A. Graham, A. Hasani, E. W. Alton, G. P. Martin, C. Marriott, M. E. Hodson, S. W. Clarke and D. M. Geddes, No added benefit from nebulized amiloride in patients with cystic fibrosis, *Eur. Respir. J.*, 1993, **6**(9), 1243–1248.

41. M. R. Knowles, N. L. Church, W. E. Waltner, J. R. Yankaskas, P. Gilligan, M. King, L. J. Edwards, R. W. Helms and R. C. Boucher, A pilot study of aerosolized amiloride for the treatment of lung disease in cystic fibrosis, *N. Engl. J. Med.*, 1990, **322**(17), 1189–1194.

42. G. Pons, M. C. Marchand, P. d'Athis, E. Sauvage, C. Foucard, P. Chaumet-Riffaud, A. Sautegeau, J. Navarro and G. Lenoir, French

multicenter randomized double-blind placebo-controlled trial on nebulized amiloride in cystic fibrosis patients. The Amiloride-AFLM Collaborative Study Group, *Pediatr. Pulmonol.*, 2000, **30**(1), 25–31.

43. P. G. Noone, J. A. Regnis, X. Liu, K. L. Brouwer, M. Robinson, L. Edwards and M. R. Knowles, Airway deposition and clearance and systemic pharmacokinetics of amiloride following aerosolization with an ultrasonic nebulizer to normal airways, *Chest*, 1997, **112**(5), 1283–1290.

44. M. A. Perazella, Drug-induced hyperkalemia: old culprits and new offenders, *Am. J. Med.*, 2000, **109**(4), 307–314.

45. H. Danahay, H. Atherton, D. Paisley, M. Bradley, P. Gedeck, C. Howsham, N. Devereux, M. Pipet, M. Vyas, G. Bhalay, N. Smith, S. Collingwood and M. Gosling, The identification of a novel structural class of ENaC blockers by high-throughput screening, *Pediatric Pulmonology*, 2008, **43**(S31), 296 (Abstract).

46. K. Gatfield, Assay development and triage strategy for the screening of ENaC Blockers. 9th Annual Ion Channel Retreat, 2011.

47. J. H. Li, C. J. Cragoe Jr. and B. Lindemann, Structure-activity relationship of amiloride analogs as blockers of epithelial Na channels: II. Side-chain modifications, *J. Membr. Biol.*, 1985, **83**, 45–56.

48. C. A. Venanzi, C. Plant and T. J. Venanzi, Molecular recognition of amiloride analogues: a molecular electrostatic potential analysis. 1. Pyrazine ring modification, *J. Med. Chem.*, 1992, **35**, 1643–1649.

49. A. W. Cuthbert and G. M. Fanelli, Effects of some pyrazinecarboxamides on sodium transport in frog skin, *Br. J. Pharm*, 1978, **63**, 139–149.

50. T. R. Kleyman and E. J. Cragoe Jr., Amiloride and its analogs as tools in the study of ion transport, *J. Membr. Biol.*, 1988, **105**, 1–21.

51. T. Hofmann, M. J. Stutts, A. Ziersch, C. Rückes, W. M. Weber, M. R. Knowles, H. Lindemann and R. C. Boucher, Effects of topically delivered benzamil and amiloride on nasal potential difference in cystic fibrosis, *Am. J. Respir. Crit. Care. Med.*, 1998, **157**(6 Pt 1), 1844–1849.

52. H. C. Rodgers and A. J. Knox, The effect of topical benzamil and amiloride on nasal potential difference in cystic fibrosis, *Eur. Respir. J.*, 1999, **14**(3), 693–696.

53. A. J. Hirsh, B. F. Molino, J. Zhang, N. Astakhova, W. B. Geiss, B. J. Sargent, B. D. Swenson, A. Usyatinsky, M. J. Wyle, R. C. Boucher, R. T. Smith, A. Zamurs and R. J. Johnson, Design, Synthesis, and Structure-Activity Relationships of Novel 2-Substituted Pyrazinoylguanidine Epithelial Sodium Channel Blockers: Drugs for Cystic Fibrosis and Chronic Bronchitis, *J. Med. Chem.*, 2006, **49**, 4098–4115.

54. A. J. Hirsh, J. Zhang, A. Zamurs, J. Fleegle, W. R. Thelin, R. A. Caldwell, J. R. Sabater, W. M. Abraham, M. Donowitz, B. Cha, K. B. Johnson, J. A. St George, M. R. Johnson and R. C. Boucher, Pharmacological properties of N-(3,5-diamino-6-chloropyrazine-2-carbonyl)-N′-4-[4-(2,3-dihydroxypropoxy)-phenyl]butyl-guanidine methanesulfonate (552-02), a novel epithelial sodium channel blocker with potential clinical efficacy for cystic fibrosis lung disease, *J. Pharmacol. Exp. Ther.*, 2008, **325**(1), 77–88.

55. J. L. Gorlaski, R. C. Boucher and B. Button, Osmolytes and ion transport modulators: new strategies for airway surface rehydration, *Curr. Opin. Pharmacol.*, 2012, **10**, 294–299.

56. A. J. Hirsh, J. George, W. Thelin, K. Stapleton, B. Montgomery, R. Boucher and M. R. Johnson, GS-9411: a potential aerosol pharmacotherapy for CF, *Am. J. Respir. Crit. Care. Med.*, 2009, **179**, A1195.

57. T. G. O'Riordan, K. H. Donn, P. Hodsman, J. H. Ansede, T. Newcomb, S. A. Lewis, W. D. Flitter, V. S. White, M. R. Johnson, A. B. Montgomery, D. G. Warnock and R. C. Boucher, Acute Hyperkalemia Associated with Inhalation of a Potent ENaC Antagonist: Phase 1 Trial of GS-9411, *J. Aerosol Med. Pulm. Drug Deliv.*, 2014, **27**(3), 200–208.

58. S. P. Collingwood and N. Smith, Pyrazine compounds and their preparation, pharmaceutical compositions and use in the treatment of diseases mediated by blockade of an epithelial sodium channel, such as inflammatory and allergic conditions, *PCT Int. Appl.*, 2007, WO 2007071396 A2.

59. R. A. Fairhurst, R. Butler, P. Oakley, S. P. Collingwood, N. Smith, E. Stanley and M. I. Rodriguez Perez, Pyrazine-amide derivatives as epithelial sodium channel blockers and their preparation, pharmaceutical compositions and use in the treatment of diseases, *PCT Int. Appl.*, 2008, WO 2008135557 A1.

60. G. Bhalay, E. Budd, G. D. C. Bloomfield, S. P. Collingwood, A. Dunstan, L. Edwards, P. Gedeck, C. Howsham, P. Hunt, T. A. Hunt, P. Oakley and N. Smith, Preparation of cyclic guanidino carbonylpyrazines as epithelial sodium channel (ENaC) blockers, *PCT Int. Appl.*, 2009, WO 2009074575 A2.

61. T. Hunt, H. C. Atherton-Watson, J. Axford, S. P. Collingwood, K. J. Coote, B. Cox, S. Czarnecki, H. Danahay, N. Devereux, C. Howsham, P. Hunt, V. Paddock, D. Paisley and A. Young, Discovery of a novel chemotype of potent human ENaC blockers using a bioisostere approach. Part 1: Quaternary amines, *Bioorg. Med. Chem. Lett.*, 2012a, **22**(2), 929–932.

62. T. Hunt, H. C. Atherton-Watson, S. P. Collingwood, K. J. Coote, S. Czarnecki, H. Danahay, C. Howsham, P. Hunt, D. Paisley and A. Young, Discovery of a novel chemotype of potent human ENaC blockers using a bioisostere approach. Part 2: α-Branched quaternary amines, *Bioorg. Med. Chem. Lett.*, 2012b, **22**(8), 2877–2879.

63. D. I. Laine, T. Li and H. Yan, Preparation of benzimidazole derivatives as epithelial sodium channels (ENaC) blockers, Pct Int. Appl (2011), WO 2011079087 A10.

64. Z. Tong, B. Illek, V. J. Bhagwandin, G. M. Verghese and G. H. Caughey, Prostasin, a membrane-anchored serine peptidase, regulates sodium currents in JME/CF15 cells, a cystic fibrosis airway epithelial cell line, *Am. J. Physiol. Lung Cell Mol. Physiol.*, 2004, **287**(5), L928–935.

65. K. Coote, H. C. Atherton-Watson, R. Sugar, A. Young, A. MacKenzie-Beevor, M. Gosling, G. Bhalay, G. Bloomfield, A. Dunstan, R. J. Bridges, J. R. Sabater, W. M. Abraham, D. Tully, R. Pacoma, A. Schumacher, J. Harris and H. Danahay, Camostat attenuates airway epithelial sodium

channel function in vivo through the inhibition of a channel-activating protease, *J. Pharmacol. Exp. Ther.*, 2009, **329**(2), 764–774.

66. K. J. Coote, H. Atherton, A. Young, R. Sugar, R. Burrows, N. J. Smith, J. M. Schlaeppi, P. J. Groot-Kormelink, M. Gosling and H. Danahay, The guinea-pig tracheal potential difference as an in vivo model for the study of epithelial sodium channel function in the airways, *Br. J. Pharmacol.*, 2008, **155**(7), 1025–1033.

67. S. M. Rowe, G. Reeves, H. Hathorne, G. M. Solomon, S. Abbi, D. Renard, R. Lock, P. Zhou, H. Danahay, J. P. Clancy and D. A. Waltz, Reduced sodium transport with nasal administration of the prostasin inhibitor camostat in subjects with cystic fibrosis, *Chest*, **144**(1), 200–207.

68. D. C. Tully, A. Vidal, A. K. Chatterjee, J. A. Williams, M. J. Roberts, H. M. Petrassi, G. Spraggon, B. Bursulaya, R. Pacoma, A. Shipway, A. M. Schumacher, H. Danahay and J. L. Harris, Discovery of inhibitors of the channel-activating protease prostasin (CAP1/PRSS8) utilizing structure-based design, *Bioorg. Med. Chem. Lett.*, 2008, **18**(22), 5895–5899.

69. R. A. Caldwell, R. C. Boucher and M. J. Stutts, Neutrophil elastase activates near-silent epithelial Na+ channels and increases airway epithelial Na+ transport, *Am. J. Physiol Lung Cell Mol. Physiol.*, 2005, **288**(5), L813–819.

CHAPTER 8

The Therapeutic Potential of Small-molecule Modulators of the Cystic Fibrosis Transmembrane Conductance Regulator (CFTR) Cl⁻ Channel

JIA LIU,[a] GERTA CAMI-KOBECI,[b] YITING WANG,[a] PISSARED KHUITUAN,[a,c] ZHIWEI CAI,[a] HONGYU LI,[a,d] STEPHEN M. HUSBANDS[b] AND DAVID N. SHEPPARD*[a]

[a] School of Physiology and Pharmacology, University of Bristol, Medical Sciences Building, University Walk, Bristol BS8 1TD, UK; [b] Department of Pharmacy and Pharmacology, University of Bath, Claverton Down, Bath BA2 7AY, UK; [c] Center of Calcium and Bone Research, Department of Physiology, Faculty of Science, Mahidol University, Bangkok 10400, Thailand; [d] Department of Applied Sciences, London South Bank University, London SE1 0AA, UK
*Email: D.N.Sheppard@bristol.ac.uk

8.1 Introduction

The cystic fibrosis transmembrane conductance regulator (CFTR)[1] is an ATP-binding cassette (ABC) transporter.[2] ABC transporters are found in all organisms, where they perform diverse physiological roles. The vast majority of ABC transporters pump assorted substrates across cell membranes either

RSC Drug Discovery Series No. 39
Ion Channel Drug Discovery
Edited by Brian Cox and Martin Gosling
© The Royal Society of Chemistry 2015
Published by the Royal Society of Chemistry, www.rsc.org

into (importers) or out of (exporters) cells.[2] Some ABC transporters (*e.g.* the sulfonylurea receptor; SUR1 (ABCC8; pancreas) and SUR2 (ABCC9; heart and muscle)) regulate associated ion channels.[3,4] But, only CFTR (ABCC7) forms an ion channel.[5,6]

CFTR is principally expressed in the apical membrane of epithelia throughout the body lining ducts and tubes.[7] It controls the quantity and composition of epithelial secretions by (i) forming a small-conductance (6–10 pS) anion-selective channel with complex regulation[8,9] and (ii) regulating the function of other ion channels and transporters.[10,11] Thus, CFTR plays a pivotal role in epithelial physiology that influences development, growth, innate immunity and fecundity. The importance of CFTR is dramatically highlighted by three common diseases: cystic fibrosis (CF), secretory diarrhoea and autosomal dominant polycystic kidney disease (ADPKD). CF is caused by loss-of-function mutations in CFTR,[7] whereas secretory diarrhoea and ADPKD are associated with unphysiological CFTR activity.[12,13] However, these are not the only diseases linked to CFTR. Some forms of male infertility, pancreatitis and bronchiectasis are termed CFTR-related disorders because they are caused by mutations that preserve partial CFTR function,[14] while some reproductive disorders involve unphysiological CFTR activity.[15]

Here, we selectively review the therapeutic potential of small-molecule CFTR modulators in the treatment of CF, secretory diarrhoea and ADPKD. We discuss how CFTR's physiological role and malfunction in disease have informed drug development. We explore the structure and function of CFTR to understand the mechanism of action of small-molecule CFTR modulators. Finally, we consider some of the challenges to be overcome before small-molecule CFTR modulators realise their promise as therapeutics.

8.2 The Pathophysiology of CFTR

8.2.1 The Physiology of CFTR

CFTR plays a crucial role in regulating the quantity and composition of epithelial secretions throughout the body. However, examination of the physiology of individual epithelia reveals tissue-specific differences in CFTR function. A variety of factors contribute to this diversity including tissue architecture, the distribution of proteins that interact with CFTR to modulate its expression and function and the role of CFTR as both an anion channel and transport regulator.

Ducts and tubes in the pancreas, intestine, hepatobiliary system and reproductive tissues transport a variety of protein-rich cargoes. For this purpose, these ducts and tubes are lined by epithelia, which secrete fluid and electrolytes to lubricate their surfaces (Figure 8.1).[7,16] The process involves the active transcellular movement of Cl^- and the passive paracellular movement of Na^+ and water. A series of ion channels and transporters in the basolateral membrane accumulate actively Cl^- inside epithelial cells.

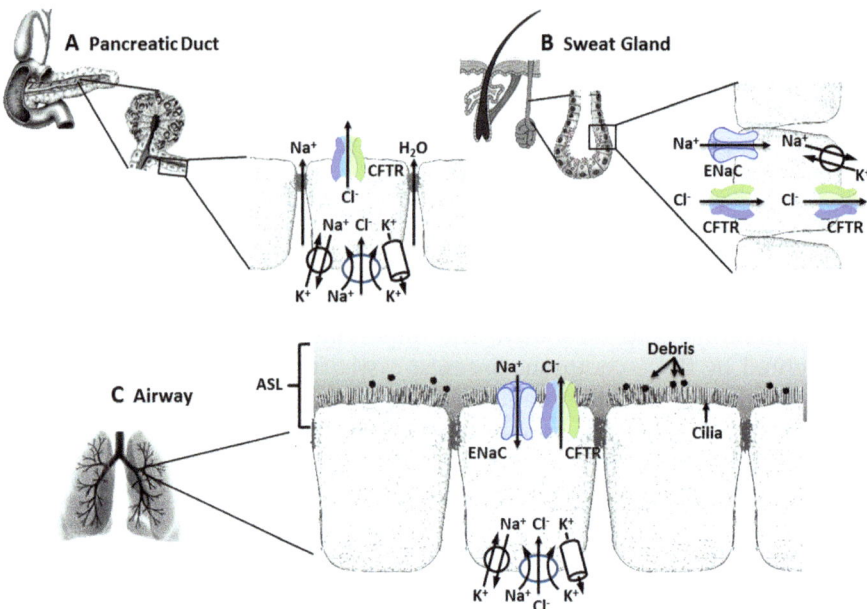

Figure 8.1   Role of CFTR in transepithelial fluid and electrolyte movements. (A) CFTR-driven fluid secretion lubricates ducts and tubes. (B) CFTR-mediated Cl⁻ absorption in sweat duct epithelia. (C) Regulation of ASL by CFTR. See text for further information.

However, CFTR plays the pivotal role by providing an apical membrane pathway for Cl⁻ to exit passively from cells down a favourable electrochemical gradient and a key point at which to regulate transepithelial fluid and electrolyte movements (Figure 8.1). By contrast, in sweat duct epithelia, the coordinated activity of CFTR and the epithelial Na⁺ channel (ENaC) reabsorb salt across a water-impermeable epithelium to generate a hypotonic fluid in the duct lumen that emerges onto skin as sweat (Figure 8.1). Interestingly, CFTR is located in both the apical and basolateral membranes of sweat duct epithelia where it provides pathways for passive Cl⁻ transport across the epithelium following active Na⁺ transport.[7,17]

Like sweat duct epithelia, airway epithelia possess both CFTR and ENaC channels in their apical membrane (Figure 8.1). They also possess a similar repertoire of basolateral membrane ion channels and transporters to those found in secretory epithelia. Thus, airway epithelia either absorb or secrete fluid and electrolytes depending on the prevailing electrochemical gradient and neurohumoral signals.[7,16,18] The coordinated activity of CFTR and ENaC in airway epithelia generates the thin layer of airway surface liquid (ASL) above which lies a mucus gel that traps debris in inhaled air (Figure 8.1). Beating cilia protruding from the apical membrane of airway epithelial cells propel mucus up the airways to remove debris from the lungs (termed mucociliary clearance). CFTR is also expressed in serous cells of submucosal glands in the proximal airways. CFTR-driven fluid and electrolyte secretion by these cells flushes mucins and antimicrobial factors from submucosal glands onto the surface epithelium where they play important roles in mucociliary clearance and host defence.[19]

In the kidney, CFTR is differentially expressed along the length of the nephron,[20] where it serves several functions. First, CFTR acts as a regulated anion channel to transport Cl^- across the apical membrane of epithelial cells in different nephron segments.[21] Second, some data suggest that CFTR co-assembles with the renal outer medullar potassium (ROMK) channel (Kir1.1) to form the ATP-sensitive K^+ channel that recycles K^+ across the apical membrane of epithelial cells in the thick ascending limb of the loop of Henle during salt reabsorption.[22] Third, CFTR stabilises cubulin expression during receptor-mediated endocytosis in the proximal tubule.[23] Thus, CFTR plays wide-ranging roles in epithelial physiology. Loss of CFTR function or unphysiological CFTR activity has profound consequences for epithelial ion transport and human health. Below, we consider CF, secretory diarrhoea and ADPKD.

8.2.2 Cystic Fibrosis

The genetic disease CF is an important medical problem, accounting for most cases of severe chronic obstructive lung disease and exocrine pancreatic dysfunction in Caucasian children and young adults.[7,24] Current therapies for CF are directed against disease symptoms, not its root cause CFTR malfunction. They include physiotherapy, mucolytic drugs and antibiotics to treat lung disease, and pancreatic enzyme replacement therapy and supplemental nutrition to overcome gastrointestinal dysfunction.[7,24] Together with specialist patient care in CF centres, improvements in symptomatic therapy have increased average life expectancy to around 40 years in North America and some European countries.[24]

The wide-ranging manifestations of CF, including chronic lung disease, exocrine pancreatic insufficiency, meconium ileus, male infertility and salty sweat are caused by defective CFTR-mediated fluid and electrolyte movements across epithelia.[7] CFTR malfunction in secretory epithelia prevents their lubrication leading to the stasis of protein-rich cargoes, the obstruction and ultimately destruction of ducts and tubes in organs such as the pancreas, intestine, hepatobiliary system and reproductive tissues. By contrast, loss of CFTR function in CF sweat ducts leads to salty sweat, a hallmark used clinically to diagnose the disease.[17]

The major cause of morbidity and mortality in CF is progressive chronic lung disease. CF lung disease is caused by defective epithelial ion transport, which impairs mucociliary clearance leading to mucus accumulation, airway obstruction and persistent bacterial infections.[7,18,25] In general, there are two mechanisms by which CF mutations cause harm. First, the mutant protein is not delivered to the apical membrane of epithelial cells. Second, the mutant protein is present at the apical membrane, but its function is altered. Both mechanisms are exemplified by the most common CF mutation, F508del, the deletion of a phenylalanine residue at position 508 of the CFTR protein sequence, which is located in the first nucleotide-binding domain (NBD1). About 70% of CF patients carry two copies of the F508del mutation and 90% at least one. The prevalence of the F508del mutation argues strongly that it is the primary target for therapy development.

The molecular basis for the F508del defect is protein misfolding, which causes retention of the mutant protein in the endoplasmic reticulum and its subsequent degradation by the proteasome.[26–28] The vast majority of F508del-CFTR therefore fails to reach its correct cellular location. Worse, the little that reaches the apical membrane is unstable, residing there for <4 h compared with >24 h for wild-type CFTR.[29] The F508del-CFTR mutation also disrupts severely CFTR channel gating (Figure 8.2).[30] For wild-type CFTR, channel openings occur frequently and are separated by short periods of channel closure (Figure 8.2). By contrast, F508del-CFTR channel openings occur infrequently and are separated by prolonged periods of channel closure (Figure 8.2).

In airway epithelia, loss of CFTR-driven transepithelial fluid and electrolyte transport has three main consequences. First, defective HCO_3^- secretion prevents secreted mucins expanding to form mucus-gel networks; instead they remain aggregated and difficult to transport.[31] Second, CFTR malfunction in submucosal glands disrupts the secretion of fluid, mucins and anti-microbial factors onto the surface epithelium.[19] Third, CFTR malfunction in the surface epithelium depletes ASL.[18] This latter defect highlights the dual roles of CFTR as a Cl^- channel and transport regulator. Loss of CFTR Cl^- channel function prevents fluid and electrolyte secretion onto the surface epithelium, whereas dysregulation of ENaC by CFTR leads to excessive absorption of Na^+ and fluid, dehydrating ASL.[18] Loss of ASL

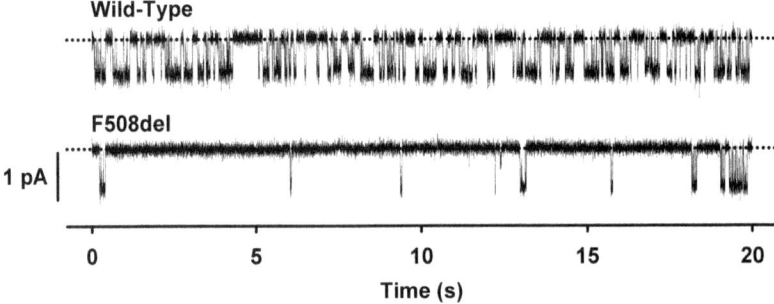

Figure 8.2   F508del-CFTR disrupts CFTR channel gating. Representative recordings of wild-type and F508del-CFTR Cl^- channels in excised inside-out membrane patches from cells expressing recombinant human CFTR. ATP (1 mM) and PKA (75 nM) were continuously present in the intracellular solution, voltage was −50 mV, there was a large Cl^- concentration gradient across the membrane patch ($[Cl^-]_{int}$, 147 mM; $[Cl^-]_{ext}$, 10 mM) and temperature was 37 °C. Dotted lines indicate where channels are closed and downward deflections of the traces correspond to channel openings.

impairs mucociliary clearance because cilia are unable to beat effectively on dehydrated airway surfaces. Mucus accumulation causes airway obstruction. Moreover, defective mucus transport results in persistent infection of CF airways by microbial pathogens (*e.g. Pseudomonas aeruginosa*).[7] Bacterial infections trigger inflammatory responses in CF airways that are excessive in magnitude and protracted in duration. Together, infection and inflammation play a central role in the destruction of lung tissue in CF, leading to bronchiectasis, emphysema and end-stage lung disease.

8.2.3 Secretory Diarrhoea

The commonest cause of secretory diarrhoea is infection of the small intestine by enterotoxin-producing bacteria most notably *Vibrio cholerae* and *Escherichia coli*.[12] These bacteria do not damage the intestinal epithelium. Instead, they cause disease by modifying profoundly transepithelial ion transport in the small intestine. For example, it is not unknown for cholera patients to expel more stools than their own body weight.[32] Contamination of water supplies with enterotoxin-producing bacteria is a major global health problem, causing the death of several million lives annually, mostly young children in underdeveloped countries.[33] However, these organisms, especially *Escherichia coli*, also cause disease in developed countries, particularly among vulnerable young and elderly populations.

Cholera toxin, the major toxin secreted by *Vibrio cholerae* and *Escherichia coli* heat labile toxin have identical cellular mechanisms of action, targeting adenylyl cyclase in intestinal epithelial cells to activate the cAMP signalling cascade leading to the phosphorylation of target proteins by protein kinase A (PKA).[12] By contrast, *Escherichia coli* heat stable toxin activates guanylate cyclase and hence, cGMP-dependent protein kinase (PKG) in intestinal epithelial cells.[34] Activation of the cAMP signaling cascade in the small intestine stimulates CFTR-mediated transepithelial Cl^- secretion by crypt epithelial cells.[12] It also inhibits electroneutral salt reabsorption by villus epithelial cells in the small intestine and stimulates K^+ secretion by crypt epithelial cells in the colon.[12] Like PKA, PKG phosphorylates CFTR leading to the activation of CFTR-mediated transepithelial Cl^- secretion by crypt epithelial cells in the small intestine.[35] Because Na^+-coupled glucose absorption by villus epithelial cells is unaffected by PKA and PKG, oral rehydration therapy (ORT) with isosmolar sodium glucose solutions is the treatment of choice for secretory diarrhoea in all countries.[36] Although ORT effectively rehydrates afflicted patients, it does not lessen the diarrhoea.[12] For this reason, pharmacological interventions to prevent diarrhoea (*e.g.* small-molecule CFTR inhibitors) would be a welcome adjuvant to ORT.

8.2.4 Autosomal Dominant Polycystic Kidney Disease

ADPKD is the most common single gene disorder to affect kidney function, with an incidence of approximately 1 in 1000 live births in all ethnic groups.[37] The disease is caused by mutations in the polycystin proteins (polycystin-1 and polycystin-2) that initiate a cascade of events, which result in the formation of multiple epithelial cysts containing a fluid-filled cavity surrounded by a single layer of immature renal epithelial cells.[37–39] The development and growth of cysts in ADPKD is accelerated by activation of the cAMP signal transduction cascade, which stimulates the proliferation of ADPKD epithelial cells and fluid accumulation within the cyst lumen powered by CFTR-mediated transepithelial Cl^- secretion.[13,40] The insidious formation and growth of multiple ADPKD cysts progressively destroys kidney function leading to severe renal failure.[37,38] At present, there are no cures for ADPKD, which accounts for approximately 7–10% of patients requiring kidney transplantation and dialysis.[38] Among the many approaches now under consideration to develop rational new therapies for ADPKD,[37] one strategy is to prevent fluid accumulation within ADPKD cysts using small-molecule CFTR inhibitors.

8.3 CFTR Structure and Function

CFTR has a modular design composed of two membrane-spanning domains (MSDs), two nucleotide-binding domains (NBDs) and a unique regulatory domain (RD).[8,9] The MSDs are composed of six transmembrane segments linked by short extracellular, but long intracellular loops. The NBDs possess conserved ATP-binding motifs (ABC signature (LSGGQ) sequence and Walker motifs). The RD is distinguished by multiple consensus phosphorylation sites and many charged residues.

Initial efforts to investigate CFTR structure and function utilised functional studies of CFTR constructs bearing site-directed mutations. This work demonstrated that CFTR is a small conductance anion-selective channel with complex regulation. The MSDs form an anion-selective pore, but instead of multiple transmembrane segments governing conduction and permeation, the sixth transmembrane segment (M6) plays a dominant role.[8,41] Phosphorylation of the RD by PKA at multiple consensus sites is a prerequisite for CFTR activity.[8,42] ATP binding and hydrolysis at the NBDs control the opening and closing of the CFTR pore.[8,42] But, detailed mechanistic insight has for a long time been lacking.

The elucidation of crystal structures of isolated NBDs (*e.g.* CFTR NBD1)[43] and intact ABC transporters, especially Sav1866, the multidrug transporter of *Staphylococcus aureus*,[44] revolutionised understanding of CFTR structure and function. They revealed that the NBDs function as a head-to-tail dimer with two ATP-binding sites located at the dimer interface (Figure 8.3). One ATP-binding site is formed by the Walker A and B motifs of NBD1 and the LSGGQ motif of NBD2 (termed site 1), while the other is formed by the Walker A and B motifs of NBD2 and the LSGGQ motif of NBD1 (termed site 2) (Figure 8.3).[43] Structural studies also illuminated the impact of CF mutations on CFTR function. For example, F508, the site of the commonest CF mutation, is located remote from the NBD dimer interface on the surface

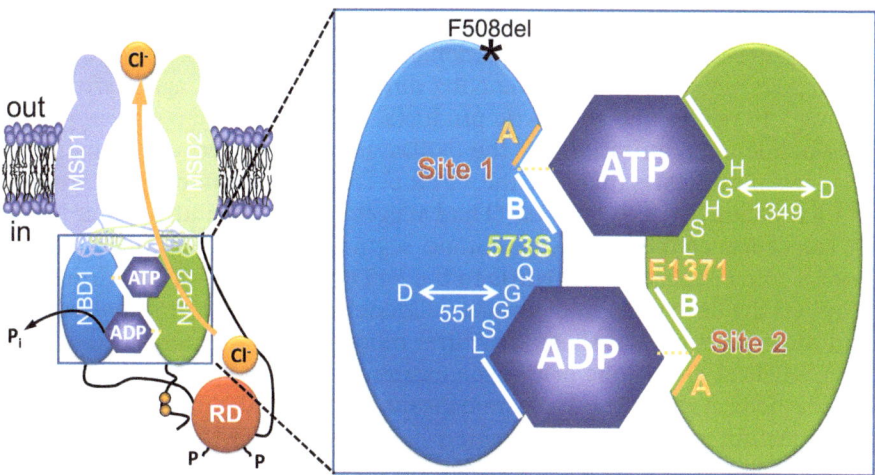

Figure 8.3 ATP-driven NBD dimerisation opens the CFTR pore. The schematic model shows an activated CFTR Cl⁻ channel. ATP-driven NBD dimerisation opens the CFTR pore. The magnified NBDs show that each ATP-binding site is formed by the Walker A and B motifs (labeled A and B, respectively) of one NBD and the LSGGQ motif of the other NBD. ATP-binding site 2 contains a canonical LSGGQ motif, whereas ATP-binding site 1 contains a non-canonical LSGGQ motif (LSHGH). ATP-binding site 2 also contains a catalytic base (E1371) at the distal end of the Walker B motif, but this residue is absent in site 1 (S573). The location of the CF mutations F508del (surface of NBD1 opposite ICL4), G551D (site 2) and G1349D (site 1) are shown. Abbreviations: MSD, membrane-spanning domain; NBD, nucleotide-binding domain; P, phosphorylation of the RD; Pᵢ, inorganic phosphate; RD, regulatory domain. In and Out denote the intra- and extracellular sides of the membrane, respectively. See text for further information.

Reproduced, with permission, from Chen *et al.*,[135] © S. Karger AG, Basel (2006), and Hwang and Sheppard,[136] © John Wiley & Sons Ltd. (2009).

of NBD1 in a prime position to interact with the MSDs (Figure 8.3).[43] By contrast, other CF mutations (*e.g.* G551D and G1349D) affect critical residues in the two ATP-binding sites (Figure 8.3), providing explanations for the deleterious effects of these mutations on CFTR channel gating.[45,46]

Vergani *et al.*[47] integrated biochemical (*e.g.* Aleksandrov *et al.*[48]) and functional (*e.g.* Zeltwanger *et al.*[49]) data with the results of structural studies of ABC transporters to propose the ATP-driven NBD dimerisation model of CFTR channel gating. In this model, CFTR channel gating is controlled by ATP binding and hydrolysis at site 2 driving cycles of NBD dimer assembly and disassembly. In support of this model, R555 (NBD1) and T1246 (NBD2), two residues on either side of ATP-binding site 2, are energetically coupled in open, but not closed channels, arguing that the NBDs change their conformation during channel gating.[50] However, subsequent studies revealed that ATP hydrolysis at ATP-binding site 2 leads to partial, not full, separation of the NBD dimer.[51,52] Thus, during the gating cycle, ATP-binding site 1 remains tightly closed around a bound ATP molecule, whereas ATP-binding site 2 undergoes conformational changes, closing following binding of an ATP molecule to form the NBD dimer, but opening again after ATP cleavage to release the hydrolytic products.

Dynamic reorganization of the NBDs in turn power conformational changes of the MSDs that gate the CFTR pore to control anion flow. Studies of CF mutants (*e.g.* Cotton *et al.*[53]) had suggested that the intracellular loops (ICLs) which connect transmembrane segments within the MSDs play a crucial role in coupling the NBDs to the CFTR pore. However, confirmation required the development of structural models of CFTR by Serohijos *et al.*[54] and Mornon *et al.*[55] In these structural models, each ICL consisted of two long α-helical extensions of transmembrane segments with an intervening short α-helix at its most cytoplasmic location orientated parallel to the plane of the membrane. Because this short α-helix interacted with the NBDs, it was termed the coupling helix.[54,55] Several important conclusions about NBD–MSD communication are suggested by the structural models of Serohijos *et al.*[54] and Mornon *et al.*[55] First, the ICLs communicate with both the same and the opposite NBD (*e.g.* ICL1 (MSD1) with NBD1 and ICL2 (MSD1) with NBD2). Second, the coupling helices of ICL1 and ICL3 interact directly with ATP-binding sites 1 and 2, respectively. Third, the coupling helix of ICL4 interacts with F508. Thus, the F508del mutation affects a critical interface in the CFTR Cl⁻ channel.

Electron crystallography of two-dimensional CFTR crystals suggested that the CFTR pore resembles an asymmetric hour-glass, with a deep wide intracellular vestibule and a shallow extracellular vestibule separated by a constriction.[56] This image of the CFTR pore demonstrates striking agreement with predictions from functional studies.[8,41] Building on these studies, Norimatsu *et al.*[57] used cysteine scanning mutagenesis to test the predicted orientation of amino acid side-chains in M3, M6, M9 and M12 from a Sav1866-based molecular model of the CFTR pore. The authors' data suggest that the constriction in the CFTR pore is located towards the extracellular end of the channel with the boundary between the outer vestibule and the constriction located near T338 (M6) and I1131 (M12) and that between the constriction and the inner vestibule located near S341 (M6) and T1134 (M12). Functional evidence argues that the pore constriction determines the anion selectivity of the CFTR,[58] exhibiting characteristics of a lyotropic series (*i.e.* anions with the lowest free energy of hydration enter first and bind strongest).[59] In support of the degraded ABC transporter model of CFTR function,[60] the constriction is also a plausible location of the gate, which controls anion flow through the CFTR pore.[61] However, other studies argue that the gate might occupy a more cytoplasmic location within the CFTR pore.[62] Identifying the location of this gate is crucial to understand the CFTR gating pathway, the sequence of conformation changes initiated by ATP binding to the NBDs that lead to Cl⁻ flow through the CFTR pore. In turn, this knowledge as well as information about the architecture of the CFTR pore is vital to understand the mechanism of action of small-molecule CFTR modulators.

8.4 Restoration of CFTR Function

To target the root cause of CF, future therapies should (i) overcome the F508del-CFTR processing defect and traffic the mutant protein to the apical membrane;[26] (ii) extend the residence time of F508del-CFTR at the apical membrane[29] and attenuate channel "rundown" (*e.g.* Schultz *et al.*[63]) and (iii) rescue the defective channel gating of F508del-CFTR.[30] Thus, small molecules with two, possibly three, types of activity are required to suppress the deficits in F508del-CFTR expression and function.

Some CFTR modulators (*e.g.* genistein)[64] were identified using conventional assays of CFTR function. However, by far the most successful strategy is high-throughput screening (HTS).[65] HTS exploits a reliable, sensitive, cost-effective assay to screen libraries of chemically diverse small-molecules to identify lead compounds for medicinal chemistry optimisation. For further information about the use of HTS to discover small-molecule CFTR modulators, see Verkman and Galietta.[65] Below, we discuss selectively, small molecules that rescue the expression and function of F508del-CFTR.

8.4.1 CFTR Correctors

Small molecules that overcome the processing defect of F508del-CFTR to rescue its cell surface expression are termed CFTR correctors.[65,66] CFTR correctors might interact directly with CFTR, by acting as substrate mimics or active site inhibitors. Alternatively, they might target one or more of the many CFTR interacting proteins that orchestrate and control CFTR processing, its delivery to, and expression at the apical membrane. This latter group of CFTR correctors is termed proteostasis regulators because they aim to treat CF by manipulating the concentration, conformation, quaternary structure and/or location of CFTR.[67]

The Cystic Fibrosis Foundation (Bethesda, USA), a nonprofit, donor-supported organization, developed a new model for supporting drug development (termed Venture Philanthropy) when it funded Aurora Biosciences then Vertex Pharmaceuticals to develop small-molecule therapies that target directly CF mutants. The first CFTR corrector identified by Vertex Pharmaceuticals was the quinazoline VRT-325, which restored CFTR function to CF bronchial epithelia expressing F508del-CFTR (F508del-CFBE) to ~ 10% that of human bronchial epithelia expressing wild-type CFTR (HBE),[68] an amount similar to that achieved by the bisaminomethylbithiazole CFTR corrector corr-4a.[69] Subsequently, Vertex Pharmaceuticals developed VX-809 (Figure 8.4), a potent, selective and orally bioavailable CFTR corrector that partially restores CFTR expression and function to F508del-CFBE.[70a] Treatment of cells expressing recombinant F508del-CFTR with VX-809 (3 μM) for 24 h abrogated the protease-sensitivity of mutant protein and accelerated its conversion from the immature, core-glycosylated form (band B) to the mature, fully-glycosylated form (band C).[70a] These data suggest that VX-809 promotes the correct folding and intracellular transport of a fraction of F508del-CFTR

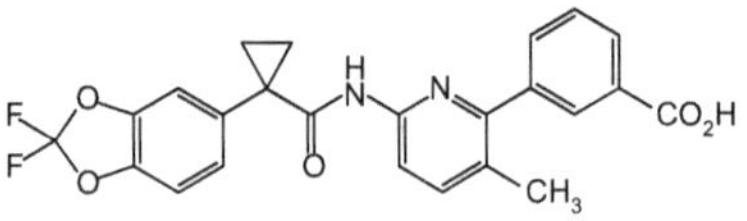

VX-809

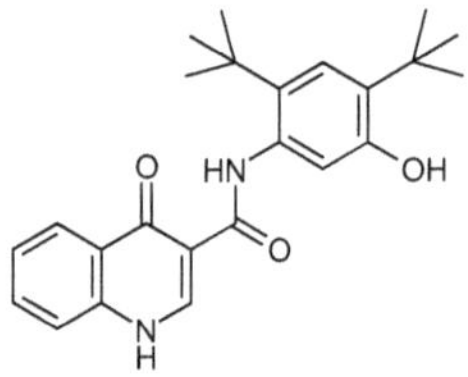

VX-770

Figure 8.4 Chemical structures of clinically-tested CFTR correctors and potentiators. Abbreviations: VX-809 (Lumacaftor), 3-{6-{[1-(2,2-difluoro-1,3-benzodioxol-5-yl)cyclopropanecarbonyl]amino}-3-methylpyridin-2-yl}benzoic acid; VX-770 (Ivacaftor), N-(2,4-di-tert-butyl-5-hydroxyphenyl)-4-oxo-1,4-dihydroquinoline-3-carboxamide.

protein. Following treatment with VX-809 (3 µM for 24–48 h), Van Goor et al.[70a] reported that F508del-CFTR Cl⁻ channels exhibited the same open probability (P_o) as wild-type, whereas Kopeikin et al.[70b] found that VX-809 by itself did not improve channel function. Finally, when compared with other CFTR correctors VX-809 demonstrated improved selectivity and greater efficacy, achieving 14% HBE function restored to F508del-CFBE.[70a]

Based on the success of preclinical studies, VX-809 was tested in 89 adult CF patients homozygous for F508del-CFTR in a 28 day randomised, double-blind placebo-controlled (phase IIa) clinical trial.[71] Although VX-809 was well tolerated by CF patients and caused a modest reduction in sweat Cl⁻ concentration, it was without effect on patient-reported outcomes, lung function (measured by forced expiratory volume in one second, FEV_1) and nasal potential difference (NPD; a measure of CFTR function in the nasal epithelium).[71]

A potential explanation for the lack of clinical efficacy of VX-809 is provided by two recent studies.[72,73] Mendoza et al.[72] identified an apparent efficacy ceiling among CFTR correctors that likely interact directly with F508del-CFTR: none had potency <1 µM nor efficacy >15% wild-type CFTR function. To elucidate the mechanistic basis of this efficacy ceiling, the authors identified amino acid positions statistically coupled to position 508 in ABC transporter sequences and evaluated the impact of mutations at these positions on NBD1 folding and CFTR protein maturation.[72] Consistent with previous results (e.g. Serohijos et al.[54] and Thibodeau et al.[74]), F508del altered both NBD1 folding and the interaction of NBD1 with ICL4. Mutations that corrected either defect alone partially restored F508del-CFTR expression and function. Only combinations of mutations that rescued NBD1 folding and stabilised the interaction of NBD1 with ICL4 restored F508del-CFTR expression and function to wild-type levels.[72] Identical conclusions were obtained by Rabeh *et al.*[73] investigating the thermodynamics of NBD1 folding and the stability of the NBD1:ICL4 domain interface. Taken together, the results of these studies provide a framework to identify efficacious CFTR correctors that restore wild-type expression and function to F508del-CFTR.

8.4.2 CFTR Potentiators

Small molecules that repair the F508del-CFTR gating defect are termed CFTR potentiators (not openers or activators because they do not open silent CFTR Cl$^-$ channels).[65] They interact directly with CFTR to enhance channel gating following PKA-dependent phosphorylation. Some CFTR potentiators increase the frequency of channel opening (*e.g.* dichlorofluorescein) (Z. Cai and D. N. Sheppard, unpublished observation), other CFTR potentiators prolong the duration of channel openings (*e.g.* phloxine B)[75] and yet others increase both the frequency and duration of channel openings (*e.g.* UC$_{CF}$-853)[76] (Y. Wang, Z. Cai and D. N. Sheppard, unpublished observation) (Figure 8.5).

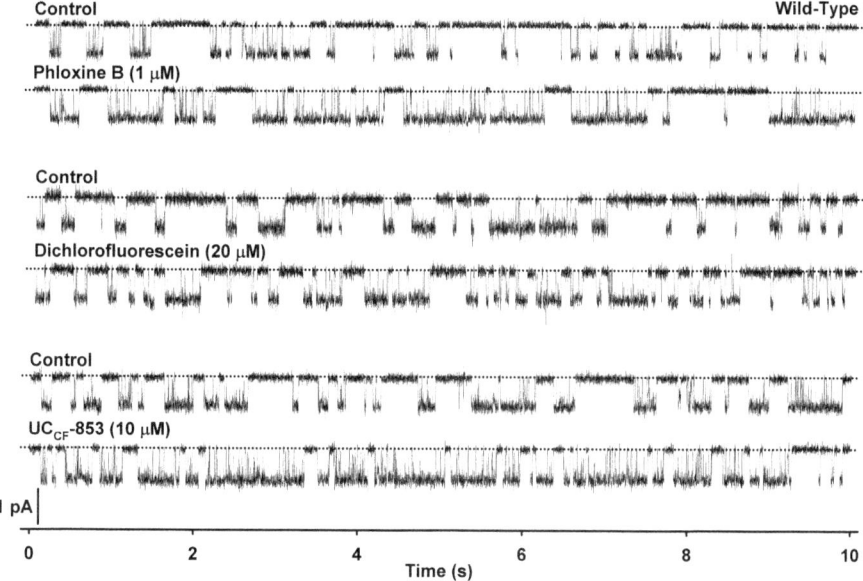

Figure 8.5 CFTR potentiators enhance channel gating. Representative recordings show the effects of dichlorofluorescein (20 μM), phloxine B (1 μM) and UC$_{CF}$-853 (10 μM) on the single-channel activity of wild-type CFTR in excised inside-out membrane patches from cells expressing recombinant human CFTR. With the exception that ATP (0.3 mM) was used to support channel activity, experimental conditions were the same as those described in Figure 8.2. Dotted lines indicate where channels are closed and downward deflections of the traces correspond to channel openings.

The CFTR potentiator that has received greatest attention is the isoflavone genistein.[64,77] In the presence of a maximally effective concentration of the cAMP agonist forskolin (10 μM), genistein (50 μM) robustly enhanced F508del-CFTR activity following its delivery to the cell surface by low-temperature incubation (*e.g.* Hwang *et al.*[78]). Genistein augments strongly F508del-CFTR channel activity by accelerating the rate and prolonging the duration of channel openings.[77] Using the ATP-driven NBD dimerisation model of CFTR channel gating,[47,50] Ai *et al.*[79] speculated that genistein might bind at the NBD dimer interface and promote channel opening by lowering the free energy of the transition state. The authors also suggested that genistein might slow the rate of channel closure by stabilising the NBD dimer in its ATP bound configuration.[79] Consistent with these ideas, Moran *et al.*[80] used virtual ligand docking with a molecular model of the NBD dimer to suggest that genistein binds at the dimer interface (Figure 8.6). The genistein-binding site is distinct from the two ATP-binding sites of CFTR.[81,82] It is composed of sequences from both NBD1 (Walker A, Walker B and LSGGQ) and NBD2 (LSGGQ), with those from NBD1 forming a cavity in which genistein docks.[80] Because of its efficacy *in vitro*, Illek *et al.*[83] tested genistein on CF patients bearing the G551D mutation, a relatively common

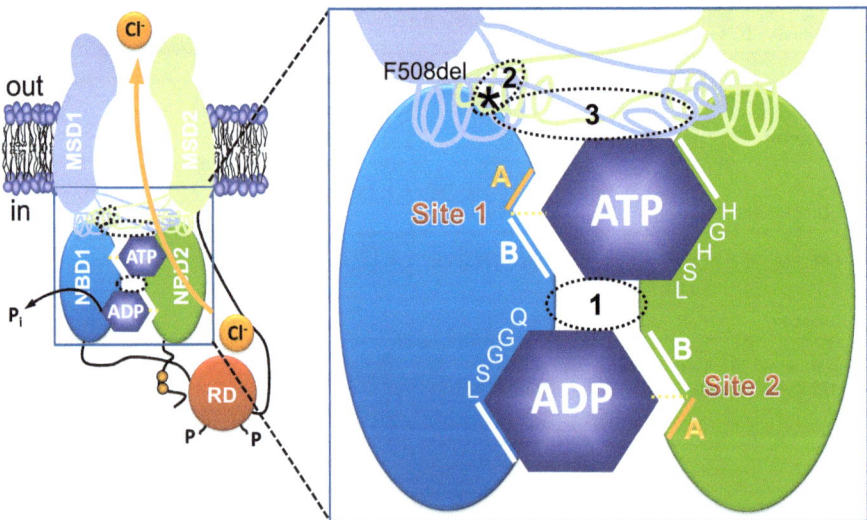

Figure 8.6   Location of some drug-binding sites on CFTR. Simplified model of an open CFTR Cl⁻ channel illustrating the general position of some drug-binding sites. Some small molecules bind at the NBD dimer interface at a location distinct from the two ATP-binding sites (labeled 1). Other small molecules dock with a cavity formed by deletion of F508 at the interface of NBD1 and ICL4 (labeled 2). Yet other small molecules bind at the interface of the NBDs and MSDs involving ICL 1, 2 and 4 (labeled 3). For further information, see text and Figure 8.3.
Modified with permission from Hwang and Sheppard,[136] © John Wiley & Sons Ltd. (2009), and Sheppard,[137] © Cell Press (2011).

CFTR mutation that is without effect on protein processing, but severely disrupts channel gating.[46] Of note, application of genistein (50 μM) to the nasal epithelium of CF patients with the G551D mutation restored 15% of wild-type CFTR function (measured by NPD).[83] However, genistein's lack of potency and selectivity precludes its application in the clinic.

In 2012, ivacaftor (Kalydeco™; VX-770) (Figure 8.4), a potent, selective and orally bioavailable CFTR potentiator developed by Vertex Pharmaceuticals, became the first small-molecule targeting CFTR to be approved for patient use. It was developed from a chemical scaffold identified as a CFTR potentiator in the second round of HTS conducted by Vertex Pharmaceuticals to identify small-molecule CFTR modulators.[68,84] By increasing the frequency and duration of channel openings, ivacaftor (1 μM) restored F508del-CFTR channel activity to wild-type levels (measured by P_o). Treatment of F508del-CFBE with ivacaftor (10 μM) increased CFTR-mediated transepithelial ion transport (measured by short-circuit current (I_{sc})) to about half those of HBE.[84] Consistent with these data, ivacaftor (10 μM) increased ASL volume and ciliary beat frequency of F508del-CFBE to about half those of HBE.[84] These data suggest that ivacaftor restores partially CFTR function and rescues transepithelial ion transport by F508del-CFBE.

In its initial (phase II) clinical trial, ivacaftor was tested in 39 adult CF patients carrying the G551D-CFTR mutation.[85] The patients took ivacaftor orally in 14 or 28 day blinded, randomised, placebo-controlled trials. Ivacaftor was well tolerated by CF patients.[85] More importantly, however, in CF patients receiving ivacaftor (150 mg twice daily) sweat Cl⁻ concentration decreased by ∼40 mmol/l to approach the normal range (<60 mmol/l).[85] At this dose, lung function (measured by FEV_1) increased by 9%.[85] A similar magnitude increase in lung function (measured by FEV_1) was observed in the phase III clinical trial of ivacaftor, which involved 161 adult CF patients with the G551D-CFTR mutation.[86] For three reasons, the results of this clinical trial were highly significant for CF patient care. First, lung function increased rapidly within 2 weeks of commencing therapy and was sustained through the 48 weeks of the trial.[86] Second, in comparison with placebo-treated patients, ivacaftor-treated patients were 55% less likely to experience a pulmonary exacerbation over the 48 week period, suggesting that ivacaftor improved lung defence against bacterial infections.[86] Third, ivacaftor-treated patients gained significant weight (∼2.7 kg) over the trial period, suggesting that ivacaftor improves intestinal function resulting in better absorption of nutrients.[86] When ivacaftor was tested in CF children aged 6–11 years with at least one copy of the G551D mutation, very similar results were obtained. Lung function improved, weight was gained and sweat Cl⁻ concentration fell.[87] Based on the results of these trials, the FDA and EMEA approved ivacaftor for use in CF patients aged 6 years and older with one copy of the G551D mutation.

Finally, recent work suggests that ivacaftor might potentiate CFTR channel gating by a mechanism distinct from that of CFTR potentiators, which enhance ATP-dependent channel gating.[79] Using purified, reconstituted re-combinant CFTR protein, Eckford *et al.*[88] demonstrated that ivacaftor opens CFTR channels following PKA-dependent phosphorylation, in the absence of ATP. These and other data[88] suggest that ivacaftor enhances ATP-independent CFTR channel gating,[89] such as that exhibited by the G551D-CFTR mutant.[46] They further suggest that CFTR channel gating can be potentiated by multiple mechanisms. However, the observation that iva-caftor potentiates multiple CF mutants,[90] including some that do not disrupt CFTR channel gating,[91a] argues that it and related small molecules will have widespread utility in the future treatment of CF patients.

8.4.3   CFTR Corrector-potentiators

Because F508del-CFTR disrupts both the expression and function of CFTR,[26,30] the ideal small-molecule therapy for CF patients bearing this mutation would combine the properties of CFTR correctors with those of CFTR potentiators. Consistent with this idea, Van Goor *et al.*[68] demonstrated that the effects of the CFTR potentiator VRT-532 on F508del-CFBE were synergistic with those of the CFTR corrector VRT-325. Together VRT-325 and VRT-532 generated levels of CFTR-mediated transepithelial ion transport in F508del-CFBE >20% those in HBE.[68] Similar results were observed using the CFTR corrector VX-809 and the CFTR potentiator ivacaftor.[70a] Moreover, in clinical trials both small molecules together restored significant levels of lung function to some CF patients homozygous for F508del-CFTR,[91b] whereas by themselves neither VX-809 nor ivacaftor improved the lung function of these CF patients.[71,92] These data argue that dual-acting small molecules, possessing corrector and potentiator activity (so-called CFTR corrector-potentiators) have significant therapeutic potential. But, could such molecules be discovered or designed?

The first two CFTR corrector-potentiators identified were discovered serendipitously. First, Wang *et al.*[93] demonstrated that the CFTR poten-tiator VRT-532, by itself, rescues the misprocessing of F508del-CFTR. Building on these data, Wellhauser *et al.*[94] showed that VRT-532 reduces the ATPase activity of F508del-CFTR and abrogates its protease suscepti-bility. The authors' data suggest that VRT-532 binds directly to F508del-CFTR to induce a change in conformation that improves its stability.[94] They also suggest that VRT-532 enhances channel gating by slowing ATP hydrolysis at ATP-binding site 2.[94] Second, Pedemonte *et al.*[95] discovered that the aminoarylthiazole corr-2b, one of the first CFTR correctors iden-tified by HTS,[69] exhibits dual activity as both a CFTR corrector and a CFTR potentiator. When compared with small molecules that act as CFTR correctors alone, (*e.g.* the bisaminomethylbithiazole corr-4a), corr-2b generated double the amount of forskolin-stimulated CFTR Cl$^-$ current (I_{FSK}) relative to the total CFTR Cl$^-$ current measured in the presence of forskolin and the CFTR potentiator genistein (I_{TOT}) (corr-4a: I_{FSK}/I_{TOT} ~40%; corr-2b: I_{FSK}/I_{TOT} ~80%).[95] Interestingly, aminoarylthiazoles are atypical CFTR potentiators; they do not act acutely, but instead require protein synthesis to exert their effects.[95]

In search of dual-acting small molecules to rescue F508del-CFTR, Verkman and colleagues screened ∼110,000 drug-like chemicals with their HTS assay.[96] From this screen, Phuan *et al.*[96] identified cyanoquinolines, a chemical scaffold with independent CFTR corrector and potentiator activity. One cyanoquinoline, CoPo-22 trafficked F508del-CFTR to the cell surface with comparable potency to corr-4a and potentiated F508del-CFTR channel activity with potency similar to genistein.[96] Structure-activity relationship studies revealed that the corrector and potentiator activities of cyanoquinolines could be dissociated from each other.[96] Molecular modeling suggested that cyanoquinolines require both a flexible diamine tether and a short bridge between the cyanoquinoline and arylamide moieties for optimal F508del-CFTR corrector and potentiator activities.[97] Phuan *et al.*[96] speculated that separation of corrector and potentiator activities might be achieved by distinct binding sites on F508del-CFTR and/or CFTR-interacting proteins. By contrast, Knapp *et al.*[97] posited the imaginative idea of a single binding site that alters its conformation during the processing of F508del-CFTR and its delivery to the cell surface. Finally, Kalid *et al.*[98] adopted *in silico* structure-based screening to search for CFTR correctors, potentiators and dual-acting small molecules. Based on analyses of homology modeling, three binding sites for small molecules were identified on F508del-CFTR: (i) the NBD1:NBD2 interface, (ii) the NBD1:ICL4 cavity formed by deletion of F508 and (iii) the interface of NBD1 and NBD2 with ICL1, 2 and 4 (Figure 8.6).[98] From 100,000 chemicals screened *in silico* and 496 selected for functional testing, two CFTR corrector-potentiators were identified. EPX-108380 docked at the NBD1:NBD2 interface, while EPX-107979 docked at the NBD1:NBD2:ICL1, 2 and 4 interface.[98] This study demonstrates convincingly the utility of *in silico* structure-based screening to search for CFTR modulators.

8.4.4 Towards the Therapeutic Application of CFTR Correctors and Potentiators

The application of HTS to CFTR drug discovery has led to the identification of abundant small molecules that rescue the expression and function of CF mutants.[65] The pressing challenge now is to transform discoveries in the laboratory into effective therapies for CF patients. The pioneering efforts of Vertex Pharmaceuticals especially the success of ivacaftor demonstrate convincingly the value of targeting directly CFTR with small molecules. However, they urge further prompt progress to develop CFTR correctors and potentiators for CF patients with the F508del mutation, the largest group of CF patients.

The amount of function that must be restored to individual CF mutants to achieve therapeutic benefit is currently unknown. Some attempts have been made to address this fundamental question by analysing published data on the relationship between the genotype of CF patients, their clinical phenotype and the effects of CF mutants on the CFTR Cl$^-$ channel. For example, Van Goor *et al.*[68] speculated that restoration of 5–30% of wild-type CFTR function to CF patients bearing the F508del mutation would be of therapeutic benefit. Using biochemical and functional data and the relationship $I^{CFTR} = N \times i \times P_o$ (I^{CFTR}, CFTR-mediated apical membrane Cl$^-$ current; N, the number of CFTR Cl$^-$ channels in the apical membrane; i, the amount of current that flows through an open CFTR Cl$^-$ channel and P_o, a measure of its activity), it is possible to predict how much function different therapeutic strategies restore to CF mutants *in vitro*.[99] By contrast, it is not possible to quantify CFTR function *in vivo* using currently available research tools (*e.g.* NPD), while measures of lung function (*e.g.* FEV$_1$) lack sensitivity during the crucial early stages of lung disease. These considerations argue that improvements in disease biomarkers and therapeutic end-points are required to determine how much CFTR function is required for therapeutic benefit.

A further obstacle for CF drug development is the use of animal models to demonstrate therapeutic proof of principle. The failure of mouse models of CF to recapitulate the clinical features of CF[100] limits their utility in preclinical studies of experimental therapeutics. Furthermore, the lack of effects of CFTR potentiators on murine CFTR,[101,102] argues that these agents should not be tested in CF mice. To overcome these difficulties, Vertex Pharmaceuticals employed well-differentiated primary cultures of human airway epithelia grown at an air-liquid interface for therapeutic proof of principle studies.[84] The success of this approach argues that primary cultures of human airway epithelia are the model system of choice for therapeutic proof of principle studies. However, the emergence of animal models that emulate the clinical features of CF (*e.g.* F508del-CFTR pigs),[103] argues that testing small-molecule therapies in animal models should now be reevaluated.

Finally, the prospects of designing rationally better CFTR correctors and potentiators beckons. The binding sites of some CFTR correctors, potentiators and dual-acting molecules are beginning to be revealed.[98] Precise mapping of these sites will require further understanding of CFTR structure and how it is perturbed by CF mutations. Advances in the structural biology of CFTR (*e.g.* Rosenberg *et al.*[104]) and modeling CFTR using the structures of related ABC transporters and knowledge of CFTR function (*e.g.* Norimatsu *et al.*[57] and Mornon *et al.*[105]) argue that rational design might be enhanced. Such therapeutics will likely have improved efficacy and specificity.

8.5   Inhibition of CFTR Function

Secretory diarrhoea, ADPKD and some reproductive disorders are characterised by unphysiological CFTR-mediated transepithelial fluid and electrolyte movements.[12,13,15] The cellular mechanism of CFTR-mediated transepithelial ion transport suggests a number of targets for pharmacological intervention to prevent unphysiological fluid and electrolyte movements. These targets include the basolateral membrane ion channels and

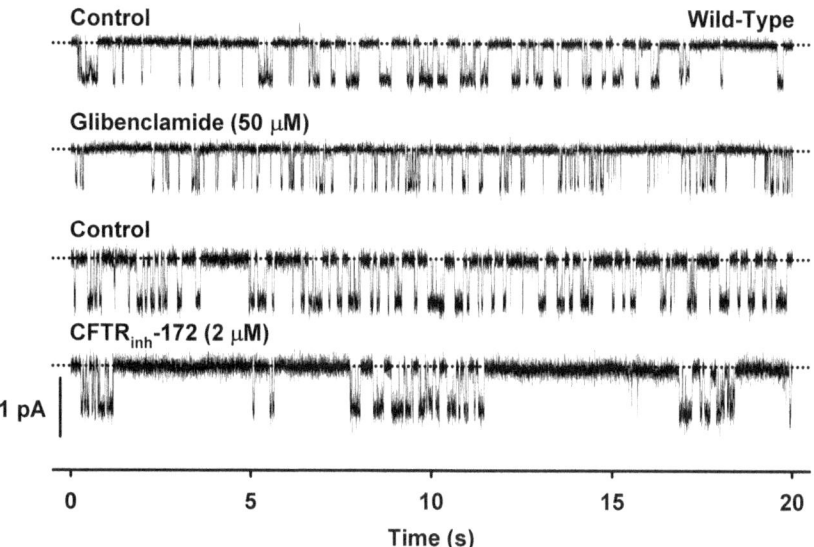

Figure 8.7 Blockade of CFTR by open-channel and allosteric inhibitors. Representative recordings show the effects of glibenclamide (50 μM) and CFTR$_{inh}$-172 (2 μM) on the single-channel activity of wild-type CFTR in excised inside-out membrane patches from cells expressing recombinant human CFTR. With the exception that ATP (0.3 mM) was used to support channel activity, experimental conditions were the same as those described in Figure 8.2. Dotted lines indicate where channels are closed and downward deflections of the traces correspond to channel openings.

transporters that accumulate Cl$^-$ within epithelial cells, signaling molecules that regulate CFTR activity and CFTR, itself. There are two general mechanisms by which small molecules inhibit directly the CFTR Cl$^-$ channel: allosteric inhibition (*e.g.* CFTR$_{inh}$-172)[106,107] and open-channel blockade (*e.g.* glibenclamide).[108] Agents that act as allosteric inhibitors bind to CFTR and interfere with channel gating to hinder Cl$^-$ entry into the channel pore (Figure 8.7). By contrast, ligands that act as open-channel blockers bind within the open channel pore to obstruct transmembrane Cl$^-$ flow (Figure 8.7).

Using conventional assays of CFTR function, many chemically diverse small-molecule CFTR inhibitor have been identified (for review, see Hwang & Sheppard[77] and Schultz *et al.*[109]). However, few of these agents are high-affinity CFTR inhibitors and none demonstrate specificity. Allosteric inhibitors interact with other targets within cells at concentrations similar to those that inhibit CFTR, while open-channel blockers of CFTR invariably block other types of Cl$^-$ channels.

8.5.1 CFTR Inhibitors and Secretory Diarrhoea

In search of lead compounds to develop therapeutically-active CFTR inhibitors for secretory diarrhoea, Verkman and colleagues deployed the HTS

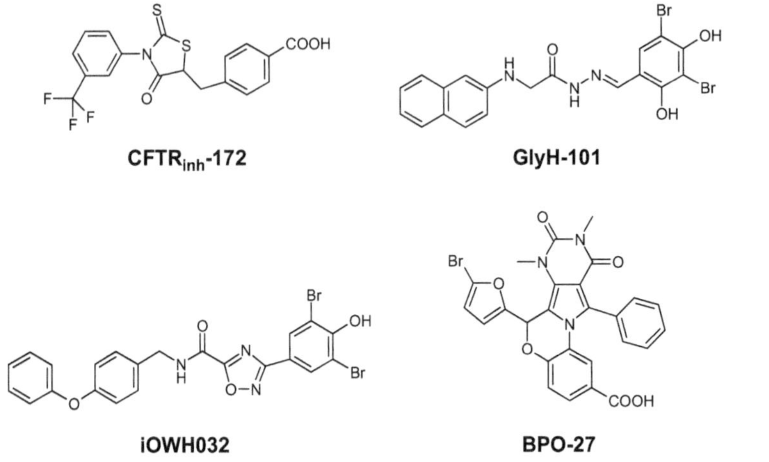

Figure 8.8 Chemical structures of some CFTR inhibitors. Abbreviations: CFTR$_{inh}$-172, 3-[(3-trifluoromethyl)phenyl]-5-[(4-carboxyphenyl)methylene]-2-thioxo-4-thiazolidinone; GlyH-101, N-(2-naphthalenyl)-[(3,5-dibromo-2,4-dihydroxyphenyl)methylene]glycine hydrazide; iOWH032, 3-(3,5-dibromo-4-hydroxy-phenyl)-N-[(4-phenoxyphenyl)methyl]-1,2,4-oxadiazole-5-carboxamide; BPO-27, 6-(5-bromofuran-2-yl)-7,9-dimethyl-8,10-dioxo-11-phenyl-7,8,9,10-tetrahydro-6*H*-benzo[*b*]pyrimido[4',5':3,4]-pyrrolo[1,2-*d*][1,4]oxazine-2-carboxylic acid.

assay used with great success to identify CFTR potentiators.[110] By screening 50,000 compounds, Ma *et al.*[106] identified the thiazolidinone CFTR$_{inh}$-172 (Figure 8.8). CFTR$_{inh}$-172 has several highly desirable properties as a CFTR inhibitor. First, potency: CFTR$_{inh}$-172 reversibly inhibited CFTR Cl$^-$ currents with half-maximal inhibitory potency (K$_i$) of ~300 nM, an increase in potency of almost 500-fold compared with the widely used CFTR inhibitor glibenclamide.[106] Second, specificity: CFTR$_{inh}$-172 was without effect on several epithelial ion channels and transporters, including Ca^{2+}-activated and volume-regulated Cl$^-$ channels and the ABC transporter P-glycoprotein.[106] Third, efficacy: CFTR$_{inh}$-172 inhibited cholera toxin-induced fluid secretion in the small intestine of mice, highlighting its value as a lead compound for drug discovery.[106] Although CFTR$_{inh}$-172 has the drawbacks of limited water solubility (~20 μM) and reduced potency in intact cells and tissues (K$_i$ ~5 μM),[111] it has become a very widely used research tool. To date, it remains the most selective CFTR inhibitor available.

Early studies to identify the mechanism of action of CFTR$_{inh}$-172 demonstrated that it interfered with channel gating without altering single-channel conductance.[112] These data suggested that CFTR$_{inh}$-172 might act as an allosteric inhibitor of CFTR. However, its mechanism of action is distinct from other allosteric inhibitors (*e.g.* genistein)[113,114] that interfere with the control of CFTR gating by ATP-driven NBD dimerisation by competing with ATP for binding to ATP-binding site 2. First, Kopeikin *et al.*[107] demonstrated that CFTR inhibition by CFTR$_{inh}$-172 occurs from either the open or closed state and does not involve disassembly of the NBD dimer, but instead is reminiscent of the inactivation mechanism of voltage-gated cation channels. Second, Taddei *et al.*[112] found that the CF mutation F508del has increased sensitivity to CFTR$_{inh}$-172 compared to wild-type CFTR, while Caci *et al.*[115] showed that mutation of R347, a non-pore lining residue located at the intracellular end of M6, attenuated markedly the inhibitory potency of CFTR$_{inh}$-172. Taken together, these data raise the possibility that CFTR$_{inh}$-172 might interfere with coupling between the NBD dimer and the channel pore.

Following the identification of CFTR$_{inh}$-172, Verkman and colleagues screened 100,000 small molecules selected for chemical diversity and drug-like properties.[111] Among the four new classes of CFTR inhibitors identified by the authors, the glycine hydrazides attracted special attention because of their unusual mechanism of action. Previous work had demonstrated that open-channel blockers inhibit CFTR by occluding the deep wide intracellular vestibule of the CFTR pore and preventing outwardly directed Cl$^-$ flow through the channel.[77] By contrast, Muanprasat *et al.*[111] revealed that glycine hydrazides, such as GlyH-101 (Figure 8.8), are open-channel blockers that occlude the extracellular vestibule of the CFTR Cl$^-$ channel. When GlyH-101 binds within the extracellular vestibule, Cl$^-$ flow from the extra- to the intracellular side of the membrane is more strongly attenuated than that in the opposite direction. Recently, Norimatsu *et al.*[116] used induced-fit virtual ligand docking and site-directed mutagenesis to identify the GlyH-101-binding site. The authors' data argue that GlyH-101 occludes the CFTR pore constriction with its hydrophobic naphthalene tail binding in the vicinity of F342 and its negatively charged benzene ring interacting with F337 and T338. These data provide a compelling explanation why GlyH-101 is such an effective open-channel blocker of CFTR.

Like CFTR$_{inh}$-172,[106] GlyH-101 strongly inhibited cholera toxin-induced fluid secretion in the small intestine of mice.[111] However, in marked contrast to CFTR$_{inh}$-172, which must be administered intraperitoneally,[106] GlyH-101 was active when added directly to the lumen of the small intestine.[111] These data argue that GlyH-101 and related compounds might be developed into a non-absorbable drug therapy for secretory diarrhoea. Towards this goal, Sonawane *et al.*[117] synthesized a series of malonic acid dihydrazides (MalH) linked to polar moieties and polyethylene glycol (PEG)-coupled butyric acid hydrazides [Gly-(PEG)$_n$] and demonstrated the therapeutic potential of these small molecules for secretory diarrhoea using a mouse model of cholera toxin-induced fluid secretion. To improve the potency and efficacy of their MalH-based CFTR inhibitors, Verkman and colleagues adopted two approaches. First, chemical conjugation with lectins to form MalH-lectins.[118] Second, formation of divalent, macromolecular MalH conjugates (MalH-PEG-MalH).[119] Of note, the small molecules developed by Verkman and colleagues[118,119] possess a number of important properties that make them excellent lead compounds for secretory diarrhoea therapy development. These include high water solubility; low toxicity; potent inhibition of

CFTR and cholera toxin-induced intestinal fluid secretion; efficacy when delivered directly into the intestine; minimum absorption by intestinal epithelial cells, but stable binding to the glycocalyx covering intestinal epithelial cells to resist drug washout.

The data of Sonawane *et al.*[119] suggest that non-absorbable divalent polyethylene glycol-malonic acid hydrazides are likely to be especially valuable in the development of rational new therapies for enterotoxin-induced secretory diarrhoea. However, the cost of such therapies might preclude their widespread use in developing countries, where treatment costs should be around US$1 per patient per day.[120] Nevertheless, One World Health (San Francisco, California, USA) in collaboration with BioFocus (Saffron Walden, Essex, UK) have developed glycine hydrazide derivatives where the labile acyl hydrazone is substituted with a stable heterocycle (*e.g.* iOWH032; Figure 8.8).[120] Based on the success of preclinical studies, clinical trials of iOWH032 began in 2011.[120]

8.5.2 CFTR Inhibitors and ADPKD

Hanaoka & Guggino[40] first explored the therapeutic potential of CFTR inhibitors in ADPKD. The authors demonstrated that diphenylamine-2-carboxylate (DPC)[121] and glibenclamide,[108] two non-specific CFTR inhibitors, diminished ADPKD cyst growth *in vitro*, whereas 4,4'-diisothiocyanatostilbene-2,2'-disulfonic acid (DIDS), which inhibits other types of epithelial Cl^- channels, but not CFTR, when added to the outside of cells[109] was without effect. Building on these data Li *et al.*[122] used Madin Darby canine kidney (MDCK) cells as a model system to explore the effects of ion transport inhibitors on renal cyst formation and growth. The data demonstrate that renal cyst growth is retarded by: (i) small molecules that inhibit directly CFTR by allosteric and open-channel blockade mechanisms; (ii) agents that act indirectly by inhibiting PKA-dependent phosphorylation of CFTR and (iii) chemicals that interfere with the basolateral membrane ion channels and transporters responsible for Cl^- accumulation within epithelial cells.[122] Two important conclusions can be drawn from the data of Li *et al.*[122] First, the pharmacology of renal cyst growth resembles that of the CFTR Cl^- channel. Second, inhibition of cyst growth by CFTR blockers is correlated with blockade of cAMP-stimulated transepithelial Cl^- secretion, not cell proliferation.

As for secretory diarrhoea, the therapeutic potential of CFTR inhibitors in ADPKD has been strengthened significantly by the identification of CFTR inhibitors by HTS. Building on the observation that $CFTR_{inh}$-172 retarded cyst formation and growth,[122] Yang *et al.*[123] screened 32 CFTR inhibitors related to $CFTR_{inh}$-172 and GlyH-101 for their ability to block cyst formation and growth. Through these studies, Yang *et al.*[123] identified tetrazolo-$CFTR_{inh}$-172 and Ph-GlyH-101, two small molecules that efficaciously inhibit cyst formation and enlargement, without affecting MDCK cell proliferation or inducing apoptosis. When tested using a neonatal kidney organ culture model of cyst formation and growth, both tetrazolo-CFTR$_{inh}$-172 and Ph-GlyH-101 reduced markedly the formation of cysts.[123] Based on these data, Yang *et al.*[123] investigated the effects of tetrazolo-CFTR$_{inh}$-172 and Ph-GlyH-101 on cyst formation and kidney enlargement *in vivo* using polycystin-1 knockout mice. Of special note, both CFTR inhibitors reduced cyst formation and decreased kidney size in polycystin-1 knockout mice, demonstrating convincingly proof of concept for the use of CFTR inhibitors in ADPKD.[123]

Despite their effectiveness as inhibitors of renal cyst growth,[122,123] thiazolidinones, such as CFTR$_{inh}$-172, have reduced potency in intact cells because their negative charge hinders small-molecule accumulation within the cytoplasm.[124] In search of new CFTR inhibitors with increased potency, Verkman and colleagues screened more than 100,000 small synthetic and natural compounds using their HTS assay.[124] Following electrophysiological evaluation of active compounds, Verkman and colleagues[124] identified a new class of CFTR inhibitors, pyrimido-pyrrolo-quinoxalinedione (PPQ) compounds. In contrast to other CFTR inhibitors, PPQs are uncharged at physiological pH.[124] The most potent ligand within this chemical class, PPQ-102 inhibited CFTR Cl$^-$ currents by a voltage-independent mechanism that increases channel closed-time without altering current flow through open channels.[124] These data suggest that PPQ-102 is an allosteric inhibitor of CFTR. However, further experiments are required to determine whether its mode of action is similar to CFTR$_{inh}$-172 or genistein.

When tested in polarised epithelia, PPQ-102 inhibited CFTR-mediated transepithelial Cl$^-$ secretion in Fischer rat thyroid (FRT) epithelia expressing recombinant human CFTR with IC$_{50}$ ~ 90 nM and airway and intestinal epithelia expressing native CFTR with IC$_{50}$ < 1 μM.[124] To begin to explore the therapeutic potential of PPQ-102 in ADPKD, Tradtrantip *et al.*[124] used their neonatal kidney organ culture model of cyst formation and growth. In this assay, PPQ-102 was more effective than either tetrazolo-CFTR$_{inh}$-172 or Ph-GlyH-101 at preventing cyst formation and growth.[123,124] Because PPQ-102 had poor metabolic stability, Snyder *et al.*[125] synthesized PPQ and benzopyrimido-pyrrolo-oxazine-dione (BPO) analogues. Like PPQ-102, the most potent small-molecule identified, BPO-27 (Figure 8.8), was uncharged at physiological pH.[125] However, it exhibited >10-fold metabolic stability than PPQ-102 when tested in hepatic microsomes.[125] Strikingly, BPO-27 inhibited CFTR-mediated transepithelial Cl$^-$ secretion in FRT epithelia expressing recombinant human CFTR with IC$_{50}$ ~ 8 nM and prevented cyst formation and growth in cultured neonatal kidneys with IC$_{50}$ ~ 100 nM, a 5-fold improvement over PPQ-102.[124,125] These data argue persuasively that BPO-27 is an excellent lead compound for testing in animal models of ADPKD.

8.5.3 Towards the Therapeutic Application of CFTR Inhibitors

The evaluation of the therapeutic potential of CFTR inhibitors has reached an exciting stage. There is understanding of their mechanism of action, efficacious lead compounds have been tested in animal models and clinical trials of one CFTR inhibitor have been initiated for secretory diarrhoea. But, what of the future? First, it will be important to establish the specificity of lead compounds by testing these small molecules on a battery of targets, including ion channels and transporters commonly found in epithelial tissues. Little information is currently available about the specificity of BPO-27,[125] $CFTR_{inh}$-172 appears quite selective for CFTR (Ma et al.[106] but, see Kelly et al.[126]), whereas GlyH-101 is less specific, inhibiting the Ca^{2+}-activated Cl^- channel TMEM16A,[127] SLC26 transporters[128] and various cation channels in cardiac myocytes.[129] Second, it is unknown whether CFTR inhibitors, by themselves, will be an effective therapy for ADPKD. The observation that neither tetrazolo-$CFTR_{inh}$-172 nor Ph-GlyH-101 affected cell proliferation[123] raises the possibility that CFTR inhibitors might need to be given in combination with a small molecule that inhibits cAMP-stimulated cell proliferation, a key element in the pathogenesis of ADPKD.[40] Third, cyst growth occurs slowly over decades in ADPKD patients.[37] There is therefore concern that long-term treatment of ADPKD patients with CFTR inhibitors might lead to the development of CF-like symptoms. The toxicity of CFTR inhibitors will therefore need to be evaluated rigorously prior to their use in the clinic. Finally, small-molecule CFTR inhibitors work by inhibiting the activity of individual CFTR Cl^- channels. However, the magnitude of CFTR-mediated Cl^- secretion across the apical membrane of an epithelium is determined both by the number of CFTR channels in the apical membrane and the activity of individual channels (see Section 8.4.4 Towards the therapeutic application of CFTR correctors and potentiators). Based on the effects of the F508del-CFTR mutation on renal cyst formation and growth *in vivo* and *in vitro*,[130–132] an alternative strategy to retard cyst formation and growth in ADPKD would be to eliminate the apical membrane expression of CFTR. In support of this idea, pioglitazone, a peroxisome proliferator-activated receptor-γ (PPARγ) agonist, sharply attenuated CFTR-mediated transepithelial Cl^- secretion by decreasing CFTR mRNA levels *in vitro*[133] and abrogated cyst growth *in vivo*.[134] It will therefore be interesting to further evaluate the therapeutic potential of pioglitazone and related agents using polycystin-1 knockout mice.

8.6   Conculsion

The success of ivacaftor in clinical trials demonstrates convincingly that small molecules that target directly CFTR have significant therapeutic potential. In the case of CF, personalized medicines might be developed to precisely target the genetic defects harboured by individuals afflicted by the disease. Early intervention with these medicines is expected to improve significantly the life expectancy and quality of life of CF patients. In the case of secretory diarrhoea and ADPKD, preclinical studies have demonstrated convincingly the therapeutic promise of CFTR inhibitors. The challenge now is to successfully translate these results into benefits for patients in the clinic. Finally, the success of ivacaftor is a paradigm for the creation of therapeutics targeting the root cause of other genetic diseases.

Acknowledgements

We thank former laboratory colleagues for stimulating discussions. DNS's laboratory is supported by the Cystic Fibrosis Trust and the Engineering and Physical Sciences Research Council [grant no. EP/J00961X/1]. SMH's laboratory is supported by the Cystic Fibrosis Trust, the National Institute on Drug Abuse [grant nos. DA07315 and DA023281] and the Medical Research Council [grant no. G0801501]. JL was supported by scholarships from the University of Bristol and the Overseas Research Students Awards Scheme of Universities UK and PK by the Strategic Scholarships Fellowships Frontier Research Networks, Office of the Higher Education Commission of Thailand.

References

1. J. R. Riordan, J. M. Rommens, B.-S. Kerem, N. Alon, R. Rozmahel, Z. Grzelczak, J. Zielenski, S. Lok, N. Plavsic, J.-L. Chou, M. L. Drumm, M. C. Iannuzzi, F. S. Collins and L.-C. Tsui, *Science*, 1989, **245**, 1066.
2. I. B. Holland, S. P. C. Cole, K. Kuchler and C. F. Higgins, *ABC Proteins: from bacteria to man*, Academic Press, London, 2003.
3. N. Inagaki, T. Gonoi, J. P. Clement IV, N. Namba, J. Inazawa, G. Gonzalez, L. Aguilar-Bryan, S. Seino and J. Bryan, *Science*, 1995, **270**, 1166.
4. N. Inagaki, T. Gonoi, J. P. Clement IV, C.-Z. Wang, L. Aguilar-Bryan, J. Bryan and S. Seino, *Neuron*, 1996, **16**, 1011.
5. M. P. Anderson, R. J. Gregory, S. Thompson, D. W. Souza, S. Paul, R. C. Mulligan, A. E. Smith and M. J. Welsh, *Science*, 1991, **253**, 202.
6. C. E. Bear, C. Li, N. Kartner, R. J. Bridges, T. J. Jensen, M. Ramjeesingh and J. R. Riordan, *Cell*, 1992, **68**, 809.
7. M. J. Welsh, B. W. Ramsey, F. Accurso and G. R. Cutting, in *The Metabolic and Molecular Basis of Inherited Disease*, ed. C. R. Scriver, A. L. Beaudet, W. S. Sly and D. Valle, McGraw-Hill Inc., New York, 2001, p. 5121.
8. D. N. Sheppard and M. J. Welsh, *Physiol. Rev.*, 1999, **79**(Suppl 1), S23.
9. D. C. Gadsby, P. Vergani and L. Csanády, *Nature*, 2006, **440**, 477.
10. E. M. Schwiebert, D. J. Benos, M. E. Egan, M. J. Stutts and W. B. Guggino, *Physiol. Rev.*, 1999, **79**(Suppl 1), S145.
11. K. Kunzelmann, *News Physiol. Sci.*, 2001, **16**, 167.
12. M. Field, *J. Clin. Invest.*, 2003, **111**, 931.
13. L. P. Sullivan, D. P. Wallace and J. J. Grantham, *J. Am. Soc. Nephrol.*, 1998, **9**, 903.
14. C. Bombieri, M. Claustres, K. De Boeck, N. Derichs, J. Dodge, E. Girodon, I. Sermet, M. Schwarz, M. Tzetis, M. Wilschanski, C. Bareil,

D. Bilton, C. Castellani, H. Cuppens, G. R. Cutting, P. Drevínek, P. Farrell, J. S. Elborn, K. Jarvi, B. Kerem, E. Kerem, M. Knowles, M. Macek Jr, A. Munck, D. Radojkovic, M. Seia, D. N. Sheppard, K. W. Southern, M. Stuhrmann, E. Tullis, J. Zielenski, P. F. Pignatti and C. Ferec, *J. Cyst. Fibros.*, 2011, **10**(Suppl 2), S86.

15. H. C. Chan, Y. C. Ruan, Q. He, M. H. Chen, H. Chen, W. M. Xu, W. Y. Chen, C. Xie, X. H. Zhang and Z. Zhou, *J. Physiol.*, 2009, **587**, 2187.

16. R. A. Frizzell and J. W. Hanrahan, *Cold Spring Harb. Perspect. Med.*, 2012, **2**, a009563.

17. P. M. Quinton, *Physiology*, 2007, **22**, 212.

18. R. C. Boucher, *Annu. Rev. Med.*, 2007, **58**, 157.

19. J. J. Wine and N. S. Joo, *Proc. Am. Thorac. Soc.*, 2004, **1**, 47.

20. M. M. Morales, T. P. Carroll, T. Morita, E. M. Schwiebert, O. Devuyst, P. D. Wilson, A. G. Lopes, B. A. Stanton, H. C. Dietz, G. R. Cutting and W. B. Guggino, *Am. J. Physiol.*, 1996, **270**, F1038.

21. B. A. Stanton, *Wien. Klin. Wochenschr.*, 1997, **109**, 457.

22. A. Ruknudin, D. H. Schulze, S. K. Sullivan, W. J. Lederer and P. A. Welling, *J. Biol. Chem.*, 1998, **273**, 14165.

23. F. Jouret, A. Bernard, C. Hermans, G. Dom, S. Terryn, T. Leal, P. Lebecque, J.-J. Cassiman, B. J. Scholte, H. R. de Jonge, P. J. Courtoy and O. Devuyst, *J. Am. Soc. Nephrol.*, 2007, **18**, 707.

24. P. B. Davis, *Am. J. Respir. Crit. Care Med.*, 2006, **173**, 475.

25. S. M. Rowe, S. Miller and E. J. Sorscher, *N. Engl. J. Med.*, 2005, **352**, 1992.

26. S. H. Cheng, R. J. Gregory, J. Marshall, S. Paul, D. W. Souza, G. A. White, C. R. O'Riordan and A. E. Smith, *Cell*, 1990, **63**, 827.

27. C. L. Ward, S. Omura and R. R. Kopito, *Cell*, 1995, **83**, 121.

28. T. J. Jensen, M. A. Loo, S. Pind, D. B. Williams, A. L. Goldberg and J. R. Riordan, *Cell*, 1995, **83**, 129.

29. G. L. Lukacs, X.-B. Chang, C. Bear, N. Kartner, A. Mohamed, J. R. Riordan and S. Grinstein, *J. Biol. Chem.*, 1993, **268**, 21592.

30. W. Dalemans, P. Barbry, G. Champigny, S. Jallat, K. Dott, D. Dreyer, R. G. Crystal, A. Pavirani, J.-P. Lecocq and M. Lazdunski, *Nature*, 1991, **354**, 526.

31. P. M. Quinton, *Am. J. Physiol.*, 2010, **299**, C1222.

32. R. A. Phillips, *Fed. Proc.*, 1964, **23**, 705.

33. C. G. Victora, J. Bryce, O. Fontaine and R. Monasch, *Bull. World Health Organ.*, 2000, **78**, 1246.

34. S. M. Lohmann, A. B. Vaandrager, A. Smolenski, U. Walter and H. R. de Jonge, *Trends Biochem. Sci.*, 1997, **22**, 307.

35. A. B. Vaandrager, A. Smolenski, B. C. Tilly, A. B. Houtsmuller, E. M. E. Ehlert, A. G. M. Bot, M. Edixhoven, W. E. M. Boomaars, S. M. Lohmann and H. R. de Jonge, *Proc. Natl. Acad. Sci. U.S.A.*, 1998, **95**, 1466.

36. M. Santosham, R. S. Daum, L. Dillman, J. L. Rodriguez, S. Luque, R. Russell, M. Kourany, R. W. Ryder, A. V. Bartlett, A. Rosenberg, A. S. Benenson and R. B. Sack, *N. Engl. J. Med.*, 1982, **306**, 1070.

37. V. E. Torres, P. C. Harris and Y. Pirson, *Lancet*, 2007, **369**, 1287.
38. P. D. Wilson and B. Goilav, *Annu. Rev. Pathol. Mech. Dis.*, 2007, **2**, 341.
39. P. Delmas, *Pflügers Arch.*, 2005, **451**, 264.
40. K. Hanaoka and W. B. Guggino, *J. Am. Soc. Nephrol.*, 2000, **11**, 1179.
41. N. A. McCarty, *J. Exp. Biol.*, 2000, **203**, 1947.
42. D. C. Gadsby and A. C. Nairn, *Physiol. Rev.*, 1999, **79**(Suppl 1), S77.
43. H. A. Lewis, S. G. Buchanan, S. K. Burley, K. Conners, M. Dickey, M. Dorwart, R. Fowler, X. Gao, W. B. Guggino, W. A. Hendrickson, J. F. Hunt, M. C. Kearins, D. Lorimer, P. C. Maloney, K. W. Post, K. R. Rajashankar, M. E. Rutter, J. M. Sauder, S. Shriver, P. H. Thibodeau, P. J. Thomas, M. Zhang, X. Zhao and S. Emtage, *EMBO J.*, 2004, **23**, 282.
44. R. J. P. Dawson and K. P. Locher, *Nature*, 2006, **443**, 180.
45. Z. Cai, A. Taddei and D. N. Sheppard, *J. Biol. Chem.*, 2006, **281**, 1970.
46. S. G. Bompadre, Y. Sohma, M. Li and T.-C. Hwang, *J. Gen. Physiol.*, 2007, **129**, 285.
47. P. Vergani, A. C. Nairn and D. C. Gadsby, *J. Gen. Physiol.*, 2003, **121**, 17.
48. L. Aleksandrov, A. A. Aleksandrov, X.-B. Chang and J. R. Riordan, *J. Biol. Chem.*, 2002, **277**, 15419.
49. S. Zeltwanger, F. Wang, G.-T. Wang, K. D. Gillis and T.-C. Hwang, *J. Gen. Physiol.*, 1999, **113**, 541.
50. P. Vergani, S. W. Lockless, A. C. Nairn and D. C. Gadsby, *Nature*, 2005, **433**, 876.
51. M.-F. Tsai, M. Li and T.-C. Hwang, *J. Gen. Physiol.*, 2010, **135**, 399.
52. A. Szollosi, D. R. Muallem, L. Csanády and P. Vergani, *J. Gen. Physiol.*, 2011, **137**, 549.
53. J. F. Cotten, L. S. Ostedgaard, M. R. Carson and M. J. Welsh, *J. Biol. Chem.*, 1996, **271**, 21279.
54. A. W. R. Serohijos, T. Hegedús, A. A. Aleksandrov, L. He, L. Cui, N. V. Dokholyan and J. R. Riordan, *Proc. Natl. Acad. Sci. U.S.A.*, 2008, **105**, 3256.
55. J.-P. Mornon, P. Lehn and I. Callebaut, *Cell. Mol. Life Sci.*, 2008, **65**, 2594.
56. M. F. Rosenberg, A. B. Kamis, L. A. Aleksandrov, R. C. Ford and J. R. Riordan, *J. Biol. Chem.*, 2004, **279**, 39051.
57. Y. Norimatsu, A. Ivetac, C. Alexander, J. Kirkham, N. O'Donnell, D. C. Dawson and M. S. P. Sansom, *Biochemistry*, 2012, **51**, 2199.
58. P. Linsdell, A. Evagelidis and J. W. Hanrahan, *Biophys. J.*, 2000, **78**, 2973.
59. J. A. Tabcharani, P. Linsdell and J. W. Hanrahan, *J. Gen. Physiol.*, 1997, **110**, 341.
60. T.-Y. Chen and T.-C. Hwang, *Physiol. Rev.*, 2008, **88**, 351.
61. Y. Bai, M. Li and T.-C. Hwang, *J. Gen. Physiol.*, 2011, **138**, 495.
62. Y. El Hiani and P. Linsdell, *J. Biol. Chem.*, 2010, **285**, 32126.
63. B. D. Schultz, R. A. Frizzell and R. J. Bridges, *J. Membr. Biol.*, 1999, **170**, 51.
64. B. Illek, H. Fischer, G. F. Santos, J. H. Widdicombe, T. E. Machen and W. W. Reenstra, *Am. J. Physiol.*, 1995, **268**, C886.

65. A. S. Verkman and L. J. V. Galietta, *Nat. Rev. Drug Discov.*, 2009, **8**, 153.
66. M. D. Amaral, *Curr Drug Targets*, 2011, **12**, 683.
67. W. E. Balch, R. I. Morimoto, A. Dillin and J. W. Kelly, *Science*, 2008, **319**, 916.
68. F. Van Goor, K. S. Straley, D. Cao, J. González, S. Hadida, A. Hazlewood, J. Joubran, T. Knapp, L. R. Makings, M. Miller, T. Neuberger, E. Olson, V. Panchenko, J. Rader, A. Singh, J. H. Stack, R. Tung, P. D. J. Grootenhuis and P. Negulescu, *Am. J. Physiol.*, 2006, **290**, L1117.
69. N. Pedemonte, G. L. Lukacs, K. Du, E. Caci, O. Zegarra-Moran, L. J. V. Galietta and A. S. Verkman, *J. Clin. Invest.*, 2005, **115**, 2564.
70. (a) F. Van Goor, S. Hadida, P. D. J. Grootenhuis, B. Burton, J. H. Stack, K. S. Straley, C. J. Decker, M. Miller, J. McCartney, E. R. Olson, J. J. Wine, R. A. Frizzell, M. Ashlock and P. A. Negulescu, *Proc. Natl. Acad. Sci. U.S.A.*, 2011, **108**, 18843; (b) Z. Kopeikin, Z. Yuksek, H. Y. Yang and S. G. Bompadre, *J. Cyst. Fibros.*, 2014, DOI: 10. 1016/ J.JCF.2014.04.003.
71. J. P. Clancy, S. M. Rowe, F. J. Accurso, M. L. Aitken, R. S. Amin, M. A. Ashlock, M. Ballmann, M. P. Boyle, I. Bronsveld, P. W. Campbell, K. De Boeck, S. H. Donaldson, H. L. Dorkin, J. M. Dunitz, P. R. Durie, M. Jain, A. Leonard, K. S. McCoy, R. B. Moss, J. M. Pilewski, D. B. Rosenbluth, R. C. Rubenstein, M. S. Schechter, M. Botfield, C. L. Ordoñez, G. T. Spencer-Green, L. Vernillet, S. Wisseh, K. Yen and M. W. Konstan, *Thorax*, 2012, **67**, 12.
72. J. L. Mendoza, A. Schmidt, Q. Li, E. Nuvaga, T. Barrett, R. J. Bridges, A. P. Feranchak, C. A. Brautigam and P. J. Thomas, *Cell*, 2012, **148**, 164.
73. W. M. Rabeh, F. Bossard, H. Xu, T. Okiyoneda, M. Bagdany, C. M. Mulvihill, K. Du, S. di Bernardo, Y. Liu, L. Konermann, A. Roldan and G. L. Lukacs, *Cell*, 2012, **148**, 150.
74. P. H. Thibodeau, J. M. Richardson III, W. Wang, L. Millen, J. Watson, J. L. Mendoza, K. Du, S. Fischman, H. Senderowitz, G. L. Lukacs, K. Kirk and P. J. Thomas, *J. Biol. Chem.*, 2010, **285**, 35825.
75. Z. Cai and D. N. Sheppard, *J. Biol. Chem.*, 2002, **277**, 19546.
76. E. Caci, C. Folli, O. Zegarra-Moran, T. Ma, M. F. Springsteel, R. E. Sammelson, M. H. Nantz, M. J. Kurth, A. S. Verkman and L. J. V. Galietta, *Am. J. Physiol.*, 2003, **285**, L180.
77. T.-C. Hwang and D. N. Sheppard, *Trends Pharmacol. Sci.*, 1999, **20**, 448.
78. T.-C. Hwang, F. Wang, I. C. H. Yang and W. W. Reenstra, *Am. J. Physiol.*, 1997, **273**, C988.
79. T. Ai, S. G. Bompadre, X. Wang, S. Hu, M. Li and T.-C. Hwang, *Mol. Pharmacol.*, 2004, **65**, 1415.
80. O. Moran, L. J. V. Galietta and O. Zegarra-Moran, *Cell. Mol. Life Sci.*, 2005, **62**, 446.
81. F. Wang, S. Zeltwanger, I. C. H. Yang, A. C. Nairn and T.-C. Hwang, *J. Gen. Physiol.*, 1998, **111**, 477.
82. O. Zegarra-Moran, M. Monteverde, L. J. V. Galietta and O. Moran, *J. Biol. Chem.*, 2007, **282**, 9098.

83. B. Illek, L. Zhang, N. C. Lewis, R. B. Moss, J.-Y. Dong and H. Fischer, *Am. J. Physiol.*, 1999, **277**, C833.
84. F. Van Goor, S. Hadida, P. D. J. Grootenhuis, B. Burton, D. Cao, T. Neuberger, A. Turnbull, A. Singh, J. Joubran, A. Hazlewood, J. Zhou, J. McCartney, V. Arumugam, C. Decker, J. Yang, C. Young, E. R. Olson, J. J. Wine, R. A. Frizzell, M. Ashlock and P. Negulescu, *Proc. Natl. Acad. Sci. U.S.A.*, 2009, **106**, 18825.
85. F. J. Accurso, S. M. Rowe, J. P. Clancy, M. P. Boyle, J. M. Dunitz, P. R. Durie, S. D. Sagel, D. B. Hornick, M. W. Konstan, S. H. Donaldson, R. B. Moss, J. M. Pilewski, R. C. Rubenstein, A. Z. Uluer, M. L. Aitken, S. D. Freedman, L. M. Rose, N. Mayer-Hamblett, Q. Dong, J. Zha, A. J. Stone, E. R. Olson, C. L. Ordoñez, P. W. Campbell, M. A. Ashlock and B. W. Ramsey, *N. Engl. J. Med.*, 2010, **363**, 1991.
86. B. W. Ramsey, J. Davies, N. G. McElvaney, E. Tullis, S. C. Bell, P. Drevínek, M. Griese, E. F. McKone, C. E. Wainwright, M. W. Konstan, R. Moss, F. Ratjen, I. Sermet-Gaudelus, S. M. Rowe, Q. Dong, S. Rodriguez, K. Yen, C. Ordoñez and J. S. Elborn for the VX08-770-102 Study Group, *N. Engl. J. Med.*, 2011, **365**, 1663.
87. R. Aherns, S. Rodriguez, K. Yen and J. C. Davies, *Pediatr. Pulmonol. Suppl.*, 2011, **34**, 283.
88. P. D. W. Eckford, C. Li, M. Ramjeesingh and C. E. Bear, *J. Biol. Chem.*, 2012, **287**, 36639.
89. W. Wang, J. Wu, K. Bernard, G. Li, G. Wang, M. O. Bevensee and K. L. Kirk, *Proc. Natl. Acad. Sci. U.S.A.*, 2010, **107**, 3888.
90. H. Yu, B. Burton, C.-J. Huang, J. Worley, D. Cao, J. P. Johnson Jr, A. Urrutia, J. Joubran, S. Seepersaud, K. Sussky, B. J. Hoffman and F. Van Goor, *J. Cyst. Fibros.*, 2012, **11**, 237.
91. (a) F. Van Goor, H. Yu, B. Burton, C.-J. Huang and B. J. Hoffman, *J. Cyst. Fibros.*, 2012, **11**(Suppl 1), S31; (b) M. P. Boyle, S. Bell, M. Konstan, S. A. McColley, L. Kang and N. Patel, *Pediatr. Pulmonol. Suppl.*, 2003, **35**, 315.
92. P. A. Flume, T. G. Liou, D. S. Borowitz, H. Li, K. Yen, C. L. Ordoñez and E. Geller for the VX08-770-104 Study Group, *Chest*, 2012, **142**, 718.
93. Y. Wang, M. C. Bartlett, T. W. Loo and D. M. Clarke, *Mol. Pharmacol.*, 2006, **70**, 297.
94. L. Wellhauser, P. Kim Chiaw, S. Pasyk, C. Li, M. Ramjeesingh and C. E. Bear, *Mol. Pharmacol.*, 2009, **75**, 1430.
95. N. Pedemonte, V. Tomati, E. Sondo, E. Caci, E. Millo, A. Armirotti, G. Damonte, O. Zegarra-Moran and L. J. V. Galietta, *J. Biol. Chem.*, 2011, **286**, 15215.
96. P.-W. Phuan, B. Yang, J. M. Knapp, A. B. Wood, G. L. Lukacs, M. J. Kurth and A. S. Verkman, *Mol. Pharmacol.*, 2011, **80**, 683.
97. J. M. Knapp, A. B. Wood, P.-W. Phuan, M. W. Lodewyk, D. J. Tantillo, A. S. Verkman and M. J. Kurth, *J. Med. Chem.*, 2012, **55**, 1242.
98. O. Kalid, M. Mense, S. Fischman, A. Shitrit, H. Bihler, E. Ben-Zeev, N. Schutz, N. Pedemonte, P. J. Thomas, R. J. Bridges, D. R. Wetmore,

      Y. Marantz and H. Senderowitz, *J. Comput. Aided Mol. Des.*, 2010,
      24, 971.
 99. D. N. Sheppard and L. S. Ostedgaard, *Mol. Med. Today*, 1996, **2**, 290.
100. M. Wilke, R. M. Buijs-Offerman, J. Aarbiou, W. H. Colledge,
      D. N. Sheppard, L. Touqui, A. Bot, H. Jorna, H. R. de Jonge and
      B. J. Scholte, *J. Cyst. Fibros.*, 2011, **10**(Suppl 2), S152.
101. T. S. Scott-Ward, Z. Cai, E. S. Dawson, A. Doherty, A. C. Da Paula,
      H. Davidson, D. J. Porteous, B. J. Wainwright, M. D. Amaral,
      D. N. Sheppard and A. C. Boyd, *Proc. Natl. Acad. Sci. U.S.A.*, 2007,
      104, 16365.
102. H. de Jonge, M. Wilke, A. Bot and D. N. Sheppard, *Pediatr. Pulmonol.
      Suppl.*, 2009, **32**, 291.
103. L. S. Ostedgaard, D. K. Meyerholz, J.-H. Chen, A. A. Pezzulo, P. H. Karp,
      T. Rokhlina, S. E. Ernst, R. A. Hanfland, L. R. Reznikov, P. S. Ludwig,
      M. P. Rogan, G. J. Davis, C. L. Dohrn, C. Wohlford-Lenane, P. J. Taft,
      M. V. Rector, E. Hornick, B. S. Nassar, M. Samuel, Y. Zhang,
      S. S. Richter, A. Uc, J. Shilyansky, R. S. Prather, P. B. McCray Jr,
      J. Zabner, M. J. Welsh and D. A. Stoltz, *Sci. Transl. Med.*, 2011, **3**, 74ra24.
104. M. F. Rosenberg, L. P. O'Ryan, G. Hughes, Z. Zhao, L. A. Aleksandrov,
      J. R. Riordan and R. C. Ford, *J. Biol. Chem.*, 2011, **286**, 42647.
105. J.-P. Mornon, P. Lehn and I. Callebaut, *Cell. Mol. Life Sci.*, 2009,
      66, 3469.
106. T. Ma, J. R. Thiagarajah, H. Yang, N. D. Sonawane, C. Folli,
      L. J. V. Galietta and A. S. Verkman, *J. Clin. Invest.*, 2002, **110**, 1651.
107. Z. Kopeikin, Y. Sohma, M. Li and T.-C. Hwang, *J. Gen. Physiol.*, 2010,
      136, 659.
108. D. N. Sheppard and K. A. Robinson, *J. Physiol.*, 1997, **503**, 333.
109. B. D. Schultz, A. K. Singh, D. C. Devor and R. J. Bridges, *Physiol. Rev.*,
      1999, **79**(Suppl 1), S109.
110. H. Yang, A. A. Shelat, R. K. Guy, V. S. Gopinath, T. Ma, K. Du,
      G. L. Lukacs, A. Taddei, C. Folli, N. Pedemonte, L. J. V. Galietta and
      A. S. Verkman, *J. Biol. Chem.*, 2003, **278**, 35079.
111. C. Muanprasat, N. D. Sonawane, D. Salinas, A. Taddei, L. J. V. Galietta
      and A. S. Verkman, *J. Gen. Physiol.*, 2004, **124**, 125.
112. A. Taddei, C. Folli, O. Zegarra-Moran, P. Fanen, A. S. Verkman and
      L. J. V. Galietta, *FEBS Lett.*, 2004, **558**, 52.
113. K. A. Lansdell, Z. Cai, J. F. Kidd and D. N. Sheppard, *J. Physiol.*, 2000,
      524, 317.
114. R. Dérand, L. Bulteau-Pignoux and F. Becq, *J. Biol. Chem.*, 2002,
      277, 35999.
115. E. Caci, A. Caputo, A. Hinzpeter, N. Arous, P. Fanen, N. Sonawane,
      A. S. Verkman, R. Ravazzolo, O. Zegarra-Moran and L. J. V. Galietta,
      Biochem. J., 2008, **413**, 135.
116. Y. Norimatsu, A. Ivetac, C. Alexander, N. O'Donnell, L. Frye,
      M. S. P. Sansom and D. C. Dawson, *Mol. Pharmacol.*, 2012, **82**, 1042.

117. N. D. Sonawane, J. Hu, C. Muanprasat and A. S. Verkman, *FASEB J.*, 2006, **20**, 130.

118. N. D. Sonawane, D. Zhao, O. Zegarra-Moran, L. J. V. Galietta and A. S. Verkman, *Gastroenterology*, 2007, **132**, 1234.

119. N. D. Sonawane, D. Zhao, O. Zegarra-Moran, L. J. V. Galietta and A. S. Verkman, *Chem. Biol.*, 2008, **15**, 718.

120. E. L. de Hostos, R. K. M. Choy and T. Nguyen, *Future Med. Chem.*, 2011, **3**, 1317.

121. N. A. McCarty, S. McDonough, B. N. Cohen, J. R. Riordan, N. Davidson and H. A. Lester, *J. Gen. Physiol.*, 1993, **102**, 1.

122. H. Li, I. A. Findlay and D. N. Sheppard, *Kidney Int.*, 2004, **66**, 1926.

123. B. Yang, N. D. Sonawane, D. Zhao, S. Somlo and A. S. Verkman, *J. Am. Soc. Nephrol.*, 2008, **19**, 1300.

124. L. Tradtrantip, N. D. Sonawane, W. Namkung and A. S. Verkman, *J. Med. Chem.*, 2009, **52**, 6447.

125. D. S. Snyder, L. Tradtrantip, C. Yao, M. J. Kurth and A. S. Verkman, *J. Med. Chem.*, 2011, **54**, 5468.

126. M. Kelly, S. Trudel, F. Brouillard, F. Bouillaud, J. Colas, T. Nguyen-Khoa, M. Ollero, A. Edelman and J. Fritsch, *J. Pharmacol. Exp. Ther.*, 2010, **333**, 60.

127. A. Caputo, E. Caci, L. Ferrera, N. Pedemonte, C. Barsanti, E. Sondo, U. Pfeffer, R. Ravazzolo, O. Zegarra-Moran and L. J. V. Galietta, *Science*, 2008, **322**, 590.

128. A. K. Stewart, B. E. Shmukler, D. H. Vandorpe, F. Reimold, J. F. Heneghan, M. Nakakuki, A. Akhavein, S. Ko, H. Ishiguro and S. L. Alper, *Am. J. Physiol.*, 2011, **301**, C289.

129. P. P. Barman, S. C. M. Choisy, H. C. Gadeberg, J. C. Hancox and A. F. James, *Biochem. Biophys. Res. Commun.*, 2011, **408**, 12.

130. D. A. O'Sullivan, V. E. Torres, P. A. Gabow, S. N. Thibodeau, B. F. King and E. J. Bergstralh, *Am. J. Kidney Dis.*, 1998, **32**, 976.

131. N. Xu, J. F. Glockner, S. Rossetti, D. Babovic-Vuksanovic, P. C. Harris and V. E. Torres, *J. Nephrol.*, 2006, **19**, 529.

132. H. Li, W. Yang, F. Mendes, M. D. Amaral and D. N. Sheppard, *Am. J. Physiol.*, 2012, **303**, F1176.

133. C. Nofziger, K. K. Brown, C. D. Smith, W. Harrington, D. Murray, J. Bisi, T. T. Ashton, F. P. Maurio, K. Kalsi, T. A. West, D. Baines and B. L. Blazer-Yost, *Am. J. Physiol.*, 2009, **297**, F55.

134. B. L. Blazer-Yost, J. Haydon, T. Eggleston-Gulyas, J.-H. Chen, X. Wang, V. Gattone and V. E. Torres, *PPAR Res*, 2010, **2010**, 274376.

135. J.-H. Chen, Z. Cai, H. Li and D. N. Sheppard, in *Cystic Fibrosis in the 21st Century*, ed. A. Bush, E. W. F. W. Alton, J. C. Davies, U. Griesenbach and A. Jaffe, Karger, Basel, 2006, p. 38.

136. T.-C. Hwang and D. N. Sheppard, *J. Physiol.*, 2009, **587**, 2151.

137. D. N. Sheppard, *Chem. Biol.*, 2011, **18**, 145.

CHAPTER 9

TRPV1 Antagonism: From Research to Clinic

MARK S. NASH,*[a] J. MARTIN VERKUYL*[b] AND
GURDIP BHALAY*[b]

[a] Novartis Institutes for Biomedical Research, Forum 1, Novartis Campus,
CH-4056 Basel, Switzerland; [b] Novartis Institutes for Biomedical Research,
Wimblehurst Road, Horsham, West Sussex, RH12 5AB, UK
*Email: mark.nash@novartis.com; gurdip.bhalay@novartis.com;
martin.verkuijl@novartis.com

9.1 Introduction

TRPV1 (transient receptor potential vanilloid 1) is a member of the large TRP
ion channel superfamily and has a central role in peripheral nociception and
neurogenic inflammation. It has a highly specific distribution, being predominantly associated with polymodal nociceptors, and is an established
target for analgesic intervention. TRPV1 was first identified as the receptor
for capsaicin, the active constituent of hot chilli peppers, in the laboratory of
David Julius by expressing a sensory ganglion cDNA library in non-neuronal
cells and screening for capsaicin responsiveness.[1] It was immediately recognized that TRPV1 is also activated by noxious heat.[1] The mammalian
TRPV family consists of six family members with two subgroups TRPV1 to 4
and TRPV5 and 6, distinguished based on sequence homology, their response to temperature and calcium selectivity.[2–4] The six transmembrane
regions of TRPV1 form a tetrameric, non-selective cation ion channel,
with selectivity for Ca^{2+} ions about 10 times greater than for Na^+.

RSC Drug Discovery Series No. 39
Ion Channel Drug Discovery
Edited by Brian Cox and Martin Gosling
© The Royal Society of Chemistry 2015
Published by the Royal Society of Chemistry, www.rsc.org

The cytosolic loops and large N- and C-terminal regions contain regulatory domains for protein kinase C (PKC), protein kinase A (PKA), Ca^{2+}/calmodulin-dependent protein kinase (CaMKII), ATP and phosphatidylinositol 4,5-bisphosphate (PIP_2) (reviewed by Tominaga and Tominaga, 2005;[5] Urban *et al.*, 2011[6]). The C terminal loop contains the classical TRP-like amino acid motif and the N-terminal loop ankryin domains which are involved in binding of ligands such as ATP and calmodulin and influences channel behaviour.[2,4,7]

TRPV1 is highly expressed in small to medium sized neurons of the dorsal root, trigeminal and nodose ganglia which form the myelinated Aδ and unmyelinated C nociceptive nerve fibres. To a lesser extent TRPV1 expression has also been found in the brain, and in non-neuronal cells of the skin, bladder and vascular system (reviewed Fernandes *et al.*, 2012[8]). TRPV1 is activated by a range of different compounds; in particular, several substances released during inflammation are endogenous activators of TRPV1, leukotriene B4, and compounds of the arachidonic pathway such as anandamide, N-arachidonoyl-dopamine (NADA) and N-oleoyldopamine. Also the mild acidification associated with inflamed tissues can directly activate TRPV1. Other compounds of the inflammatory soup such as bradykinin, nerve growth factor (NGF), ATP and prostaglandin E2 can sensitize TRPV1.

9.2 Preclinical Perspectives on TRPV1

9.2.1 Expression in Disease Models

The role of TRPV1 in disease is supported by data showing that it is both sensitized and up-regulated in a variety pre-clinical pain models. Sensitization of TRPV1 can occur through multiple pathways, including those associated with phospholipase C, adenylate cyclase, tyrosine kinase receptor activation,[9–13] leading to an enhancement of function by lowering the threshold of activation for both heat and acid (*e.g.* Nicholas *et al.*, 1999[14]) or through increased insertion of TRPV1 into the plasma membrane.[15] In this context TRPV1 can be viewed as a key co-incidence detector in the peripheral terminals of nociceptive neurons (Figure 9.1). This role has made the pursuit of TRPV1 antagonists one of the most heavily investigated areas in pain research since antagonists could potentially block the response to a variety of well-established pain signals.

The expression of TRPV1 in pre-clinical disease models is also highly regulated. Thus, increased expression of TRPV1 in sensory ganglia has been reported in models of peripheral inflammation,[14,16,17] osteoarthritis,[18] nerve injury,[19,20] bone cancer,[21,22] lung inflammation[23,24] and visceral hypersensitivity.[25–28] Neurotrophic factors, in particular NGF, are likely to mediate the increased expression of TRPV1 and release of NGF following nerve damage and inflammation has been shown to increase TRPV1 expression through the p38 MAP kinase pathway.[14,29] The nature of the increased

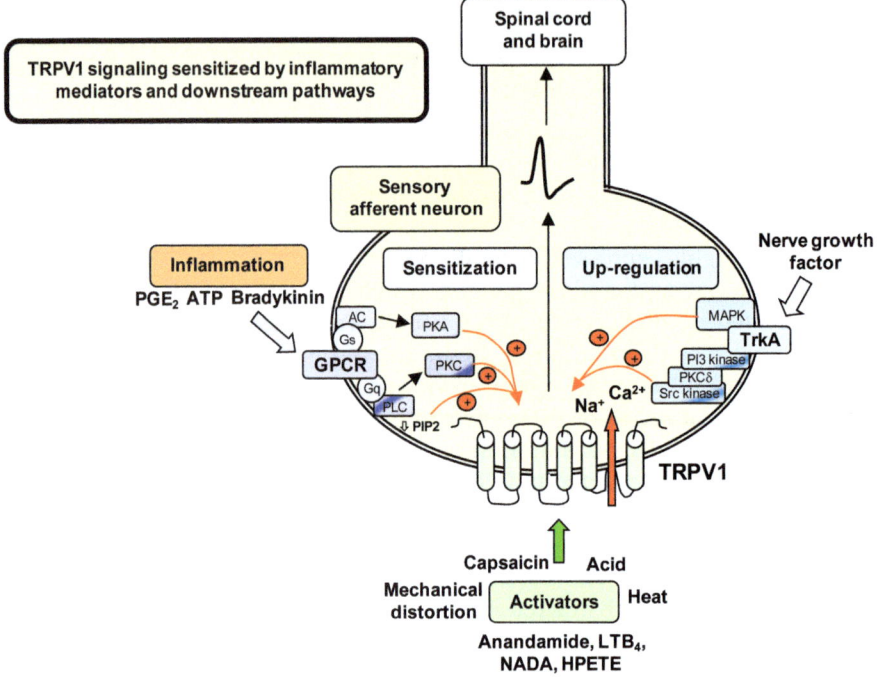

Figure 9.1 TRPV1 as a molecular coincidence detector in the peripheral terminals of nociceptive neurons.

expression of TRPV1 in the sensory neurons is complex with both an elevated expression in those nociceptors already expressing TRPV1 as well as apparent *de novo* expression in both small non-peptidergic IB4-positive C-fibres and medium diameter (Aδ, Aβ) sensory neurons, which greatly enhances the overall sensitivity of the innervated tissue to TRPV1 activation.[17,30,31]

9.2.2 Knockout Phenotype

Details of the effect of TRPV1 knockout in mice were first described by Julius *et al.*[32] TRPV1 null mice were shown to be viable and fertile with no overt phenotypic difference with wild-type littermates and no alterations in the development of the small-diameter nociceptive neurons in which TRPV1 is predominantly expressed were observed. However, cultured primary sensory neurons and afferent nociceptor fiber responses in a skin-nerve preparation were insensitive to capsaicin. TRPV1 -/- mice were also unresponsive to capsaicin-induced pain and taste aversion. These data provided the final confirmation after the cloning of the channel that TRPV1 is indeed the capsaicin receptor.[1] Similar deficits were observed in *in vitro* responses to protons (pH 5.0) and heat responsiveness was also greatly reduced from a threshold of ∼43 °C to 55 °C with a marked drop in the number of C-fibers activated by noxious heat. These *in vitro* findings were recapitulated in the TRPV1 -/- mice, which showed increases in latency of response to noxious heat in tail immersion, hot plate and Hargreaves tests. In contrast, mechanical responses were unaffected by the lack of TRPV1 expression. Thermal hyperalgesia was also observed to be reduced following an inflammatory challenge but not a mechanical challenge. In contrast, responses after sciatic nerve injury (partial ligation) were apparently normal indicating some interesting differences in the development of hyperalgesia in inflammatory or neuropathic settings. It is noteworthy that these effects on noxious heat perception first observed in knockout mice alluded to findings that would be seen later in humans with TRPV1 antagonists (see below).

Subsequent to the identification of altered thermosensation in the TRPV1 knockout mice, effects on bladder function were studied.[33] Although no gross differences with wild-type mice in anatomy or daily bladder function were observed in the TRPV1 -/- mice abnormal urodynamic responses were detected. Increased small volume urine spotting and decreased non-voiding contractions (NVCs) during cystometry were identified in the knockout mice. Moreover, reflex voiding to bladder filling was impaired and there was a consequent increase in bladder volume. TRPV1 is expressed in the bladder by both the afferent terminals innervating the musculature as well as the urothelial cells. The authors were subsequently able to show that stretch-evoked responses in isolated bladder and cultured urothelial cells were decreased in tissue from knockout mouse suggesting that TRPV1 mediates a mechanosensory role in bladder. This is in marked contrast to the normal somatic mechanical responses in knockout mice suggesting that TRPV1 is not inherently mechanosensitive but may form a mechanosensory complex in bladder urothelium. These early findings in knockout mouse likely explain why several companies have been clinically targeting over-active bladder with a TRPV1 antagonist (*e.g.* Charrua *et al.*, 2009;[34] Round *et al.*, 2011[35]).

Although no influence of TRPV1 loss on body temperature was found in the original characterization of the knockout mice,[32] inhibition of the well described phenomenon of capsaicin-induced hypothermia was observed.[36] Since capsaicin mediates this effect *via* activation of receptors in the thermoregulatory centers of the pre-optic hypothalamus it is perhaps surprising that there was no impact on body temperature. A subsequent study identified changes in the daily body temperature rhythm in TRPV1 knockout mice with a modest elevation in the daytime maximum.[37] Moreover, Iida *et al.* (2005) demonstrated that TRPV1 knockout mice have an attenuated fever-induced hyperthermia.[38] These data would prelude the subsequent clinical observations of antagonist-induced increases in body temperature.

Since these initial publications on the knockout mouse there have been a number of studies linking TRPV1 expression to a diverse range of functions. These will not be discussed in detail here but are summarized in Table 9.1.

Table 9.1 Summary of findings in TRPV1 KO mice.

System	Observation	Refs
Sensory perception	Attenuation of thermal perception. Normal mechanical and cold perception	32
Inflammatory pain	Reduced thermal hypersensitivity after noxious chemical or inflammatory stimuli	32, 39–46
	Attenuated PKC-induced acute nociception. No effect on formalin-induced pain	
	Attenuation of inflammation and pain following intra-articular injection of CFA into the knee joint	
	Partial reduction in bradykinin-induced nocifensive responses	
	Reduced heat hyperalgesia following intraplantar skin incision	
	Normal mechanical sensitivity	
	Decreased H_2O_2-induced heat-, but not mechanical-, hyperalgesia	
Neuropathic pain	No effect on mechanical or thermal hyperalgesia in PNL model	32, 39, 47
	Enhanced induction of mechanical hyperalgesia in diabetes- and cisplatin-induced neuropathy	
	Reduced cisplatin-induced heat hyperalgesia	
Visceral	Reduced activation of gastroesophageal vagal afferents by distension and acid	48–55
	Decreased jejunal distension- and acid-evoked afferent discharge	
	Protection from oleoylethanolamide-induced visceral pain	
	Attenuated visceral nociception and inflammation-induced colonic mechanical hypersensitivity	
	Reduced DSS-induced colitis. Enhanced DNBS-induced colitis	
	Reduced mustard oil-induced referred allodynia	
Bladder	Reduced NVCs and stretch-evoked contraction	33, 56–58
	Increased bladder capacity	
	Distension-induced afferent activation attenuated	
	Decreased cystitis-induced bladder mechanical hyper-reactivity	
	Attenuation of NGF-induced increases in voiding contractions	
Thermoregulation	Elevation of daytime maximal body temperature	37, 38, 59, 60
	Normal adaptive heat responses	
	Attenuation of bacterial LPS-induced fever response	
	Decreased fasting-induced hypothermia	
	Altered thermoregulatory properties	
CNS	Reduced anxiety, conditioned fear, and hippocampal long-term potentiation	61
Lung	Involved in capsaicin, acid and bradykinin-induced bronchopulmonary C-fibre activation	62, 63
	Reduced endotoxin induced inflammation	

Table 9.1 (*Continued*)

System	Observation	Refs
Skin	Delayed hair follicle cycling	64–68
	Enhanced response to oxazalone-induced allergic contact dermatitis	
	Loss of protective effect	
	Increased cutaneous thickening and fibrosis following bleomycin treatment	
	Increased skin carcinogenicity	
	Attenuated formaldehyde-induced dermatitis	
Heart	Normal cardiovascular profile	69–71
	Reduced pressure overload cardiac hypertrophy. Reduced fibrosis, tissue remodeling and inflammation	
	Impaired cardiac recovery following ischaemia	
Other	Altered behavioral responses to ethanol consumption	72–75
	Increased locomotor activity in young -/- mice, decreased in aged -/- mice and associated obesity	
	Protection from diet-induced obesity	
	Loss of TRPV1-mediated protection in endotoxin-induced sepsis	

Abbreviations: PKC, Protein Kinase C; CFA, Complete Freund's Adjuvant; NVCs, Non-voiding Contractions; DSS, Dextran Sodium Sulfate; DNBS, Dinitrobenzenesulfonic Acid; NGF, Nerve Growth Factor; LPS, lipopolysaccharide.

9.2.3 TRPV1 Antagonism

As a result of the significant efforts of multiple pharmaceutical companies a range of potent and selective TRPV1 antagonists have now been identified (see Section 9.4). The classical antagonists block the three main modes of TRPV1 activation (vanilloid, proton and heat) with similar potency. The efficacy of this class of TRPV1 antagonist has been well established in many different pre-clinical models of inflammatory and neuropathic pain as well as on different types of pain (*i.e.* allodynia and hyperalgesia) and to different noxious stimuli (*e.g.* thermal, mechanical, *etc.*). Moreover, this activity is broadly similar across different chemical classes (Section 9.4) giving confidence that inhibition of TRPV1 has analgesic properties, at least in rodents. The collective pre-clinical pharmacology data gleaned from the various selective TRPV1 antagonists is summarized in Table 9.2. The literature is vast in the area so only selected references are included and the reader is referred to the numerous excellent reviews and primary literature on TRPV1 antagonism for more detail on the individual compounds.[76–83]

9.2.3.1 Site of Action

From the earliest studies on the action of inhibitory capsaicin analogues there was evidence that the anti-nociceptive effect maybe mediated *via*

Table 9.2 Summary of pre-clinical findings on TRPV1 antagonism.

Pre-clinical model	Effect of TRPV1 antagonism	Examples
Intra-plantar CFA injection	Decreased thermal and mechanical hyperalgesia and allodynia	84–86
Formalin-injection	Reduced paw flinches in both early and late phase response	87, 88
Intra-plantar carageenan-injection	Reduced thermal hyperalgesia No effect on oedema	85
Iodoacetate-induced joint pain (OA model)	Decreased weight-bearing differences	89
Bone cancer pain	Reduced spontaneous and evoked pain	89
Partial nerve ligation (Seltzer model)	Reversed mechanical hyperalgesia and tactile allodynia	55, 84
Spinal nerve ligation (Chung model)	Weak reversal of mechanical allodynia	85
Chronic sciatic nerve constriction (Bennett model)	Weak reversal of mechanical allodynia	85
Skin incision (Brennan model)	Rapid and sustained reversal of thermal hyperalgesia Modest and delayed reversal of mechanical allodynia	85, 89
Inflammatory visceral hypersensitivity	Attenuated distension-induced pain and visceral hypersensitivity; reduced mustard oil-induced referred allodynia	55, 90–93
Bladder hyperactivity	Decreased bladder contractions and amplitude (spinal injury) Increased micturition threshold volume and decreased bladder contraction amplitude (refill model) Inhibit acid- and LPS-induced detrusor overactivity	34, 94, 95
Acute nociception	Normal	85
Lung	Partially inhibition of tussive response to PGE2 and BK Reduced endotoxin-induced inflammation Inhibition of airway hyperresponsiveness in ovalbumin sensitized guinea pig, but not in rats and mice	96–98
Body temperature	Increased	99

Abbreviations: CFA, Complete Freund's Adjuvant; LPS, lipopolysaccharide; PGE2, Prostaglandin E2; BK, Bradykinin.

blockade of receptors at both the distal afferent terminals as well as within the spinal cord.[100] Selective binding sites for capsaicin and resiniferatoxin (RTx) were shown to exist in the spinal cord and capsaicin to activate neurotransmitter/neuropeptide release from primary afferent neurons when applied directly to the spinal cord as well as in the periphery.[101–103] Following its cloning, TRPV1 immunoreactivity was detected in peripheral nociceptor afferent nerve fibres as well as lumbar spinal cord dorsal horn, trigeminal nucleus caudalis, sacral spinal cord dorsal horn, and nucleus of the solitary tract and area postrema of the caudal medulla.[104] The therapeutic action of TRPV1 antagonists could therefore be as a consequence of inhibiting TRPV1 expressed at multiple levels in the pain pathway both within the spinal cord as well as the periphery.

Several studies have attempted to dissect out the contribution of TRPV1 expressed at different sites to induction of pain in different pre-clinical models.[105–107] This has taken on a new focus as the pharmaceutical industry explores ways to enhance the analgesic efficacy of TRPV1 antagonists whilst limiting the side effects. What is clear is that TRPV1 antagonists delivered both locally or intrathecally as well as those engineered to be peripherally restricted are able in some settings to provide analgesia in pre-clinical models.[85,105–109] What is currently less clear is to what degree the potential clinical efficacy of TRPV1 inhibition would be compromised by acting at only selected sites and to what extent the ability to block different modalities of pain would be affected. Thus, for example, Cui *et al.* (2006) compared a CNS-penetrant and a peripherally-restricted TRPV1 antagonist and showed that, whilst thermal hyperalgesia following Complete Freund's Adjuvant (CFA) injection was blocked to the same extent by the compounds, the ability of a non-brain penetrant compound to reverse mechanical allodynia in the same model was reduced.[105] Kanai *et al.* (2007) also demonstrated that the TRPV1 antagonists BCTC and SB-366791 delivered intrathecally reversed CFA-induced mechanical- and thermal-hyperalgesia, but when delivered *via* intraplantar injection only reversed thermal-hyperalgesia.[108] Moreover, peripheral administration of the inhibitor iodoresiniferatoxin was shown to inhibit thermal hyperalgesia in inflammatory and neuropathic models but not mechanical.[110] However, McGaraughty *et al.* (2006) showed that intraplantar injection of A-889425 reduced mechanical allodynia induced by CFA injection.[106]

A recent study by Kim *et al.* (2012) has added further complexity.[111] The authors showed that mechanical hypersensitivity after peripheral nerve injury was attenuated in TRPV1 knockout mice but not in mice where TRPV1-expressing peripheral neurons had been ablated by prior treatment with RTx. This was revealed to be due to the involvement of TRPV1 on post-synaptic GABAergic neurons, which reduce inter-neuronal feedback inhibition to provide analgesia.[111]

9.3 Clinical Perspectives on TRPV1

Despite extensive pre-clinical data demonstrating the therapeutic effect of blocking TRPV1 function in pre-clinical models of pain, bladder dysfunction and airways hyper-reactivity there is at present a paucity of data to clearly demonstrate its role in human disease. Activation of TRPV1 clearly causes burning pain in humans and can subsequently elicit a prolonged hyperalgesia that is indicative of a sensitization of pain pathways.[112] The sustained exposure to agonists like capsaicin or RTx also provides analgesia after the initial painful response.[113] Indeed, a number of agonist approaches are being tested clinically with a transdermal patch from NeurogesX being the most advanced in Phase III trials for treatment of neuropathic pain following post-herpetic neuralgia.[114] Moreover, RTx instillation into the bladder has been used clinically in the management of bladder overactivity.[115,116] Although these data could be interpreted as being *via* a functional inhibition of TRPV1 since the channel is well known to undergo desensitization,[117,118] this conclusion is complicated by the fact that high concentrations of agonists inhibit the activity of sensory afferents and cause denervation.[119] Currently, the main evidence for a role of TRPV1 in human disease is altered expression and genetic associations.

9.3.1 TRPV1 Expression and Human Disease

Expression analysis of human tissue samples has implicated TRPV1 in a range of diseases. Most notably, increased expression of TRPV1 relative to healthy controls has been reported in patient biopsy samples for a range of visceral disorders, which may relate to the greater number of TRPV1-positive sensory afferents innervating the viscera compared to somatic tissues.[120,121] Thus, Yiangou *et al.* (2001) reported increased TRPV1 immunoreactivity in colonic nerve fibres in inflamed biopsies from inflammatory bowel disease (IBD) patients that correlated with reported abdominal pain.[122] The same group also provided evidence for increased TRPV1 immunopositive nerve fibres in esophageal mucosal biopsies from gastroesophageal reflux disease (GERD) patients with erosive esophagitis compared to healthy controls.[123] This was confirmed by Shieh *et al.* (2010) in a study that interestingly showed that in contrast to patients with active erosive esophagitis, TRPV1 levels in asymptomatic GERD patients were comparable to controls suggesting that active inflammation may be required for sustained elevated expression.[124] This study also found that the elevated TRPV1 expression correlated with levels of NGF in line with pre-clinical work showing the induction of TRPV1 by neurotrophic factors. Intriguingly, TRPV1 mRNA levels were found to be elevated in the mucosal biopsies indicative of a local induction of expression. Dömötör *et al.* (2007) have also reported increased TRPV1 levels in mucosal biopsies from *H. pylori*-negative *versus* *H. pylori*-positive gastritis patients, although the authors argued that in this context enhanced TRPV1 expression was a protective response rather than disease contributing.[125]

Increased TRPV1 levels have also been found in visceral diseases where inflammation is generally not a major feature such as fecal urgency with rectal hypersensitivity,[126] irritable bowel syndrome (IBS)[127] and non-erosive reflux disease (NERD).[128] As for the inflammatory gastrointestinal diseases, TRPV1 expression correlated with pain scores[126,127] and neurotrophic factor expression.[128] In these functional disorders there is presumed to be prolonged visceral pain or hypersensitivity resulting from a previous insult such as acute enteritis, for example in post-infectious IBS.[129] In support of this a recent study of IBD patients in remission observed increases in TRPV1-immunoreactive fibres in biopsies from patients with abdominal pain compared to controls and asymptomatic quiescent IBD patients.[130] The hypersensitivity in functional bowel disorders is generally accepted to be due to sensitization of the sensory afferents through, for example, alterations in expression of nociceptive channels such as TRPV1.[131]

TRPV1 expression in human bladder tissue has been detected in the sensory C- and Aδ-fibres innervating the bladder musculature and sub-urothelial space as well as in the urothelium itself.[132,133] Increased TRPV1 protein levels has been observed in bladder biopsies from patients with spinal neurogenic detrusor overactivity (NDO) due to, for example, multiple sclerosis or spinal cord injury,[134,135] refractory idiopathic detrusor over-activity[136] and sensory urgency in women.[137] As discussed previously, TRPV1 agonists, particularly the ultrapotent capsaicin analog RTx, have shown efficacy in overactive bladder conditions *via* defunctionalization of the sensory afferents innervating the bladder.[134,135] Unsurprisingly the efficacy of RTx in NDO correlated with the reduction in TRPV1 levels. More intriguing is the fact that in patients with idiopathic DO refractory to anti-muscarinic drugs, RTx was most efficacious in those that had elevated initial TRPV1 levels suggestive of a role for TRPV1 over-activity in at least a subset of patients with this disease.[136]

In further support of a key role for TRPV1 in visceral tissues, studies have also reported increased TRPV1 levels associated with vulvodynia[138] and chronic pelvic pain.[139,140] Increased TRPV1 has also been observed in a number of oral pain conditions, for example in pulpal nerve fibers innervating molars with significant dental caries[141] and sensitized hypomineralized molars.[142] These reports showing a potential role for TRPV1 in dental pain may explain why a number of pharmaceutical companies chose to clinically test their TRPV1 antagonists in molar extraction trials (Section 9.4.3). An interesting study from the group of Praveen Anand in collaboration with GlaxoSmithKline correlated increased TRPV1 expression in lingual biopsies from Burning Mouth Syndrome (BMS) patients with reported pain and sensitivity to oral capsaicin application.[143] This research also reported a parallel increase in NGF expression again mirroring the pre-clinical data to show that NGF induces TRPV1 expression. Since TRPV1 activation is well-established to elicit burning pain this data would strongly support the testing of TRPV1 antagonists in this chronic, intractable pain condition affecting middle-aged women. Also of note is the potential role TRPV1 plays in cough. Groneberg *et al.* (2004) reported increased expression of TRPV1 in nerve-fibres innervating the lung mucosal biopsies from patients with chronic cough due to, for example, asthma or GERD.[144] This increased expression, presumably associated with vagal neurons mediating cough reflexes, was also found to correlate with a heightened sensitivity to a capsaicin tussive test.

There is significantly less literature providing clinical evidence for increased TRPV1 expression in somatic pain conditions, in contrast to the wealth of data from pre-clinical species. Facer *et al.* (2007) reported that peripheral nerve injury results in increased TRPV1 levels in both injured and spared nerves.[145] Skin hypersensitivity during recovery from nerve damage was also associated with elevated TRPV1 in skin biopsies. Perhaps surprisingly levels of TRPV1 were found to be decreased in diabetic neuropathy patient biopsy samples, which might be explained by denervation in this patient population. A potential role for increased TRPV1 in aging and UV-damaged skin has recently been identified.[146,147]

9.3.2   Genetic Associations

The existence of polymorphisms in the TRPV1 channel has been known for some time and some have been shown *in vitro* to affect TRPV1 expression or function.[148] However, their potential association with human diseases has not been easy to elucidate with the initial, paradoxical evidence of an increased cold pain sensation in female European Americans with the TRPV1 V585V allele apparently not replicated in subsequent studies.[149,150] More recently however there are some intriguing associations of genetic variants of the channel with increased risk for a number of clinical conditions. Two studies have identified associations of TRPV1 variants with lung disorders. Cantero-Recasens *et al.* (2010) reported less severe symptoms such as wheezing and cough in chronic asthmatic children with the TRPV1-I585V variant, which was shown to have reduced responsiveness to capsaicin and heat *in vitro*[151] Recently, Smit *et al.* (2012) reported a significant association with six single nucleotide polymorphisms (snps) in TRPV1 in a large, multi-center study of cough (nocturnal, usual and chronic) in asthmatics and non-asthmatics.[152] Snps were also found to be associated with exacerbation of cough following exposure to environmental irritants or with current smoking.[152] Carreño *et al.* (2011) demonstrated an association of TRPV1 snps with susceptibility to migraine in the Spanish population.[153] The TRPV1-I585V variant was also recently studied in OA patients and a lower risk of symptomatic knee OA compared with healthy controls found to be associated with the TRPV1 585 Ile–Ile genotype.[154]

An association of the G315C polymorphism with functional dyspepsia (FD) is intriguing given the large amount of clinical data linking TRPV1 expression with visceral pain.[155] Despite being one of the most common GI disorders affecting up to 25% of the population, the origins of FD remain unknown but visceral hypersensitivity is likely to underlie the associated pain. Interestingly FD patients show greater sensitivity to capsaicin and dyspeptic symptoms are alleviated by chronic ingestion of chili peppers to desensitize TRPV1.[156,157] The study by Tahara *et al.* (2009) found a reduced risk for FD with homozygous TRPV1 315C and also demonstrated that healthy volunteers homozygous for 315C reported less severe symptoms following rapid ingestion of 200 ml of carbonated cold water.[155] However, the interpretation of this data is complicated by the fact that the 315C variant is associated with increased TRPV1 protein levels and greater maximal capsaicin responses meaning independent confirmation of the association with FD is desirable.[148]

A recent article from the German Research Network on Neuropathic Pain (DFNS; www.neuro.med.tu-muenchen.de/dfns) elaborated on research to assess the impact of TRP channel polymorphisms in neuropathic pain patients characterized with a standardized quantitative sensory testing (QST) battery and somatosensory phenotyping.[158] In this work no association with TRPV1 polymorphisms was observed for the incidence of neuropathic pain, which is supportive of earlier data that TRPV1 knock-out mice develop neuropathic pain to the same extent as wild-type controls.[32] Interestingly, however, this study found associations with two TRPV1 polymorphisms (1911A>G and 1103C>G) and differences within the chronic neuropathic pain patients for responses to QST. Thus, for example, patients expressing a TRPV1 1911A genotype (AA or AG) tended to show heat hyperalgesia compared to TRPV1 1911G homozygotes and those homozygous for the TRPV1 1103G variant showed significant cold hypothesia in a cold detection threshold test compared with heterozygous or wild type carriers.

In conclusion, this limited number of studies provides no more than a relatively weak association for the role of TRPV1 in these diseases and none would in themselves clearly identify a patient population for testing a TRPV1 antagonist. However, coupled to the knowledge of differential functional consequences of polymorphisms on TRPV1 activity and expression and the differences in somatosensory response between genotypes there may be an opportunity to enrich patient populations to maximize efficacy or reduce liability for on-target side-effects (see below) thereby improving the chances for success of clinical trials in the future.[148,158]

9.4   The Search for TRPV1 Antagonists

One of the most widely investigated exogenous chemical mediators of pain and nociception has been capsaicin, the principle pungent component of chilli peppers (*Capsicum spp.*), and the progenitor of the vanilloid chemical class of TRPV1 agonists. The name vanilloid originates from the vanillyl-like group which features in the structures of natural product TRPV1 agonists like: capsaicin (derived from chilli peppers); piperine (derived from black pepper); zingerone (obtained from ginger); and resiniferatoxin (obtained from *Euphorbia resinifera*).[159] The presence of a vanillyl group was believed to be necessary for activity but subsequent studies showed this to be not true (Scheme 9.1).

Interest in capsaicin's pungency motivated research to synthesise capsaicin analogues to provide an understanding of structure-activity relationship (SAR) and it was such a campaign and the design of conformationally constrained compounds that led to the discovery of the first competitive TRPV1 antagonist capsazepine.[160] This was achieved prior to the identification, cloning and expression of the TRPV1 ion channel,[1] and used isolated neonatal rat dorsal root ganglion (DRG) neurones with the monitoring of the compound-induced influx of radioactive $^{45}Ca^{2+}$ ions into these cultured cells.[161] Capsazepine provided a pharmacological tool for dissecting the

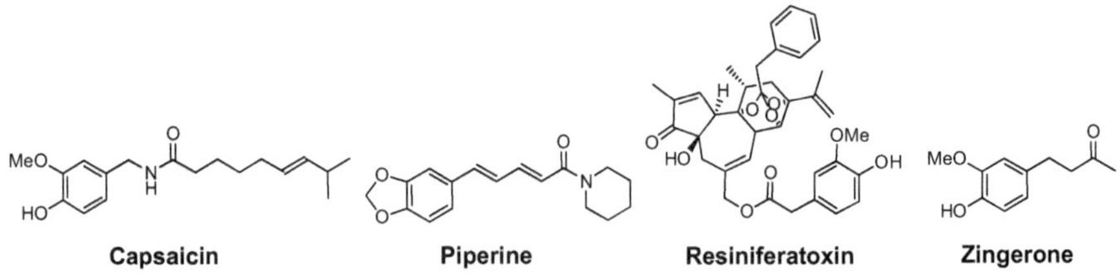

Scheme 9.1   Structures of natural product TRPV1 agonists isolated from plants.

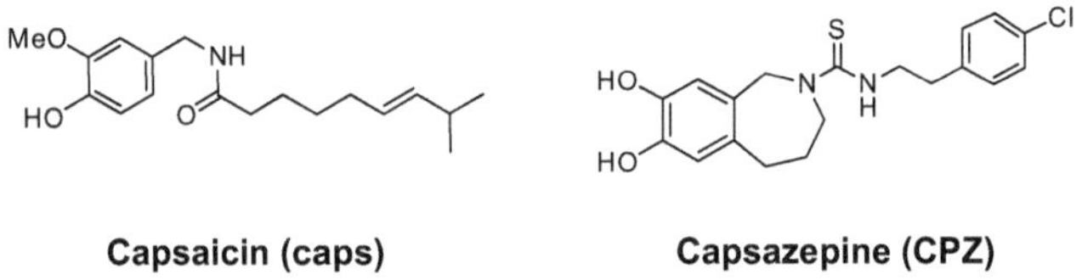

Scheme 9.2   Capsazepine the first TRPV1 antagonist.

mechanisms by which capsaicin analogues evoke their effects *via* TRPV1 (Scheme 9.2).

9.4.1   TRPV1 Agonists as Analgesics

Capsaicin-induced analgesia *via* desensitisation of TRPV1 was recognised early on for its potential to provide pain relief. A successful therapeutic based upon capsaicin's mechanism of action requires acute excitatory effects and resulting pain to be minimised, whilst promoting desensitisation of the sensory neuron. In addition, other biological effects such as apnoea, bradycardia, hypotension and hypothermia associated with TRPV1 agonism need to be minimised. Topical creams containing capsaicin have been used in the treatment of post-herpatic neuralgia and diabetic neuropathy,[162,163] as well as psoriasis, osteoarthritis and rheumatoid arthritis. Although the antinociceptive action of capsaicin has great appeal as a putative analgesic, the intensely painful response to initial exposure makes capsaicin itself unattractive, except in the most extreme cases. Several attempts to separate the antinociceptive effects of synthetic capsaicin analogues from their nociceptive and other biological actions have been published. While some divergence in the structural requirements for pungency and desensitisation was achieved, none achieved total separation. One attempt to separate capsaicin's antinociceptive properties from its hypothermic properties also failed.[164]

9.4.2   TRPV1 Antagonists as Analgesics

9.4.2.1   Capsazepine

The mechanism of TRPV1 activation by capsaicin, and the key role TRPV1 plays in peripheral nociception and neurogenic inflammation, strongly suggest that TRPV1 antagonists would provide efficient analgesics without the undesirable effects of TRPV1 agonists described in Section 9.4.1. However, early *in vivo* studies using capsazepine demonstrated that capsazepine lacked any analgesic activity in rat.[165] Data from models of acute and chronic pain concluded that capsaicin antagonists would be unlikely to provide therapeutically useful analgesics. Fortunately subsequent studies in inflammatory pain models showed capsazepine to have an antinociceptive effect in mouse,[166] and affect reversal of inflammatory mechanical hyperalgesia and neuropathic pain in the guinea pig.[167] These results renewed the interest in TRPV1 antagonists.

Further work towards understanding the basis of capsazepine's species specific analgesia came as the identification of various orthologues for mammalian TRPV1, particularly the rat,[1] human[168] and guinea pig[169] occurred, allowing capsazepine's behaviour in each of these species to be characterised (Table 9.3). Capsazepine demonstrated inhibition of capsaicin-induced activation cross-species (rat, human, guinea pig) with higher potency in the human.[170] However, TRPV1 is also activated by proton (pH $\sim$ 5.5) and noxious heat[1] (>43 °C) and it was shown that, while capsazepine successfully blocks the opening of TRPV1 by capsaicin, protons and noxious heat in the guinea pig and human, capsazepine does not block proton-induced activation of rat TRPV1 and is much less effective at inhibiting noxious heat-induced activation of TRPV1 in the rat. This suggests that inhibition of capsaicin-induced activation of TRPV1 alone is not sufficient for analgesia. The cause of this variance in activity between species appears to be the responsibility of only a few amino acid residues and this was demonstrated by the ability to give human-like activity on rat TRPV1 by point mutation of only three amino acid residues to the human orthologue equivalent.[171]

Capsazepine also inhibits nicotinic acetylcholine receptors, voltage-gated calcium channels, activated cation channels and the cold-sensitive channel TRPM8. Although capsazepine is not the ideal TRPV1 antagonist its great impact was in providing motivation to drug hunters to discover improved antagonists.

9.4.2.2 Further Vanilloid-derived Analogues

In a few cases it has been possible to convert a TRPV1 agonist to an antagonist through the addition of an iodine atom to the agonist's structure.

Table 9.3 IC_{50} values for capsazepine following capsaicin or pH activation of TRPV1 orthologues.

Activation Method	IC_{50} (μM) for capsazepine		
	Rat	Guinea Pig	Human
Caps	0.480	0.324	0.190
pH	>10	0.355	0.180

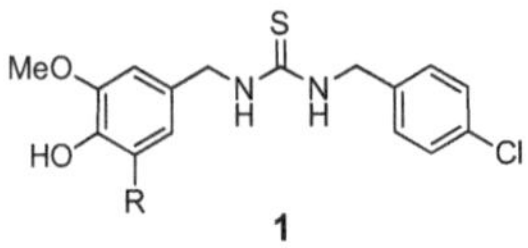

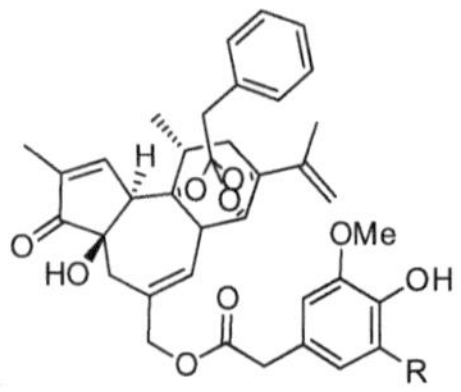

R=H 1 (agonist)   R=H Resiniferatoxin (agonist)

R=I IBTU (antagonist)   R=I Iodo-Resiniferatoxin (antagonist)

Scheme 9.3   Agonist to antagonist switching by iodination.

This switching is observed by comparing resiniferatoxin with its 5-iodo analogue (I-RTx)[172] which is a highly potent TRPV1 antagonist and comparing N-(4-chlorobenzyl)-N'-(4-hydroxy-3-methoxybenzyl)thiourea (**1**) and its antagonist N-(4-chlorobenzyl)-N'-(4-hydroxy-3-iodo-5-methoxybenzyl)thiourea (IBTU) (Scheme 9.3).[173]

9.4.3   First Generation TRPV1 Antagonists

Many pharmaceutical companies dedicated intense efforts towards developing TRPV1 antagonists as analgesics and numerous publications appeared revealing some structurally similar chemotypes. The following section aims to overview this vast research area by focusing on the available SAR studies leading to the advanced clinical candidates. This is a greatly simplified account of the enormous medicinal chemistry efforts to identify drug-like TRPV1 antagonists and the dedicated reader is directed toward the primary literature for more detail. The preclinical pharmacology will also be described only in brief since this has been covered extensively by other recent reviews to which the reader is referred.[76–83] Our focus will rather be on the recently released clinical data for the different advanced compounds.

9.4.3.1   *GSK*

At GSK **2** (Scheme 9.4) was identified though a high throughput screening (HTS) campaign as one of the most potent compounds in a collection of ureas,[174] which encouraged medicinal chemists to explore SAR of this hit series. Hit expansion through parallel synthesis of urea analogues (>700 compounds made) revealed the positional requirement of the Br-atom, urea N-substitution, linker length and of a N-ethyl group to maintain optimal TRPV1 antagonism. Although **2** was a potent TRPV1 antagonist effective against capsaicin, noxious heat, and low pH mediated activation of TRPV1 across species (human, rat and guinea pig) it suffered from high intrinsic clearance as determined using *in vitro* rat and human liver microsomes making it unsuitable for further progression. Attempts to rectify this targeted a dealkylation of the N-ethyl group as a pathway

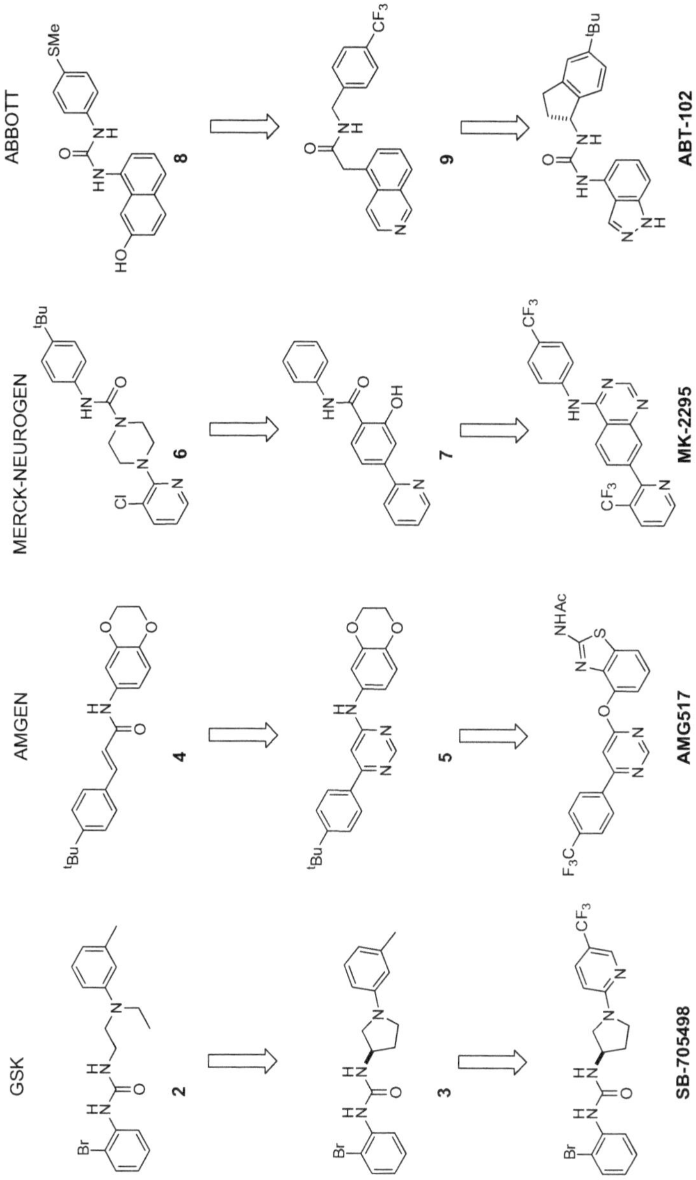

Scheme 9.4 Origin of the first generation of TRPV1 antagonist clinical candidates.

contributing to high intrinsic clearance. To ameliorate this, medicinal chemists invoked a rigidification strategy involving the N-ethyl and linker regions to form a pyrrolidine ring. The emerging compound **3** was associated with a reduction in intrinsic clearance relative to previous compounds. However, all the analogues had low aqueous solubility and attention was focussed on the pyrrolidine N-aryl group to rectify this. It was found that replacement of the N-aryl with an optimally substituted N-pyridyl group improved solubility and these features can be seen in the chemotype of the clinical candidate (SB-705498).[175]

Gunthorpe *et al.* (2007) extensively characterized the *in vitro* properties of SB-705498, revealing it to be a competitive and potent inhibitor of capsaicin- and heat-induced activation of TRPV1.[176] A slightly reduced potency for inhibition of acid-induced activation of TRPV1 was observed (79 nM *vs.* 25 nM for capsaicin at human TRPV1) and a low potency (1.6 μM) block of TRPM8 channels. SB-7054978 attenuated capsaicin-induced hyperalgesia and CFA-induced allodynia (Rami *et al.*, 2006).[175] SB-7054978 was also found to be efficacious at reversing neuropathic pain in a guinea pig partial ligation model.[177] As indicated previously (Section 9.2.3) there is good evidence from studies on TRPV1 expression that this channel may be involved in visceral pain conditions. Following neonatal maternal separation rats develop a mast cell driven visceral hypersensitivity to subsequent acute water avoidance stress that is frequently used to model the stress-induced effects in irritable bowel syndrome. SB-705498 was shown to reverse the exaggerated visceromotor responses to colorectal distension observed in these rats.[90] The potential for the therapeutic use of SB-705498 for migraine was studied preclinically by examining the role of TRPV1 in the trigeminovascular sensory system.[178] SB-705498 inhibited the activation of second-order neurons in the trigeminal nucleus caudalis of cats by electrical and mechanical stimulation of the dura mater and the facial skin. Sensitization of these responses by application of an inflammatory soup was also reversed by SB-705498. The authors conclude that SB-705498 may be useful for the treatment of the cutaneous trigeminal allodynia seen in some forms of migraine. SB-705498 was also shown to be efficacious at reversing airways hyper-responsiveness in ovalbumin-sensitized guinea pigs suggesting the potential use of this compound to manage symptoms of inflammatory respiratory diseases.[98]

GSK were the first to publish clinical data on a TRPV1 antagonist. The compound SB-705498 was shown to be safe with no adverse events up to 400 mg when administered orally and significantly "No AEs or ECG changes consistent with pyrexia were reported."[179] Using a range of sensory tests the authors were then able to show that SB-705498 had pharmacodynamic effects in treated healthy volunteers compared to placebo. Thus, at 400 mg SB-705498 significantly increased heat pain thresholds to a temperature ramp delivered to the arm or leg using a peltier device. The area of "flare response" to capsaicin application or UVB irradiation was also significantly reduced in healthy volunteers treated with SB-705498 compared to placebo. The thermal hyperalgesia measured by heat pain thresholds and tolerance were both found to be increased in skin sensitized by UVB irradiation. Surprisingly, no effect of SB-705498 was observed on the capsaicin-induced hyperalgesia. The efficacy of SB-705498 was then clinically tested in a dental pain paradigm at 400 mg and 1000 mg doses.[180] No significant effect of the drug on pain scores following tooth extraction was observed, which was attributed to the failure to achieve the anticipated exposures for efficacy. A second study with SB-705498 has recently been published assessing its effect on seasonal allergic rhinitis when delivered intranasally.[181] SB-705498 delivered by this route was first shown to pharmacologically block TRPV1 by significantly reducing the pain or smart and sensation of heat induced by an intranasal capsaicin challenge in the allergic rhinitis patients. Unfortunately, SB-705498 was then shown to be ineffective at reducing the same symptoms in patients following allergen challenge.

9.4.3.2 Amgen

The origins of Amgen's clinical compound can be traced back to cinnamide chemotypes like **4** (Scheme 9.4).[182] These were discovered through in-house screening of their corporate compound collection, and **4** was established as a novel lead molecule able to block TRPV1 channel activation *in vitro* when stimulated by activators like capsaicin, pH, heat and anandimide as well as reversing thermal and mechanical hyperalgesia in a rat model of inflammatory pain. Extensive synthesis was performed to find analogues with improved oral bioavailability and other pharmaceutical properties. The initial phase of the investigation examined the effect of restricting the s-cis conformer of cinnamide **4**. Constraining the cinnamide by incorporating the β-position of the cinnamide and the amide carbonyl into a variety of six-membered aromatic rings resulted in the construction of the 4-aminopyrimidine core of compound **5**. However, the 4-aminopyrimidine derivatives exhibited low exposure in rat, due to a high first pass metabolism which was rectified by changing the NH-linking group to an O-link. A significant improvement in the pharmacokinetic (PK) profile was observed through an increase in C_{max} and exposure. Further optimisation to improve potency and increase metabolic stability was achieved through modifications to the benzodioxane ring, culminating in the identification of the clinical candidate (AMG 517) which entered Phase I trials by September 2004.

AMG 517 was shown to be a potent and competitive antagonist on capsaicin, proton and heat activation of human TRPV1 with IC_{50} values of 0.76, 0.62 and 1.3 nM, respectively.[183] It showed good cross-species reactivity and selectivity over other TRP channels. Oral administration of AMG 517 to rats reversed the thermal hyperalgesia 24 hr following an intraplantar CFA injection. Efficacy for reversal of carrageenan-induced and surgical incision-induced thermal hyperalgesia was also reported. AMG 517 induced significant body temperature increases in rats, dogs and monkeys, which attenuated on repeat dosing and was suppressed by the antipyretic, acetaminophen.[183] The hyperthermia-induced by AMG 517 in rats was shown to be more pronounced than a second TRPV1 antagonist with increases in body temperature observed before inhibition of capsaicin-induced mechanical hyperalgesia.[55]

Clinical data published in 2008 on AMG 517 by Amgen alerted the field to a major hurdle to the development of TRPV1 antagonists[184] and significantly influenced subsequent research in this area (reviewed by Gavva, 2008;[185] Romanovsky *et al.*, 2009[186]). During Phase 1 trials AMG 517 was observed to induce a reversible dose and exposure related increase in body temperature. This was reported to attenuate on repeat dosing within 2–4 days. However, during a trial to assess analgesia following molar extraction pronounced hyperthermia was observed in all patients with the body temperature of one patient exceeding 40 °C. This patient had a >39 °C temperature sustained over 3 days that was not fully reversed by anti-pyretic agents. The potential reasons for this could have been individual susceptibility to TRPV1 antagonists or some interaction with the local inflammation induced by dental extraction. Given the extremely long half-life of AMG 517, in the range of 13 to 23 days, such a finding of sustained hyperthermia apparently led to the termination of development of this compound.

9.4.3.3   Merck-Neurogen

Compounds featuring a central piperazine **6** with pendant aryl groups have been patented[187–189] or published[84,190–193] by a number of companies (Scheme 9.4). The earliest examples were first reported in the patent literature by Neurogen as non-vanilloid TRPV1 antagonists. As well as providing tools to investigate preclincal pharmacology these compounds have been used as starting points with recognised liabilities requiring optimisation to improve solubility, oral exposure and metabolic stability. The ensuing SAR studies showed that the central piperazine of **6** could be replaced with an aryl-ring thereby converting the urea to an amide **7** with a minimal loss in potency. Further elaboration off the central aryl-ring with hydroxyl and amino groups provided compounds capable of forming intramolecular H-bonds with the carbonyl enforcing planarity of the benzamide unit. The pseudo-ring was replaced with an aminoquinazoline, a modification that provided a new chemical series and is present in the Merck-Neurogen clinical compound (MK-2295).

At the American Society Clinical Pharmacology Therapeutics (ASCPT) meeting in 2009 Merck presented some limited pre-clinical data describing the properties of MK-2295.[194] The IC_{50} for inhibition of capsaicin-induced activation of human TRPV1 was reported as approximately 0.4–6 nM with a modest 13-fold selectivity over TRPV3 and no activity at TRPV4, TRPA1 and TRPM8. Increased response latencies on the hot-plate test were observed in mice and tail withdrawal in rhesus monkeys.

MK-2295 was licensed by Merck from Neurogen for the potential treatment of pain and cough purportedly achieving proof of concept in a dental pain trial in 2007 (Neurogen Press Release, 2006). However, the presentation by Merck scientists in 2009 of clinical data on MK-2295 alerted the field to a second potential hurdle for the development of TRPV1 antagonists: altered thermal perception and an increased risk of burn injury.[194] In Phase I trials MK-2295 was observed to induce an increase in heat pain perception thresholds of approximately 4 °C that did not tolerate and at the highest dose reported (8 mg) reversed slowly over 2 days washout. Warmth perception thresholds were also increased from around 35 °C to 45 °C. Using a series of "real world" scenarios the potential risks associated with these changes in thermo-sensation were then assessed. Under normal circumstances hand withdrawal from a 49 °C waterbath, a temperature that causes irreversible skin damage after 4 mins,[195] is rapid. After dosing with MK-2295 withdrawal times were significantly increased with some volunteers able to withstand 2.5 minutes exposure, the safety cut-off for this study. Similarly, the ability to perceive 60 °C and 70 °C water as too hot to safely drink was impaired in some subjects following dosing with MK-2295. Using a capsaicin-induced dermal vasodilation paradigm and increases in core body temperature as indicators of TRPV1 target engagement, the Merck scientists were unable to establish a therapeutic window between the undesirable changes in thermal perception.[196] Clinical trials to assess MK-2295 for cough associated with upper airway diseases were planned in 2007, although no recent development of the compound has been reported.

9.4.3.4 Abbott Pharmaceuticals

Abbott's search for a TRPV1 antagonist started with an HTS campaign of the corporate compound collection and identified the 7-hydroxynapthalene **8** (Scheme 9.4) to be a potent TRPV1 antagonist as a suitable starting point for optimisation.[197,198] PK experiments demonstrated **8** was not orally bioavailable and did not show *in vivo* activity in animal models of inflammatory pain. The 7-hydroxy group turned out to be necessary to maintain potency but also presented a PK-liability by contributing to rapid metabolism. This prompted a search for a replacement of the 7-hydroxynapthyl group leading to the identification of the 5-isoquinoline ring system exemplified by **9**. Further SAR studies on the right hand side of the molecule demonstrated the superior potency of substituted benzyl groups compared to phenyl groups. Lead compound **9** possessed good selectivity and *in vitro* potency; however, it showed modest efficacy after oral dosing in models of inflammatory pain, most likely as a result of the short half-life and low volume of distribution. Medicinal chemists directed attention towards substitutions off the benzyl position. The resulting analogues had an improved PK profile, albeit at the expense of some potency; however, further analogue synthesis provided conformationally restricted examples such as the indane which when combined with a change of the core heterocyle from isoquinoline to indazole synergised to furnish the clinical compound (ABT-102) as a TRPV1 antagonist possessing activity in a wide range of preclinical pain models.

ABT-102 was shown to be a potent, selective and polymodal inhibitor of TRPV1 and to reverse heat-induced activation of nociceptive neurons in the dorsal horn of the spinal cord and to attenuate inflammatory thermal hyperalgesia by blocking TRPV1 in the periphery.[89,199] In a particularly intriguing series of studies, Abbott scientists reported that on repeat dosing the hyperthermia induced by ABT-102 was attenuated whilst, surprisingly, its analgesic potential increased over time in models of post-operative, osteoarthritic and bone cancer pain[89] This points to some disease modifying effect of TRPV1 and the authors note that the enhanced effect was observed in those pain models where central sensitization is known to play a significant role. Perhaps blocking TRPV1 dampens down the excitability of the spinal pathways to enable recovery from a self-reinforcing feed-forward loop. The implications of this data may be profound if it can be proven to also occur clinically since it may be possible to increase the therapeutic window on repeat dosing, reduce the dose with time and even treat the underlying pathology contributing to pain.

ABT-102 is a poorly water soluble, selective TRPV1 antagonist[200] and had a half-life ranging from 7 to 11 hours in single- and multi-dose Phase 1 clinical trials using oral solution and solid dispersion formulations.[201] In these trials ABT-102 showed the on-target increase in core body temperature common to all clinically tested TRPV1 antagonists with the most frequently reported adverse events being sensations of hot or cold, hot flushes, altered taste sensation, and oral hypoesthesia or dysesthesia.[201,202] Following the release of clinical data on MK-2295, Abbott conducted a comprehensive assessment of the thermosensory effects of ABT-102.[202] In these studies on healthy subjects a dose-dependent increase in core body temperature was observed, which was modest (mean 0.6 °C) at the highest dose (4 mg) reported, although an increase to 38.7 °C in one patient was observed suggesting significant inter-patient variability. On repeat dosing (*b.i.d.*) the temperature increases were no longer significant indicating that a similar tolerance to the hyperthermia is observed in humans as rodents.[85,202] Increases of approximately 4 °C were observed in cutaneous and oral heat pain thresholds, which did not show tachyphylaxis, whilst no differences in cold pain detection thresholds were observed. Using a modified version of the oral liquid test (55 °C) described by Merck scientists and a similar water bath test (49 °C), ABT-102 was found to significantly impair normal thermosensory responses. The analgesic efficacy of ABT-102 was also compared to opioid and coxib analgesics using laser evoked potentials (LEP) in UVB irritated skin.[203] The LEP amplitude on UVB skin was attenuated to the same extent by ABT-102 (2 mg), tramadol (100 mg) and etoricoxib (90 mg) but at 6 mg ABT-102 was statistically superior to the other analgesics. Subjective pain VAS scores were consistent with the LEP data. There has been no recent development reported for ABT-102.

9.4.3.5   AstraZeneca

Details of an extensive clinical program with the TRPV1 antagonist AZD1386 were recently released by AstraZeneca at the 13th World Congress on Pain

(2010).[204–207] Structural information around this compound is currently unavailable and its *in vitro* profile has apparently not been disclosed. In preclinical studies in rats AZD1386 reversed the thermal hyperalgesia associated with both inflammatory (carrageenan- and CFA-models) and neuropathic (spinal nerve ligation model) pain.[204] No tolerance to the analgesic efficacy was observed and at similar doses to those providing analgesia no hyperthermia was detected.

In a Phase I healthy volunteer trial oral AZD1386 (95 mg) increased mean body temperature by ≤0.33 °C with the highest individual body temperature recorded as 38.1 °C.[205] Significant reductions in intradermal capsaicin-induced pain and flare were induced by AZD1386 and heat pain thresholds were increased by 4.8 °C. A single oral solution of AZD1386 provided a rapid and short-lived analgesia following mandibular third molar extraction.[206] The maximal pain intensity difference from drug intake for AZD1386 relative to placebo was approximately half of that of the positive comparator 500 mg naproxen (32% for placebo *vs.* 48% for AZD1386 *vs.* 63% naproxen). Body temperature increases were equivalent to those in the Phase I trial (<0.4 °C) and so appeared unaffected by the molar extraction. In a 4-week study of 241 patients with knee osteoarthritis (OA), AZD1386 at 30 mg and 90 mg had no analgesic effect using the WOMAC pain subscale and no patient reported improvement.[207] Whilst this is seemingly a negative result for the use of TRPV1 antagonists for OA pain it is noteworthy that a main inclusion criteria included unsatisfactory pain relief to NSAIDs/coxibs suggesting that some patients did not have the classic inflammatory pain associated with OA. There were no serious AEs observed in this trial; however, dysgeusia, hot flushes, oral hypoaesthesia and feeling cold were reported. A notable reported finding, given the changes in thermal perception, was burns of mild to moderate intensity in eight patients taking either 30 mg or 90 mg AZD1386.

In view of the link between TRPV1 expression and GERD, Krarup *et al.* (2011) recently published data detailing the effect of AZD1386 on oesophageal sensitivity in healthy subjects.[208] Volunteers were dosed with 30 mg or 95 mg AZD1386 and responses to various painful stimuli to the oesophagus or forearm were studied. Consistent with earlier findings heat pain thresholds were increased in subjects treated with AZD1386 compared to placebo. However, sensitivity to painful acid, electrical and mechanical stimuli in the oesophagus and painful mechanical pressure to the forearm were not attenuated by AZD1386. The authors concluded that AZD1386 and, by inference, TRPV1 antagonists show selectivity for analgesia to acute thermal pain but not to other types of painful stimuli. Such a conclusion would be consistent with the data generated in rodents (Section 9.2). What needs noting, however, is that these studies were conducted in healthy subjects with normal TRPV1 signaling. In chronic pain conditions where TRPV1 expression can be upregulated or the channel sensitized its inhibition may provide significant analgesia. Moreover, the loss of acute nociception may be undesirable as evidenced by the incidence of burns in OA patients dosed with AZD1386.

In the OA trial with AZD1386, increases in hepatic enzymes were noted in some patients in the highest dose group and it was stated that for this reason clinical development of AZD1386 was discontinued in 2010.[207]

9.4.3.6   Other

Clinical data on two further TRPV1 antagonists have been made publically available but structures are yet to be publically disclosed. XEN-D0501 is under development for overactive bladder and Phase 1 single and multiple dose clinical data was recently published.[35] XEN-D0501 was rapidly absorbed and had a half-life of approximately 3–5 hours. XEN-D0501 induced no serious or severe adverse events with a variety of mild AEs associated with body temperature changes, such as feeling cold and chills. Following multiple dosing with XEN-D0501 *b.i.d.* for 14 days the most common AEs were dysgeusia, feeling cold, headache and oral paraesthesia. XEN-D0501 also induced a dose-dependent increase in core body temperature that ranged from 37.23 to 39.07 °C (1 mg twice daily), 37.43 to 38.35 °C (2.5 mg twice daily) and 37.23 to 38.82 °C (5 mg twice daily) compared to 37.03 to 37.99 °C for placebo treated subjects. On day 14 after multiple dosing there was some, but not complete, tachyphylaxis.[35] A second compound with clinical data from Renovis/Pfizer, PF-03864086, had a terminal half-life of approximately 2 weeks in Phase 1 trials and induced a significant increase in core body temperature (average $= 1.5$ °C with 180 mg). No tachyphylaxis was observed on repeat dosing and further development appears to have been halted (TrialTrove, Citeline Inc.).

Additional TRPV1 antagonists reported to be in clinical development include GRC-6211, which is under development by Glenmark for a variety of pain conditions and urinary incontinence and licenced by Eli Lilly. No clinical data has been released on this compound. PHE-377 is under development by PharmEste and has completed Phase I trials. Proof-of-concept Phase IIa trials in diabetic neuropathy pain (DNP) and post-herpetic neuralgia (PHN) are planned. JNJ-39439335 is under development by Johnson & Johnson for the treatment of pain such as OA with little information available about this compound. PAC-14028 is under development by Pacific Pharmaceutical for the treatment of atopic dermatitis and inflammatory bowel disease (IBD). DWP-05195 (DWJ-204) is an orally-active vanilloid receptor 1 (VR1) antagonist, under development by Daewoong for the treatment of neuropathic pain and post-herpetic neuralgia (PHN).

9.4.4   Conclusions on Clinical Experience with TRPV1 Antagonists

It is difficult to draw too many firm conclusions from the available clinical data because in many cases only a limited amount of information is available and assumptions have to be made as to the rationale for dose selection by the researchers. Thus, for example, it is possible that in different studies the maximal tolerated dose, the maximal achievable exposure due to PK limitations or a dose inducing minimal off-/on-target side effects was used. Moreover, important details around the pharmacological effects of the different drugs at TRPV1 and other related sensory TRP channels, which could influence the clinical observations, may be unavailable. However, two consistent findings have been reported with the various TRPV1 antagonists tested clinically: increased body temperature and attenuated thermal perception. Beyond these on-target effects, blocking TRPV1 appears to be well tolerated with no serious adverse effects. Given the paucity of data in patients with chronic pain it is currently difficult to judge whether these will limit the clinical development of TRPV1 antagonists because there is no clear understanding of efficacy or whether there will be a therapeutic margin in a disease setting so no risk–benefit assessment can be made.

More risky conclusions from the clinical data are that all TRPV1 antagonists are not the same with apparent differences in the extent of changes in body temperature and thermal perception observed. Thus, clinically SB-705498 reportedly induced no body temperature changes and AZD1386 induced only minimal changes (0.4 °C) whilst AMG 517 administration resulted in a 38–39 °C body temperature rises at the highest dose reported.[179,184,205] Moreover, there appears to be significant variability among the responses of individuals to the on-target effects of TRPV1 blockade. For example, the data for the effect of ABT-102 on withdrawal from 49 °C water is suggestive of some subjects showing no loss of thermal sensitivity.[202] These observations suggest that it may be possible to develop antagonists that have reduced side effects or to identify patients in which these are less pronounced, and it will be interesting to follow the development of those antagonists currently in clinical trials (Figure 9.2).

Small molecule TRPV1 antagonists in early development

Company	Phase I	Phase 2	Phase 3	Comment
GSK	• **SB-705498** (topical, pruritus)			Mar 2012, intranasal for rhinitis stopped Mar 2012, listed in Ph1 for pruritus
Provesica		• **XEN-D0501** (systemic, OAB)		Jan 2011, start for overactive bladder
Johnson & Johnson	• **JNJ-39439335** (oral, OA)			Jun 2012, additional Ph1 study planned for pain associated with osteoarthritis
PharmEste	• **PHE-377** (oral, NP)			May 2011, Ph1b started in neuropathic pain
Mochida	• **MR-1817** (oral, pain)			Ph1 completed in May 2011
Pacific Pharma	• **PAC-14028** (oral, AD, UC, CD)			Dec 2010, start for atopic dermatitis, ulcerative colitis & Crohn's disease
Daewoong Pharmaceuticals	• **DWJ-204** (oral, NP)			Aug 2011, phase II trial start for neuropathic pain, phase II for postherpetic neuralgia planned

Figure 9.2 TRPV1 clinical candidates.

9.5   Second Generation TRPV1 Antagonists and Regulation of Body Temperature

The discovery that TRPV1 antagonists elicit hyperthermia was perhaps unexpected since TRPV1 null and knockdown mice as well as TRPV1 desensitised rats present with a normal body temperature,[38,209] but it is now clear from the clinical data that increased body temperature is a common effect of first generation TRPV1 antagonists. Together with the observation of impaired thermal perception these clinical findings led many pharmaceutical companies to reassess their TRPV1 programmes. However, some organizations continued and have reported a new generation of TRPV1 antagonists, which do not elevate core body temperature in preclinical species.

Early work towards avoiding hyperthermia investigated minimising central exposure but the strategy was deemed to be non-viable since TRPV1 antagonists were shown to cause hyperthermia by acting outside the brain at peripheral tissues through intra-abdominal targets. TRPV1 channels are abundant in primary DRG and nodose neurons that innervate the abdominal viscera. Studies in rat using AMG0347 revealed that this compound did not cause or elicit hyperthermia when administered into the brain or spinal cord in contrast to when it was administered systemically, which led the authors to conclude that the drug causes hyperthermia by acting outside the blood brain barrier.[210] A separate study in rats involving the intravenous, intrathecal, and intra-POA administration of the TRPV1 antagonist A-889425 corroborated this finding.[211]

To comprehend the nociceptive and thermoregulatory function of TRPV1 researchers looked for a relationship between the mode of activation of TRPV1 and effects on body temperature. This strategy was based on the knowledge that TRPV1 has independent activation mechanisms with distinct molecular domains for vanilloids, protons and heat. Using modality-specific molecules Amgen hypothesised that the propensity of an antagonist to cause hyperthermia may correlate with its potency to block a certain mode (or modes) of TRPV1 activation.[182] Amgen studied eight TRPV1 antagonists with different potencies to block the activation of the rat TRPV1 channel by heat, protons, and capsaicin and established dose–response curves for the hyperthermic effect of these drugs. They then analyzed these curves graphically to compare the ability of TRPV1 antagonists to cause hyperthermia with their potency to block TRPV1 activation in each mode. They found that blockade of TRPV1 activation in the heat mode does not contribute to the development of hyperthermia and the propensity of TRPV1 antagonists to cause hyperthermia relates most closely to their potency to block TRPV1 activation by protons. Blockade of the capsaicin mode either does not contribute to hyperthermia or makes only a limited contribution. Despite this learning it remains unclear which TRPV1 inhibitory phenotype is optimal for clinical efficacy and safety.

9.5.1 Abbott Pharmaceuticals

Abbott reported a series of ureas featuring the chromanyl moiety which appear to have evolved from the ABT-102 clinical compound (see Section 9.4.3.4). Representative compounds are depicted over two patent applications and are described as potent TRPV1 antagonists after capsaicin activation but less effective after acid activation of TRPV1.[212,213a] Effects on rat core body temperature along with potency against capsaicin stimulated human TRPV1 are disclosed for some compounds but inhibition after acid activation is not described in the patents, however selected examples like 10, 11 (Scheme 9.5) have appeared in a recent review article.

A detailed pharmacological characterization of these[213b] modality-specific antagonists has recently been published.[214] Among the compounds exemplified were A-1106625, a classical polymodal TRPV1 antagonist, and A-1165442, which showed excellent blockade (high potency, full efficacy) of capsaicin and NADA but only partially blocked pH-induced activation of rat (14.4%) and human (61.8%) TRPV1. A-1106625 increased core body temperature when dosed orally, whilst A-1165442 and compounds with a similar *in vitro* profile did not. The authors ascribed these differences to the sparing of proton-induced activation of TRPV1 in line with the work from Amgen

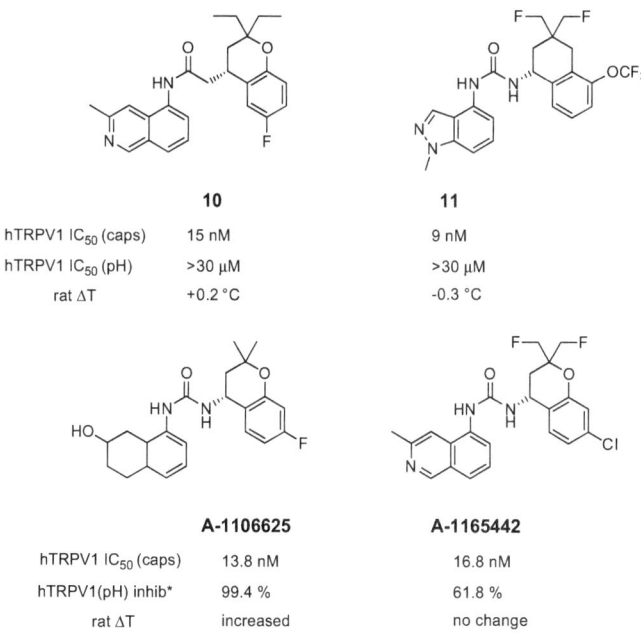

	10	**11**
hTRPV1 IC$_{50}$ (caps)	15 nM	9 nM
hTRPV1 IC$_{50}$ (pH)	>30 µM	>30 µM
rat ΔT	+0.2 °C	-0.3 °C

	A-1106625	**A-1165442**
hTRPV1 IC$_{50}$ (caps)	13.8 nM	16.8 nM
hTRPV1(pH) inhib*	99.4 %	61.8 %
rat ΔT	increased	no change

* inhibition determined at 11.25 µM concentration of test compound

Scheme 9.5 TRPV1 antagonists from Abbott Pharmaceuticals with reduced inhibition of acid activation and hyperthermic liability.

(*e.g.* Garami *et al.*, 2010[215]). It should be noted, however, that no data on inhibition of heat-induced activation of TRPV1 was provided.[214] Significantly, both compounds were able to reverse pain in an osteoarthritic knee-joint model but only A-1106625 was able to attenuate pain behaviours in a bone cancer model, suggesting that whilst the modality-specific TRPV1 antagonists can retain therapeutic efficacy it may only be in some pain paradigms.

9.5.2 Astellas Pharma

Astellas reported the TRPV1 antagonist AS1928370 with a dissociated antagonist profile in rat TRPV1.[109] AS1928370 shows little inhibitory effect upon proton-evoked TRPV1 activation but potent blockade after capsaicin activation. Rectal body temperatures were assessed in rat, with AS1928370 showing no significant effect at doses up to 10 mg/kg *p.o.* compared to a control compound BCTC. Further characterisation showed AS1928370 lowered body temperatures when dosed at 30 mg/kg *p.o.* By contrast, BCTC significantly increased rectal body temperatures at 10 to 100 mg/kg *i.p.* Analgesic effects were assessed in capsaicin-induced secondary hyperalgesia, neuropathic and inflammatory pain models. Significant effects were observed in the CFA-induced inflammatory pain model; however, the effective dose was much higher than in capsaicin and neuropathic pain models. These results suggest that blockade of TRPV1 receptors without affecting proton-mediated activation could be a promising approach to treat neuropathic pain without incurring the hyperthermia liability.

9.5.3 Gruenenthal

Gruenenthal have described aromatic carboxamides and urea derivatives in the patent literature with evidence for stimulus specific TRPV1 antagonism.[216-219] The divergence in potency at recombinant human TRPV1 receptor was acknowledged between capsaicin and acid stimuli, however effects on body temperature were not shown (representative compounds **12**, **13**, see Scheme 9.6).

9.5.4 PharmEste

PharmEste is developing PHE-377 (structure is not disclosed) for the potential oral treatment of neuropathic pain. The compound was shown to be orally bioavailable with good pharmacokinetic and safety profiles and did not elevate body temperature. Phase I clinical development is ongoing and proof of concept studies (Phase IIa) in patients are planned in two major clinical indications such as painful diabetic neuropathy and post herpetic neuralgia. A further report discloses V405,[220,221] a nicotinamide N-oxide **14** derivative which was described as a potent thermoneutral antagonist (TRPV1 IC_{50} 15nM) of capsaicin-evoked TRPV1 activation. Comparison was made

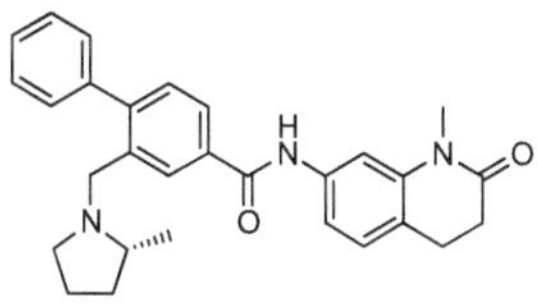

12

Astellas AS1928370

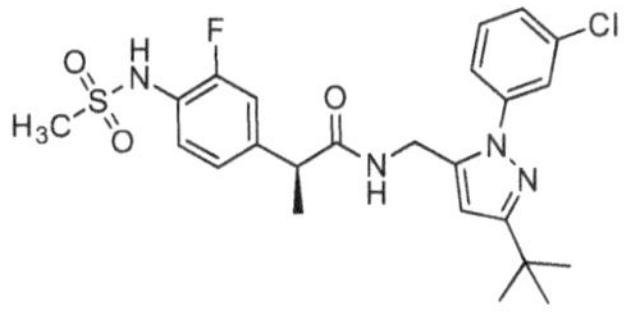

compound 17, page 141 WO 2010/127856

Gruenenthal

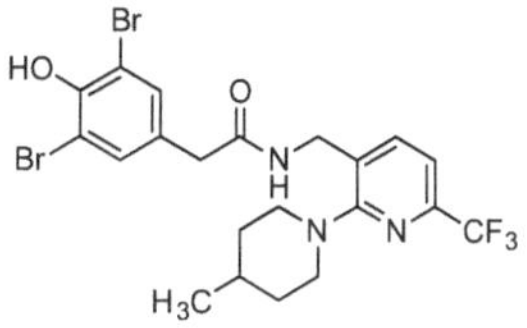

compound 23 from page 120 WO 2008/125337 Gruenenthal

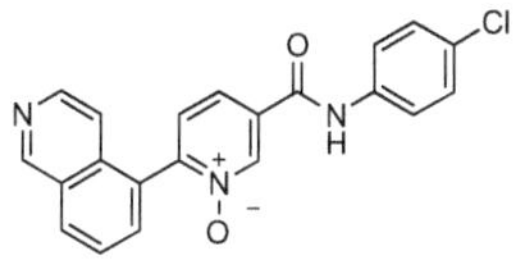

13

14

PharmEste

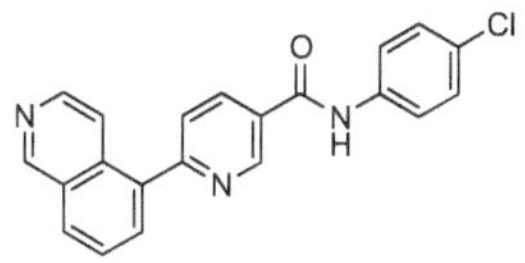

15

PharmEste

Scheme 9.6 Additional TRPV1 antagonists reported to have reduced liability for eliciting hyperthermia.

with the related non-oxidised compound **15** which increased body temperature (TRPV1 IC_{50} 0.83 nM, ΔT +0.5 °C).

9.5.5 Recent Developments

A recent publication has added further complexity to this area with the description of BCTP, a classical TRPV1 antagonist that fully inhibits the activation of rat and human channels by capsaicin, protons and heat with equivalent potency but appears to have reduced liability for hyperthermia induction.[55] This compound showed at least a 10-fold window between its analgesic efficacy in a model of visceral hypersensitivity (10 mg/kg, *p.o.*) and the induction of significant hyperthermia (100 mg/kg, *p.o.*). For reversal of capsaicin-induced mechanical hyperalgesia BCTP contrasted markedly AMG517, the clinical candidate from Amgen, for which the window was reversed, *i.e.* induction of hyperthermia appeared to occur at a lower dose than the reversal of hyperalgesia. These data clearly illustrate that not all classical TRPV1 antagonists behave the same for the different on-target effects of blocking TRPV1. Although the actual mechanism underlying the distinct properties of the two antagonists remains unclear, the authors speculated that it might be due to variations in tissue exposure, differences in ability to block an as yet unknown endogenous agonist or the temporal onset of blockade.[55] The latter is an intriguing possibility because the induction of hyperthermia appears to be more transient than the analgesic effect of BCTP suggesting some homeostatic mechanisms may be acting to restore normal body temperature. So, if the block of TRPV1 is relatively slow these may be able to compensate for any significant increase in body temperature.

9.6 Other TRP's Involved in Pain

As discussed in Section 9.4, major safety concerns of TRPV1 antagonists are increases in body temperature and the loss of noxious heat sensation. In Section 9.5 we gave an overview of modality specific antagonists that might be able to overcome the temperature-related side effects. The potential of this strategy to avoid hyperthermia whilst retaining analgesic efficacy now requires development of advanced candidates and clinical testing. Whilst our knowledge of TRPV1 is most advanced, there is a growing knowledge around the role of other related TRP channels in pain. Some of these TRP channels have been proposed to have similar analgesic effect, and might thus represent alternative drug targets. Owing to their specific biophysical and physiological characteristics, tissue distribution and regulation in response to stimuli, some of these TRP channels have been shown to play a role in certain types of pain. Like TRPV1 some of these TRP channels have effects on temperature sensation and/or thermogenesis. In the following section we discuss the pain related phenotype of knockout mice as a proxy for antagonists since in many cases good pharmacological tools are currently lacking. We also discuss what is known about the pharmacology of antagonists for these channels with respect to their analgesic properties.

9.6.1 The TRPV Family

The TRPV family consists of 6 members of which TRPV1 to TRPV4 have been implicated in pain.[222] They are expressed in DRG and trigeminal neurons, although TRPV3 is expressed only at very low levels in DRG.[223–225] TRPV3 like TRPV4 is highly expressed in keratinocytes, where its activation can lead to release of messenger molecules that activate the afferent nerve terminals in the skin.[3] All members of the TRPV1-4 subfamily are temperature sensitive spanning a temperature range from physiologically benign (TRPV4 27–34 °C, TRPV3 33–39 °C) to noxious heat (TRPV1 > 43 °C, TRPV2 >52 °C).[226,227]

9.6.1.1 TRPV2

TRPV2 seems to contribute little to pain sensation as TRPV2 adult knockout mice do not show deficits in mechanical sensitivity under normal conditions and during inflammation and nerve injury.[228] Furthermore, the temperature sensitivity of TRPV2 is debated since only heterologously expressed rodent but not human TRPV2 channels are activated by heat.[229,230] Further doubt around the temperature sensitivity was raised by the apparently normal thermosensitive phenotype of TRPV2 knockout mice, both under normal and inflamed conditions.[228] The possibility of TRPV1 substituting for the lack of TRPV2 in these studies was excluded as mice with both TRPV1 and TRPV2 genes knocked out or mice in which functional TRPV1 was ablated by agonist treatment had similar temperature sensitivity as TRPV1 knockout mice.[228] Currently no selective TRPV2 antagonists have been published in peer reviewed papers. Overall, the existing data indicates that TRPV2 does not present an alternative to TRPV1 as a target for pain.

9.6.1.2 TRPV3

Studies of the TRPV3 knockout mouse do not describe obvious alterations in pain sensation except for temperature related sensitivity. In the original papers describing the generation of TRPV3 knockout mice, Moqrich and colleagues (2005) showed that these mice have reduced responses to non-noxious and acute noxious heat, while under inflamed conditions the responses to temperature and mechanical stimuli were as for wild type control mice.[231] However, Huang *et al.* (2011) independently generated two TRPV3 knockout mice lines on different mice backgrounds that did not show alterations in temperature sensitivity over a wide range of temperatures tested.[232] Differences in the temperatures used and mouse strain might explain the discrepancy between the reports. In the study of Huang *et al.* (2011) knocking out TRPV4 in addition to TRPV3 did not uncover a temperature sensing defect, nor did pharmacologically blocking TRPV1 in these double knockout mice lead to an increased sensitivity over TRPV1 inhibition in wild type mice.[232] However, the translation of the findings in mice to humans might be complicated as TRPV3 seems to be mainly localized in keratinocytes in mice, whereas in humans it has also been found in neurons of the pain pathway mainly DRG and TG as well as in the central nervous system (CNS).[3]

In light of this discrepancy it is interesting that the endogenous lipid mediator 17-r-Resolvin D1, which has been shown to be a selective TRPV3 antagonist, has anti-nociceptive properties.[233] Local application of 17-r-Resolvin D1 reverses temperature hypersensitivity, but not mechanical hypersensitivity of inflamed tissue. Although no primary peer reviewed literature exists, several pharmaceutical companies have filed patents on TRPV3 antagonists. Hydra Biosciences has published several patents on TRPV3 antagonists showing efficacy in several different pain models.[234] In 2009 Glenmark in collaboration with Sanofi started a Phase I clinical trial with their TRPV3 antagonist (GRC 15300) for neuropathic pain, but no details has been reported (see www.glenmarkpharma.com). Possible side effects of blocking TRPV3 might arise from its interaction with the TGF-alpha/EGFR signalling complex. Through this interaction TRPV3 plays a role in keratinocyte differentiation. Indeed, knockout mice have in addition to a wavy hair phenotype, impaired barrier function of their epithelium.[235]

9.6.1.3 TRPV4

TRPV4 was originally cloned by Liedtke *et al.* (2000) and Strotmann *et al.* (2000), and was found to be activated by hyposmolarity, shear stress and heat.[236,237] Subsequent extensive characterization of the phenotype of the

TRPV4 knockout mice indicated that it is involved in a wide range of crucial physiological functions. These studies show a role in fluid homeostasis as knockout mice have altered responsiveness of the antidiuretic hormone to changes in osmotic pressure and possibly effecting maintenance of water intake.[238,239] A physiological response to water intake is a reduction of blood pressure, which is regulated by sensory neurons innervating the hepatic system. This mechanism is attenuated in TRPV4 knockout mice and additionally kidney cortical duct cells of TRPV4 knockout mice have reduced K^+ and Na^+ ion secretion.[240,241] TRPV4 knockout mice also have altered pressure responses of the bladder, which leads to defects in bladder voiding.[242] Further study of the TRPV4 knockout mice revealed that the channel plays a crucial role in vascular pressure regulation, since TRPV4 knockout mice had defects in shear stress-induced vasodilation.[233] Indeed, a selective TRPV4 agonist was found to cause profound circulatory collapse.[243]

Studies also showed that TRPV4 has a role in bone formation since protection against bone loss during limb immobilization was observed in TRPV4 knockout mice. A study showed a role for TRPV4 in hearing as TRPV4 knockout mice develop a progressive hearing loss with age.[244] Furthermore, TRPV4 might be involved in the physiological responses of the lung to stretch. Positive pressure applied during mechanical ventilation in lung injury patients can be lifesaving, however excessive pressure can aggravate lung damage and over distension by high peak inflation pressure is known to cause ventilator-induced lung injury. Deletion of TRPV4 has been suggested to offer protection through its role in activation of macrophages residing in lung alveoli.[245,246]

TRPV4 knockout mice also have defects in sensing pain and temperature. TRPV4 knockout mice showed a reduced sensitivity to pressure applied on to the tail[247] and to noxious mechanical stimuli under normal conditions[138,247] as well as during inflammation or neuropathy.[248,249] Furthermore, knockout mice also show reduced sensitivity to mild hypertonic solutions.[250]

Given that the activation temperature of TRPV4 is within the physiological range, genetic deletion of TRPV4 could result in defects in normal thermosensation. Indeed, behavioural responses to noxious heat were not different from wild type mice,[238,247] however TRPV4 knockout mice show reduced thermal hyperalgesia to non-noxious warmth applied to inflamed tissue and prolonged tail withdrawal latency to acute heating under both normal and inflammatory conditions.[232,251] Interestingly, TRPV4 knockout mice showed a preference for slightly warmer temperatures than wild-type mice possibly because of a reduced ability to discriminate between warm temperatures.[232,252] No difference in body temperature or its response to changes in ambient temperature has been reported for the TRPV4 knockout mice.[238,252] However, it is worth noting that initially no overt changes in temperature regulation were observed in TRPV1 knockout mice yet TRPV1 antagonists clearly increase body temperature (see above).

Several pharmaceutical companies have published patents for TRPV4 antagonists.[253] To date two antagonists have been described in peer reviewed publications. One antagonist, RN1734 from Renovis, partially reduced mechanical allodynia and heat hyperalgesia associated with neuropathic damage induced by the chemotherapeutic Paclitaxel.[254,255] The second antagonist, HC-67047 from Hydra Biosciences, showed efficacy at reversing the altered bladder function induced by cyclophosphamide-induced cystitis.[256] HC-67047 was also tested in models of neuropathic pain induced by Paclitaxel. As for RN1734, it partially reduced the mechanical allodynia and importantly it was able to completely reverse the neuropathic changes when co-administered with the TRPA1 antagonist HC-030031.[257]

Given the many physiological functions of TRPV4 it will be important to closely assess what on-target side effects are associated with inhibiting the channel before clinical testing. However, there may be great promise for the therapeutic value of TRPV4 modulators beyond the treatment of pain since they may be able to counter some of the debilitating phenotypes associated with gain-of-function mutations found in patients with Charcot-Marie-Tooth disease.[258,259]

9.7 Other TRP Channels

9.7.1 TRPM3

The transient receptor potential melastatin subfamily (TRPM), of which the first member was discovered as a transcript increase in melanomas,[260] now consists of eight family members.[222] Two family members have been related to sensing pain. One of them is TRPM3, which endogenously can be activated by the neuroactive steroid hormone pregnenolone sulfate[261] and by metabolites formed during sphingolipid synthesis.[262] It is mainly expressed in brain, kidney, pancreas (activation leading to insulin secretion) and vascular smooth muscle (activation inducing contraction).[263,264] It is also expressed in small diameter neurons of the DRG and the TG,[265] indicating a potential role in pain.

Interestingly in terms of temperature responses TRPM3 shares great similarity with TRPV1. Heterologously expressed TRPM3 renders human embryonic kidney (HEK) cells sensitive to temperatures above 40 °C. TRPM3 knockout mice show clear deficits in pain thermal sensation, with reduced noxious heat avoidance and inflammation-induced heat hyperalgesia, but have normal responses to noxious cold or mechanical stimuli.[265] TRPM3 knockout mice have apparently normal core body temperature regulation and in contrast to TRPV1 agonists, which cause hypothermia, injection of the endogenous TRPM3 agonist pregnenolone sulfate does not alter body temperature.[265]

Currently there are no synthetic antagonists known. Polyclonal antibodies against the third extracellular loop of TRPM3 have been generated in the laboratory of David Beech, which were shown to suppress TRPM3 function *in vitro*.[266,267] The non-steroidal anti-inflammatory drug (NSAID) mefenamic acid, a fenamate derivative used for the treatment of pain, is a selective and potent antagonist of TRPM3,[268] however direct evidence that TRPM3 antagonists alter pain perception waits for the development of potent and selective blockers.

9.7.2   TRPM8

The other TRPM family member implicated in pain is TRPM8, which was first described as a transcript increased in tumors of prostate breast, colon, lung and skin,[269] and is important for viability of these cells.[270] Subsequently, it was recognized that TRPM8 is a receptor for cold temperatures and cooling agents such as menthol and icilin.[271,272] When TRPM8 knockout mice were generated it was observed that the mice showed reduced avoidance of cold temperatures, despite the fact that TRPM8 is not the only cold sensor. Furthermore, these mice have an attenuated nocifensive behavioural response to cooling agents and reduced hypersensitivity to these agents following nerve injury and inflammation.[273–275] Nerves of TRPM8 knockout mice show normal responses to mechanical stimuli,[273] but TRPM8 dependent alterations in behavioural responses to mechanical stimuli were inconsistent.[274]

TRPM8 has received considerable attention from the pharmaceutical industry; several agonists and antagonists have been described in primary peer reviewed papers and in patents, for comprehensive reviews see DeFalco *et al.* (2011) and Preti *et al.* (2012).[276,277] The pharmacology of these has reinforced the role of TRPM8 as a cold sensor suggested by studies of the TRPM8 knockout mice. In addition, it was found that TRPM8 is involved in thermoregulation as antagonists cause a TRPM8 dependent hypothermia.[278,279] This effect is in line with the physiological and behavioural responses to agonists as administration of icilin and menthol induced hyperthermia.[280,281] However, repeated administration of a TRPM8 antagonist showed that the effect on body temperature tolerates indicating that this on-target side effect might be manageable.[279] When taken together with the observation that blocking TRPM8 mainly reduced cold sensitivity and had little evidence of affecting other pain modalities it appears that TRPM8 might have limited application as an analgesic.

9.7.3   TRPA1

In mammals TRPA1 is the only family member of the TRPA subfamily. After its initial cloning from a lung fibroblast cell line[282] it was quickly recognized to be a cold and chemical sensor expressed by nociceptive neurons.[283–285] Studies of the TRPA1 knockout mice confirmed the polymodal sensory nature of TRPA1, although its importance in cold sensing remains controversial.[286–291] TRPA1 is also thought to be involved in mechanosensitivity; however, whether TRPA1 itself is a mechanosensitive channels is also controversial.[23,292,293]

TRPA1 knockout mice have reduced sensitivity to a variety of noxious chemicals with attenuated responses to pungent plant spices such as mustard oil and garlic, environmental agents such as acrolein, formalin and diesel exhaust, and to some pharmaceutical compounds such as the volatile general anaesthetics Isoflurane.[284,286,294–297] Common to most agents that activate TRPA1 is their reactive nature containing unsaturated aldehydes, which are strong electrophiles that react with cysteine residues in the N-terminus of the channels *via* Michael Addition.[298,299]

TRPA1 knockout mice also have reduced sensitivity to endogenous products of inflammation and tissue damage, such as prostaglandin metabolites[300] and reactive oxygen species, *e.g.* H_2O_2,[301] and products of lipid peroxidation such as 4-ONE and 4-HNE.[302] A further role of TRPA1 in inflammation and related hypersensitivity is indicated by the reduced sensitivity of TRPA1 knockout mice to inflammatory mediators such as bradykinin[284,286] and PAR2,[303] which activate TRPA1 through activation of their perspective G-protein coupled receptors. Interestingly, the Mas-related G protein expressed in DRG and TGN can also regulate TRPA1, as activation of this pathway by antimalarial chloroquine drugs cause histamine-independent itch to which TRPA1 knockout mice are insensitive.[304]

The knockout mice have been tested in a variety of preclinical pain models, showing efficacy in inflammatory thermal and mechanical hypersensitivity,[286,287] visceral mechanical hypersensitivity,[305,306] hypersensitivity related to platinum-based anticancer drugs and cerulein-induced pancreatitis.[307,308] Besides a role in pain, TRPA1 knockout has been reported to attenuate cough responses and airways inflammation induced by TRPA1 agonists that could be involved in environmental pollution related asthma.[309,310]

As mentioned above, TRPA1 has been implicated in sensing cold. It is beyond the scope of this review to cover the controversy surrounding this modality. However, it is worth considering whether loss of cold sensation presents an on-target related drug development issue, in a similar manner to the loss of heat sensation presents for the development of TRPV1 antagonists. The papers reporting altered cold sensation with TRPA1 deletion observed effects at noxious cold of 0–5 °C or less. Exposure of tissue to such low temperatures might cause tissue damage, *e.g.* frost bite. However, most of the human population lives in temperate climates, where exposure to cold is seasonal and accompanied by visual changes to the outdoor environment. For TRPV1, daily life scenarios where an antagonist might be considered a risk could involve drinking excessively hot beverages or taking showers sufficiently hot to cause burns. For TRPA1 antagonists, analogous daily life scenarios involving noxious cold are difficult to envisage, so loss of cold sensation might not be such an obstacle for their future development. It is also of interest that blockade of TRPA1 with a tool antagonist A-967079 did not affect noxious cold sensation or body temperature regulation.[311] What may be a more significant consideration for the development of TRPA1 antagonists is the role this channel plays in detecting irritants and the possibility that a loss of TRPA1 signalling could alter sensing of noxious environmental chemicals and subsequent clearance from the respiratory tract due to altered cough reflexes. Improper detection and clearance of such irritants might lead to significant health risks with prolonged exposure.[312,313]

Several moderately potent but selective tool antagonists have been published. Hydra Biosciences generated the commonly used HC-030031[296,314] and ChemBridge Corporation produced a derivative of this compound (Chembridge-5861528/TCS 5861528 (Torcis Bioscience)).[315] Amgen have released data on a series of trichloro(sulfanyl)ethyl benzamides; however, data on the *in vivo* effects of these compounds have not be publically released.[316] The laboratory of Ardem Patapoutian identified AP-18, which is structurally similar to cinnamaldehyde,[317] and Abbott subsequently published the structure of A-967079 a more potent derivative of AP-18.[318] In preclinical studies these compounds confirmed the involvement of TRPA1 in pain sensation as was suggested by the phenotype of the TRPA1 knockout mice and showed an additional potential role for TRPA1 in neuropathic pain.[314]

Glenmark Pharmaceuticals is the first pharmaceutical company to move beyond the preclinical phase with their clinical candidate GRC 17536 having completed phase I trials. Phase IIa proof-of-concept studies are planned in patients with painful diabetic peripheral neuropathy and mild asthma (http://www.glenmarkpharma.com). The Phase I clinical trial for GRC 17536 showed that the compound was well tolerated up to the maximum dose tested indicating possible side effects of pharmacologically inhibiting TRPA1 are likely to be minor. The potential of TRPA1 as a target for pain has recently been further strengthened by the identification of single point, gain of function mutation in TRPA1, which was found in patients with an episodic autosomal dominant pain syndrome that is characterized by periods of debilitating upper body pain triggered by fasting, physical exercise and cold.[233]

9.8   Conclusion

In conclusion, besides TRPV1 several other TRP channels are potential candidates as analgesic pharmaceuticals. Preclinical data for TRPM3 is promising and developing an antagonist is the next critical step. For TRPV3 it will be interesting to see in which pain indications blocking its function will show efficacy. TRPV4 holds a lot of promise; an antagonist might be suited for several indications besides pain. We eagerly await the results of Phase I clinical trials for TRPA1 antagonists. With new TRP channel antagonists being developed, it will be of interest to directly compare the therapeutic efficacy of blocking the different channels in a variety of pain conditions. The potential for use of combinations or development of compounds with poly-pharmacology at the different TRP channels may also be an interesting area to explore in the future.

9.9 Important Questions Remaining in TRPV1 Research

It is a testament to the strength of the pharmaceutical industry's medicinal chemistry efforts that since the discovery of capsazepine and the subsequent cloning of TRPV1 multiple drug-like antagonists have been identified. However, despite data being publically disclosed on relatively few of the TRPV1 antagonists that have advanced to clinical phase testing, what appears significant is the lack of any clear demonstration of therapeutic efficacy in actual clinical pain conditions. This must surely stand as a great disappointment for all those scientists who have been actively involved in this field. Nonetheless, there is still hope for the progression of TRPV1 antagonists since development of some promising clinical candidates appear to have been restricted due to off-target side effects or an unusually pronounced incidence of hyperthermia and altered thermal perception. The combined data offers hope that a TRPV1 antagonist can be identified without these limitations and renewed efforts to dissect the effect of inhibiting the different modes TRPV1 activation might significantly facilitate this search.

We conclude with the observation that in spite of the long history of TRPV1 and before that capsaicin research, and with over 2700 references in PubMed identified by searching for "TRPV1" alone, there are still some important issues that need to be addressed:

1. What is the endogenous agonist responsible for mediating the effects of TRPV1 and is this the same for different types of pain?
2. Can antagonists be found that inhibit only the clinically relevant modes of TRPV1 activation?
3. For what type of clinical chronic or inflammatory pain will a TRPV1 antagonist show efficacy?
4. Are there diseases beyond pain where abnormal TRPV1 activation plays a key role?
5. Can compounds acting at other sensory TRP channels be developed that have improved therapeutic effects and/or reduced on-target side-effects over TRPV1 antagonists?

References

1. M. J. Caterina, M. A. Schumacher, M. Tominaga, T. A. Rosen, J. D. Levine and D. Julius, *Nature*, 1997, **389**, 816.
2. C. Montell, *Science's STKE*, 2005, **272**, re3.
3. A. Dhaka, V. Viswanath and A. Patapoutian, *Ann. Rev. Neurosci.*, 2006, **29**, 135.
4. B. Nilius and G. Owsianik, *Genome Biology*, 2011, **12**, 218.
5. M. Tominaga and T. Tominaga, *Pflügers Archiv: EJP*, 2005, **451**, 143.

6. L. Urban, J. P. M. White and I. Nagy, *Current Pharmaceutical Biotechnology*, 2011, **12**, 115.

7. P. V. Lishko, E. Procko, X. Jin, C. B. Phelps and R. Gaudet, *Neuron*, 2007, **54**, 905.

8. E. Fernandes, M. Fernandes and J. E. Keeble, *Br. J. Pharmacol.*, 2012, **166**, 510.

9. V. Vellani, S. Mapplebeck, A. Moriondo, J. B. Davis and P. A. McNaughton, *J. Physiol.*, 2001, **534**, 813.

10. X. Shu and L. M. Mendell, *J. Neurophysiol.*, 2001, **86**, 2931.

11. A. Varga, K. Bölcskei, E. Szöke, R. Almási, G. Czéh, J. Szolcsányi and G. Pethö, *Neuroscience*, 2006, **140**, 645.

12. T. Rohacs, B. Thyagarajan and V. Lukacs, *Mol. Neurobiol.*, 2008, **37**, 153.

13. K. Schnizler, L. P. Shutov, M. J. Van Kanegan, M. A. Merrill, B. Nichols, G. S. McKnight, S. Strack, J. W. Hell and Y. M. Usachev, *J. Neurosci.*, 2008, **28**, 4904.

14. R. S. Nicholas, J. Winter, P. Wren, R. Bergmann and C. J. Woolf, *Neuroscience*, 1999, **91**, 1425.

15. X. Zhang, J. Huang and P. A. McNaughton, *EMBO J.*, 2005, **24**, 4211.

16. H. Luo, J. Cheng, J.-S. Han and Y. Wan, *Neuroreport*, 2004, **15**, 655.

17. F. Amaya, G. Shimosato, M. Nagano, M. Ueda, S. Hashimoto, Y. Tanaka, H. Suzuki and M. Tanaka, *Eur. J. Neurosci.*, 2004, **20**, 2303.

18. J. Fernihough, C. Gentry, S. Bevan and J. Winter, *Neurosci. Lett.*, 2005, **388**, 75.

19. Y. Kanai, E. Nakazato, A. Fujiuchi, T. Hara and A. Imai, *Neuropharmacology*, 2005, **49**, 977.

20. D. Vilceanu, P. Honore, Q. H. Hogan and C. L. Stucky, *J. Pain*, 2010, **11**, 588.

21. Y. Niiyama, T. Kawamata, J. Yamamoto, K. Omote and A. Namiki, *Neuroscience*, 2007, **148**, 560.

22. M. Shinoda, A. Ogino, N. Ozaki, H. Urano, K. Hironaka, M. Yasui and Y. Sugiura, *J. Pain*, 2008, **9**, 687.

23. X. F. Zhang, J. Chen, C. R. Faltynek, R. B. Moreland and T. R. Neelands, *Eur. J. Neurosci.*, 2008, **27**, 605.

24. L.-Y. Lee and Q. Gu, *Curr. Opinion Pharmacol.*, 2009, **9**, 243.

25. P. Geppetti and M. Trevisani, *Br. J. Pharmacol.*, 2004, **141**, 1313.

26. J. Winston, M. Shenoy, D. Medley, A. Naniwadekar and P. J. Pasricha, *Gastroenterology*, 2007, **132**, 615.

27. G.-Y. Xu, J. H. Winston, M. Shenoy, H. Yin, S. Pendyala and P. J. Pasricha, *Gastroenterology*, 2007, **133**, 1282.

28. H. U. De Schepper, J. G. De Man, N. E. Ruyssers, A. Deiteren, L. Van Nassauw, J. P. Timmermans, W. Martinet, A. G. Herman, P. A. Pelckmans and B. Y. De Winter, *Am. J. Physiol. Gastrointest. Liver Physiol.*, 2008, **294**, G245.

29. R.-R. Ji, T. A. Samad, S.-X. Jin, R. Schmoll and C. J. Woolf, *Neuron*, 2002, **36**, 57.

30. N. M. Breese, A. C. George, L. E. Pauers and C. L. Stucky, *Pain*, 2005, **115**, 37.
31. L. Yu, F. Yang, H. Luo, F.-Y. Liu, J.-S. Han, G.-G. Xing and Y. Wan, *Mol. Pain*, 2008, **4**, 61.
32. M. J. Caterina, A. Leffler, A. B. Malmberg, W. J. Martin, J. Trafton, K. R. Petersen-Zeitz, M. Koltzenburg, A. I. Basbaum and D. Julius, *Science*, 2000, **288**, 306.
33. L. A. Birder, Y. Nakamura, S. Kiss, M. L. Nealen, S. Barrick, A. J. Kanai, E. Wang, G. Ruiz, W. C. De Groat, G. Apodaca, S. Watkins and M. J. Caterina, *Nat. Neurosci.*, 2002, **5**, 856.
34. A. Charrua, C. D. Cruz, S. Narayanan, L. Gharat, S. Gullapalli, F. Cruz and A. Avelino, *J. Urol.*, 2009, **181**, 379.
35. P. Round, A. Priestley and J. Robinson, *Br. J. Clin. Pharmacol.*, 2011, **72**, 921.
36. A. Jancsó-Gábor, J. Szolcsányi and N. Jancsó, *J. Physiol.*, 1970, **208**, 449.
37. Z. Szelényi, Z. Hummel, J. Szolcsányi and J. B. Davis, *Eur. J. Neurosci.*, 2004, **19**, 1421.
38. T. Iida, I. Shimizu, M. L. Nealen, A. Campbell and M. Caterina, *Neurosci. Lett.*, 2005, **378**, 28.
39. K. Bölcskei, Z. Helyes, A. Szabó, K. Sándor, K. Elekes, J. Németh, R. Almási, E. Pintér, G. Petho and J. Szolcsányi, *Pain*, 2005, **117**, 368.
40. J. Keeble, F. Russell, B. Curtis, A. Starr, E. Pinter and S. D. Brain, *Arthritis Rheum.*, 2005, **52**, 3248.
41. N. J. Barton, D. S. McQueen, D. Thomson, S. D. Gauldie, A. W. Wilson, D. M. Salter and I. P. Chessell, *Exp. Mol. Pathol.*, 2006, **81**, 166.
42. A. Szabó, Z. Helyes, K. Sándor, A. Bite, E. Pintér, J. Németh, A. Bánvölgyi, K. Bölcskei, K. Elekes and J. Szolcsányi, *J. Pharmacol. Exp. Ther.*, 2005, **314**, 111.
43. K. Katanosaka, R. K. Banik, R. Giron, T. Higashi, M. Tominaga and K. Mizumura, *Neurosci. Res.*, 2008, **62**, 168.
44. E. M. Pogatzki-Zahn, I. Shimizu, M. Caterina and S. N. Raja, *Pain*, 2005, **115**, 296.
45. R. K. Banik and T. J. Brennan, *Pain*, 2009, **141**, 41.
46. J. E. Keeble, J. Elizabeth, J. V. Bodkin, L. L. R. Wodarski, M. Davies, E. S. Fernandes, C. de Faria Coelho, F. Russell, R. Graepel, M. N. Muscara, M. Malcangio and S. D. Brain, *Pain*, 2009, **141**, 135.
47. L. E. Ta, A. J. Bieber, S. M. Carlton, C. L. Loprinzi, P. A. Low and A. J. Windebank, *Mol. Pain*, 2010, **6**, 15.
48. K. Bielefeldt and B. M. Davis, *Am. J. Physiol.*, 2008, **294**, G130.
49. W. Rong, K. Hillsley, J. B. Davis, G. Hicks, W. J. Winchester and D. Grundy, *J. Physiol.*, 2004, **560**, 867.
50. X. Wang, R. L. Miyares and G. P. Ahern, *J. Physiol.*, 2005, **564**, 541.
51. R. C. W. Jones 3rd, L. Xu and G. F. Gebhart, *J. Neurosci.*, 2005, **25**, 10981.
52. R. C. W. Jones 3rd, E. Otsuka, E. Wagstrom, C. S. Jensen, M. P. Price and G. F. Gebhart, *Gastroenterology*, 2007, **133**, 184.

53. I. Szitter, G. Pozsgai, K. Sandor, K. Elekes, A. Kemeny, A. Perkecz, J. Szolcsanyi, Z. Helyes and E. Pinter, *J. Mol. Neurosci.*, 2010, **42**, 80.

54. F. Massa, A. Sibaev, G. Marsicano, H. Blaudzun, M. Storr and B. Lutz, *J. Mol. Med.*, 2006, **84**, 142.

55. M. S. Nash, P. McIntyre, A. Groarke, E. Lilley, A. Culshaw, A. Hallett, M. Panesar, A. Fox and S. Bevan, *J. Pharmacol. Exp. Ther.*, 2012, **342**, 389.

56. D. Daly, W. Rong, R. Chess-Williams, C. Chapple and D. Grundy, *J. Physiol.*, 2007, **583**, 663.

57. Z.-Y. Wang, P. Wang, F. V. Merriam and D. E. Bjorling, *Pain*, 2008, **139**, 158.

58. B. Frias, A. Charrua, A. Avelino, M. C. Michel, F. Cruz and C. D. Cruz, *BJU Int.*, 2012, **110**, E422.

59. P. Kanizsai, A. Garami, M. Solymár, J. Szolcsányi and Z. Szelényi, *Physiol. Behav.*, 2009, **96**, 149.

60. A. Garami, E. Pakai, D. L. Oliveira, A. A. Steiner, S. P. Wanner, M. C. Almeida, V. A. Lesnikov, N. R. Gavva and A. A. Romanovsky, *J. Neurosci.*, 2011, **31**, 1721.

61. R. Marsch, E. Foeller, G. Rammes, M. Bunck, M. Kössl, F. Holsboer, W. Zieglgänsberger, R. Landgraf, B. Lutz and C. T. Wotjak, *J. Neurosci.*, 2007, **27**, 832.

62. M. Kollarik and B. J. Undem, *J. Physiol.*, 2004, **555**, 115.

63. Z. Helyes, K. Elekes, J. Németh, G. Pozsgai, K. Sándor, L. Kereskai, R. Börzsei, E. Pintér, A. Szabó and J. Szolcsányi, *Am. J. Physiol.*, 2007, **292**, L1173.

64. T. Bíró, E. Bodó, A. Telek, T. Géczy, B. Tychsen, L. Kovács and R. Paus, *J. Invest. Dermatol.*, 2006, **126**, 1909.

65. A. Bánvölgyi, L. Pálinkás, T. Berki, N. Clark, A. D. Grant, Z. Helyes, G. Pozsgai, J. Szolcsányi, S. D. Brain and E. Pintér, *J. Neuroimmunol.*, 2005, **169**, 86.

66. A. Szabó, L. Czirják, Z. Sándor, Z. Helyes, T. László, K. Elekes, T. Czömpöly, A. Starr, S. Brain, J. Szolcsányi and E. Pintér, *Arthritis Rheum.*, 2008, **58**, 292.

67. A. M. Bode, Y.-Y. Cho, D. Zheng, F. Zhu, M. E. Ericson, W.-Y. Ma, K. Yao and Z. Dong, *Cancer Research*, 2009, **69**, 905.

68. H. Usuda, T. Endo, A. Shimouchi, A. Saito, M. Tominaga, H. Yamashita, H. Nagai, N. Inagaki and H. Tanaka, *J. Pharmacol. Sci.*, 2012, **118**, 266.

69. P. Pacher, S. Bátkai and G. Kunos, *J. Physiol.*, 2004, **558**, 647.

70. C. L. Buckley and A. J. Stokes, *Channels*, 2011, **5**, 367.

71. L. Wang and D. H. Wang, *Circulation*, 2005, **112**, 3617.

72. Y. A. Blednov and R. A. Harris, *Neuropharmacology*, 2009, **56**, 814.

73. J. M. Ellingson, B. C. Silbaugh and S. M. Brasser, *Behavior Genetics*, 2009, **39**, 62.

74. S. P. Wanner, A. Garami and A. A. Romanovsky, *Aging*, 2011, **3**, 450.

75. A. L. Motter and G. P. Ahern, *FEBS Lett.*, 2008, **582**, 2257.

76. L. A. Roberts and M. Connor, *Recent Pat. CNS Drug Discovery*, 2006, **1**, 65.
77. D. C. Immke and N. R. Gavva, *Semin. Cell Dev. Biol.*, 2006, **17**, 582.
78. A. Szallasi, D. N. Cortright, C. A. Blum and S. R. Eid, *Nat. Rev. Drug Discovery*, 2007, **6**, 357.
79. M. J. Gunthorpe and B. A. Chizh, *Drug Discovery Today*, 2009, **14**, 56.
80. G. Y. Wong and N. R. Gavva, *Brain Res. Rev.*, 2009, **60**, 267.
81. C. Grimm, E. Aneiros and M. de Groot, *Channels*, 2011, **5**, 201.
82. E. Palazzo, L. Luongo, V. de Novellis, F. Rossi, I. Marabese and S. Maione, *Curr. Opinion Pharmacol.*, 2012, **12**, 9.
83. A. Szallasi and M. Sheta, *Expert Opin. Investig. Drugs*, 2012, **21**, 1351.
84. J. D. Pomonis, J. E. Harrison, L. Mark, D. R. Bristol, K. J. Valenzano and K. Walker, *J. Pharmacol. Exp. Ther.*, 2003, **306**, 387.
85. P. Honore, C. T. Wismer, J. Mikusa, C. Z. Zhu, C. Zhong, D. M. Gauvin, A. Gomtsyan, R. El Kouhen, C.-H. Lee, K. Marsh, J. P. Sullivan, C. R. Faltynek and M. F. Jarvis, *J. Pharmacol. Exp. Ther.*, 2005, **314**, 410.
86. N. R. Gavva, R. Tamir, Y. Qu, L. Klionsky, T. J. Zhang, D. Immke, J. Wang, D. Zhu, T. W. Vanderah, F. Porreca, E. M. Doherty, M. H. Norman, K. D. Wild, A. W. Bannon, J. C. Louis and J. J. Treanor, *J. Pharmacol. Exp. Ther.*, 2005, **313**, 474.
87. Y. Kanai, T. Hara and A. Imai, *J. Pharmacy Pharmacol.*, 2006, **58**, 489.
88. L. Tang, Y. Chen, Z. Chen, P. M. Blumberg, A. P. Kozikowski and Z. J. Wang, *J. Pharmacol. Exp. Ther.*, 2007, **321**, 791.
89. P. Honore, P. Chandran, G. Hernandez, D. M. Gauvin, J. P. Mikusa, C. Zhong, S. K. Joshi, J. R. Ghilardi, M. A. Sevcik, R. M. Fryer, J. A. Segreti, P. N. Banfor, K. Marsh, T. Neelands, E. Bayburt, J. F. Daanen, A. Gomtsyan, C. H. Lee, M. E. Kort, R. M. Reilly, C. S. Surowy, P. R. Kym, P. W. Mantyh, J. P. Sullivan, M. F. Jarvis and C. R. Faltynek, *Pain*, 2009, **142**, 27.
90. R. M. van den Wijngaard, T. K. Klooker, O. Welting, O. I. Stanisor, M. M. Wouters, D. van der Coelen, D. C. Bulmer, P. J. Peeters, J. Aerssens, R. de Hoogt, K. Lee, W. J. de Jonge and G. E. Boeckxstaens, *Neurogastroenterol. Motil.*, 2009, **21**, 1107.
91. A. Ravnefjord, M. Brusberg, D. Kang, U. Bauer, H. Larsson, E. Lindström and V. Martinez, *Eur. J. Pharmacol.*, 2009, **611**, 85.
92. B. J. Wiskur, K. Tyler, K. Campbell-Dittmeyer, S. R. Chaplan, A. D. Wickenden and B. Greenwood-Van Meerveld, *Methods Find. Exp. Clin. Pharmacol.*, 2010, **32**, 557.
93. A. Miranda, E. Nordstrom, A. Mannem, C. Smith, B. Banerjee and J. N. Sengupta, *Neuroscience*, 2007, **148**, 1021.
94. A. Santos-Silva, A. Charrua, C. D. Cruz, L. Gharat, A. Avelino and F. Cruz, *Auton. Neurosci.*, 2012, **166**, 35.
95. J. S. Cefalu, M. A. Guillon, L. R. Burbach, Q.-M. Zhu, D.-Q. Hu, M. J. Ho, A. P. D. W. Ford, P. A. Nunn and D. A. Cockayne, *J. Urol.*, 2009, **182**, 776.
96. M. Grace, M. A. Birrell, E. Dubuis, S. A. Maher and M. G. Belvisi, *Thorax*, 2012, **67**, 891.

97. K. C. Thomas, J. K. Roberts, C. E. Deering-Rice, E. G. Romero, R. O. Dull, J. Lee, G. S. Yost and C. A. Reilly, *Am. J. Physiol.*, 2012, **302**, L111.
98. I. Delescluse, H. Mace and J. J. Adcock, *Br. J. Pharmacol.*, 2012, **166**, 1822.
99. N. R. Gavva, A. W. Bannon, S. Surapaneni, D. N. Hovland Jr, S. G. Lehto, A. Gore, T. Juan, H. Deng, B. Han, L. Klionsky, R. Kuang, A. Le, R. Tamir, J. Wang, B. Youngblood, D. Zhu, M. H. Norman, E. Magal, J. J. Treanor and J. C. Louis, *J. Neurosci.*, 2007, **27**, 3366.
100. A. H. Dickenson and A. Dray, *Br. J. Pharmacol.*, 1991, **104**, 1045.
101. J. Winter, C. S. Walpole, S. Bevan and I. F. James, *Neuroscience*, 1993, **57**, 747.
102. A. Szallasi, C. Goso, P. M. Blumberg and S. Manzini, *J. Pharmacol. Exp. Ther.*, 1993, **267**, 728.
103. M. Rigoni, M. Trevisani, D. Gazzieri, R. Nadaletto, M. Tognetto, C. Creminon, J. B. Davis, B. Campi, S. Amadesi, P. Geppetti and S. Harrison, *Br. J. Pharmacol.*, 2003, **138**, 977.
104. M. Tominaga, M. J. Caterina, A. B. Malmberg, T. A. Rosen, H. Gilbert, K. Skinner, B. E. Raumann, A. I. Basbaum and D. Julius, *Neuron*, 1998, **21**, 531.
105. M. Cui, P. Honore, C. Zhong, D. Gauvin, J. Mikusa, G. Hernandez, P. Chandran, A. Gomtsyan, B. Brown, E. K. Bayburt, K. Marsh, B. Bianchi, H. McDonald, W. Niforatos, T. R. Neelands, R. B. Moreland, M. W. Decker, C. H. Lee, J. P. Sullivan and C. R. Faltynek, *J. Neurosci.*, 2006, **26**, 9385.
106. S. McGaraughty, K. L. Chu, C. R. Faltynek and M. F. Jarvis, *J. Neurophysiol.*, 2006, **95**, 18.
107. S. McGaraughty, K. L. Chu, B. S. Brown, C. Z. Zhu, C. Zhong, S. K. Joshi, P. Honore, C. R. Faltynek and M. F. Jarvis, *J. Neurophysiol.*, 2008, **100**, 3158.
108. Y. Kanai, T. Hara, A. Imai and A. Sakakibara, *J. Pharmacy Pharmacol.*, 2007, **59**, 733.
109. T. Watabiki, T. Kiso, T. Kuramochi, K. Yonezawa, N. Tsuji, A. Kohara, S. Kakimoto, T. Aoki and N. J. Matsuoka, *J. Pharmacol. Exp. Ther.*, 2011, **336**, 743.
110. M. D. Jhaveri, S. J. R. Elmes, D. A. Kendall and V. Chapman, *Eur. J. Neurosci.*, 2005, **22**, 361.
111. Y. H. Kim, S. K. Back, A. J. Davies, H. Jeong, H. J. Jo, G. Chung, H. S. Na, Y. C. Bae, S. J. Kim, J. S. Kim, S. J. Jung and S. B. Oh, *Neuron*, 2012, **74**, 640.
112. A. Hughes, A. Macleod, J. Growcott and I. Thomas, *Pain*, 2002, **99**, 323.
113. J. Winter, S. Bevan and E. A. Campbell, *Br. J. Anaesth.*, 1995, **75**, 157.
114. C. Noto, M. Pappagallo and A. Szallasi, *Curr. Opin. Invest. Drugs*, 2009, **10**, 702.
115. H.-C. Kuo, *J. Urol.*, 2003, **170**, 835.
116. A. Apostolidis, G. E. Gonzales and C. J. Fowler, *Eur. Urol.*, 2006, **50**, 1299.

117. G. Bhave, W. Zhu, H. Wang, D. J. Brasier, G. S. Oxford and R. W. Gereau 4th, *Neuron*, 2002, **35**, 721.
118. M. Numazaki, T. Tominaga, K. Takeuchi, N. Murayama, H. Toyooka and M. Tominaga, *Proc. Natl. Acad. Sci. U. S. A.*, 2003, **100**, 8002.
119. D. A. Simone, M. Nolano, T. Johnson, G. Wendelschafer-Crabb and W. R. Kennedy, *J. Neurosci.*, 1998, **18**, 8947.
120. S. J. Hwang, J. M. Oh and J. G. Valtschanoff, *Brain Res.*, 2005, **1047**, 261.
121. J. A. Christianson, S. L. McIlwrath, H. R. Koerber and B. M. Davis, *Neuroscience*, 2006, **140**, 247.
122. Y. Yiangou, P. Facer, N. H. Dyer, C. L. Chan, C. Knowles, N. S. Williams and P. Anand, *Lancet*, 2001, **357**, 1338.
123. P. J. Matthews, Q. Aziz, P. Facer, J. B. Davis, D. G. Thompson and P. Anand, *Eur. J. Gastroenterol. Hepatol.*, 2004, **16**, 897.
124. K. R. Shieh, C. H. Yi, T. T. Liu, H. L. Tseng, H. C. Ho, H. T. Hsieh and C. L. Chen, *Neurogastroenterol. Motil.*, 2010, **22**, 971.
125. A. Dömötör, L. Kereskay, G. Szekeres, B. Hunyady, J. Szolcsányi and G. Y. Mózsik, *Dig. Dis. Sci.*, 2007, **52**, 411.
126. C. L. H. Chan, P. Facer, J. B. Davis, G. D. Smith, J. Egerton, C. Bountra, N. S. Williams and P. Anand, *Lancet*, 2003, **361**, 385.
127. A. Akbar, Y. Yiangou, P. Facer, J. R. F. Walters, P. Anand and S Ghosh, *Gut*, 2008, **57**, 923.
128. M. P. L. Guarino, L. Cheng, J. Ma, K. Harnett, P. Biancani, A. Altomare, F. Panzera, J. Behar and M. Cicala, *Neurogastroenterol. Motil.*, 2010, **22**, 746.
129. F. Azpiroz, M. Bouin, M. Camilleri, E. A. Mayer, P. Poitras, J. Serra and R. C. Spiller, *Neurogastroenterol. Motil.*, 2007, **19**, 62.
130. A. Akbar, Y. Yiangou, P. Facer, W. G. Brydon, J. R. F. Walters, P. Anand and S. Ghosh, *Gut*, 2010, **59**, 767.
131. A. Akbar, J. R. F. Walters and S. Ghosh, *Aliment. Pharmacol. Ther.*, 2009, **30**, 423.
132. Y. Yiangou, P. Facer, A. Ford, C. Brady, O. Wiseman, C. J. Fowler and P. Anand, *BJU Int.*, 2001, **87**, 774.
133. M. Lazzeri, M. G. Vannucchi, C. Zardo, M. Spinelli, P. Beneforti, D. Turini and M.-S. Faussone-Pellegrini, *European Urology*, 2004, **46**, 792.
134. C. M. Brady, A. N. Apostolidis, M. Harper, Y. Yiangou, A. Beckett, T. S. Jacques, A. Freeman, F. Scaravilli, C. J. Fowler and P. Anand, *BJU Int.*, 2004, **93**, 770.
135. A. Apostolidis, C. M. Brady, Y. Yiangou, J. Davis, C. J. Fowler and P. Anand, *Urology*, 2005, **65**, 2400.
136. H.-T. Liu and H.-C. Kuo, *BJU Int.*, 2007, **100**, 1086.
137. L. Liu, K. J. Mansfield, I. Kristiana, K. J. Vaux, R. J. Millard and E. Burcher, *Neurourol. Urodyn.*, 2007, **26**, 433.
138. P. Tympanidis, M. A. Casula, Y. Yiangou, G. Terenghi, P. Dowd and P. Anand, *Eur. J. Pain*, 2004, **8**, 129.
139. O. B. Poli-Neto, A. A. Filho, J. C. R. Silva, H. F. Barbosa, F. J. C. D. Reis and A. A. Nogueira, *Clin. J. Pain*, 2009, **25**, 218.

140. M. G. Rocha, J. C. R. Silva, A. R. Silva, F. J. C. D. Reis, A. A. Nogueira and O. B. Poli-Neto, *Reprod. Sci.*, 2011, **18**, 511.

141. C. R. Morgan, H. D. Rodd, N. Clayton, J. B. Davis and F. M. Boissonade, *J. Orofacial Pain*, 2005, **19**, 248.

142. H. D. Rodd, C. R. Morgan, P. F. Day and F. M. Boissonade, *Eur. J. Paediatr. Dent.*, 2007, **8**, 184.

143. Z. Yilmaz, T. Renton, Y. Yiangou, J. Zakrzewska, I. P. Chessell, C. Bountra and P. Anand, *J. Clin. Neurosci.*, 2007, **14**, 864.

144. D. A. Groneberg, A. Niimi, Q. T. Dinh, B. Cosio, M. Hew, A. Fischer and K. F. Chung, *Am. J. Respir. Crit. Care Med.*, 2004, **170**, 1276.

145. P. Facer, M. A. Casula, G. D. Smith, C. D. Benham, I. P. Chessell, C. Bountra, M. Sinisi, R. Birch and P. Anand, *BMC Neurology*, 2007, **7**, 11.

146. Y. M. Lee, Y. K. Kim and J. H. Chung, *Exp. Dermatol.*, 2009, **18**, 431.

147. Y. M. Lee, S. M. Kang and J. H. Chung, *J. Dermatol. Sci.*, 2012, **65**, 81.

148. H. Xu, W. Tian, Y. Fu, T. T. Oyama, S. Anderson and D. M. Cohen, *Am. J. Physiol.*, 2007, **293**, F1865.

149. H. Kim, J. K. Neubert, A. San Miguel, K. Xu, R. K. Krishnaraju, M. J. Iadarola, D. Goldman and R. A. Dionne, *Pain*, 2004, **109**, 488.

150. H. Kim, D. P. Mittal, M. J. Iadarola and R. A. Dionne, *J. Med. Genetics*, 2006, **43**, e40.

151. G. Cantero-Recasens, J. R. Gonzalez, C. Fandos, E. Duran-Tauleria, L. A. M. Smit, F. Kauffmann, J. M. Antó and M. A. Valverde, *J. Biol. Chem.*, 2010, **285**, 27532.

152. L. A. M. Smit, M. Kogevinas, J. M. Antó, E. Bouzigon, J. R. González, N. Le Moual, H. Kromhout, A. E. Carsin, I. Pin, D. Jarvis, R. Vermeulen, C. Janson, J. Heinrich, I. Gut, M. Lathrop, M. A. Valverde, F. Demenais and F. Kauffmann, *Respiratory Res.*, 2012, **13**, 26.

153. O. Carreño, R. Corominas, J. Fernández-Morales, M. Camiña, M.-J. Sobrido, J. M. Fernández-Fernández, P. Pozo-Rosich, B. Cormand and A. Macaya, *Am. J. Med. Genet. B*, 2012, **159B**, 94.

154. A. M. Valdes, G. De Wilde, S. A. Doherty, R. J. Lories, F. L. Vaughn, L. L. Laslett, R. A. Maciewicz, A. Soni, D. J. Hart, W. Zhang, K. R. Muir, E. M. Dennison, M. Wheeler, P. Leaverton, C. Cooper, T. D. Spector, F. M. Cicuttini, V. Chapman, G. Jones, N. K. Arden and M. Doherty, *Ann. Rheum. Dis.*, 2011, **70**, 1556.

155. T. Tahara, T. Shibata, M. Nakamura, H. Yamashita, D. Yoshioka, I. Hirata and T. Arisawa, *J. Clin. Gastroenterology*, 2010, **44**, e1.

156. J. Hammer, M. Führer, L. Pipal and J. Matiasek, *Neurogastroenterol. Motil.*, 2008, **20**, 125.

157. M. Bortolotti, G. Coccia, G. Grossi and M. Miglioli, *Aliment. Pharmacol. Ther.*, 2002, **16**, 1075.

158. A. Binder, D. May, R. Baron, C. Maier, T. R. Tölle, R. D. Treede, A. Berthele, F. Faltraco, H. Flor, J. Gierthmühlen, S. Haenisch, V. Huge, W. Magerl, C. Maihöfner, H. Richter, R. Rolke, A. Scherens, N. Uçeyler, M. Ufer, G. Wasner, J. Zhu and I. Cascorbi, *PloS One*, 2011, **6**, e17387.

159. A. Szallasi and P. M. Blumberg, *Neuroscience*, 1989, **30**, 515.
160. C. S. J. Walpole, S. Bevan, G. Bovermann, J. J. Boelsterli, R. Breckenridge, J. W. Davies, G. A. Hughes, I. James and L. Oberer, *J. Med. Chem.*, 1994, **37**, 1942.
161. J. N. Wood, J. Winter, I. F. James, H. P. Rang, J. Yeats and S. Bevan, *J. Neurosci.*, 1988, **8**, 3208.
162. H. Knotova, M. Pappagallo and A. Szallasi, *Clin. J. Pain*, 2008, **24**, 142.
163. M. J. Iadarola and A. J. Mannes, *Curr. Top. Med. Chem.*, 2011, **11**, 2171.
164. A. G. Hayes, A. Oxford, M. Reynolds, A. H. Shingler, M. Skingle, C. Smith and M. B. Tyers, *Life Sci.*, 1984, **34**, 1241.
165. M. N. Perkins and E. A. Campbell, *Br. J. Pharmacol.*, 1992, **107**, 329.
166. A. R. S. Santos and J. B. Calixto, *Neurosci. Lett.*, 1997, **235**, 73.
167. K. M. Walker, L. Urban, S. J. Medhurst, S. Patel, M. Panesar, A. J. Fox and P. McIntyre, *J. Pharmacol. Exp. Ther.*, 2003, **304**, 56.
168. P. Hayes, H. J. Meadows, M. J. Gunthorpe, M. H. Harries, D. M. Duckworth, W. Cairns, D. C. Harrison, C. E. Clarke, K. Ellington, R. K. Prinjha, A. J. Barton, A. D. Medhurst, G. D. Smith, S. Topp, P. Murdock, G. J. Sanger, J. Terrett, O. Jenkins, C. D. Benham, A. D. Randall, I. S. Gloger and J. B. Davis, *Pain*, 2000, **88**, 205.
169. J. Savidge, C. Davis, K. Shah, S. Colley, E. Phillips, S. Ranasinghe, J. Winter, P. Kotsonis, H. Rang and P. McIntyre, *Neuropharmacol.*, 2002, **43**, 450.
170. P. McIntyre, L. M. McLatchie, A. Chambers, E. Phillips, M. Clarke, J. Savidge, C. Toms, M. Peacock, K. Shah, J. Winter, N. Weerasakera, M. Webb, H. P. Rang, S. Bevan and I. F. James, *Br. J. Pharmacol.*, 2001, **132**, 1084.
171. E. Phillips, A. Reeve, S. Bevan and P. McIntyre, *J. Biol. Chem.*, 2004, **279**, 17165.
172. P. Wahl, C. Foged, S. Tullin and C. Thomsen, *Mol. Pharmacol.*, 2001, **59**, 9.
173. A. Toth, P. M. Blumberg, Z. Chen and A. P. Kozikowski, *Mol. Pharmacol.*, 2004, **65**, 282.
174. H. K. Rami, M. Thompson, P. Wyman, J. C. Jerman, J. Egerton, S. Brough, A. J. Stevens, A. D. Randall, D. Smart, M. J. Gunthorpe and J. B. Davis, *Bioorg. Med. Chem. Lett.*, 2004, **14**, 3631.
175. H. K. Rami, M. Thompson, G. Stemp, S. Fell, J. C. Jerman, A. J. Stevens, D. Smart, B. Saargent, D. Sanderson, A. D. Randall, M. J. Gunthorpe and J. B. Davis, *Bioorg. Med. Chem. Lett.*, 2006, **16**, 3287.
176. M. J. Gunthorpe, S. L. Hannan, D. Smart, J. C. Jerman, S. Arpino, G. D. Smith, S. Brough, J. Wright, J. Egerton, S. C. Lappin, V. A. Holland, K. Winborn, M. Thompson, H. K. Rami, A. Randall and J. B. Davis, *J. Pharmacol. Exp. Ther.*, 2007, **321**, 1183.
177. M. Gunthorpe, presented at the SCI Conference, London, 2006.
178. G. A. Lambert, J. B. Davis, J. M. Appleby, B. A. Chizh, K. L. Hoskin and A. S. Zagami, *Naunyn Schmiedebergs Arch. Pharmacol.*, 2009, **380**, 311.

179. B. A. Chizh, M. B. O'Donnell, A. Napolitano, J. Wang, A. C. Brooke, M. C. Aylott, J. N. Bullman, E. J. Gray, R. Y. Lai, P. M. Williams and J. M. Appleby, *Pain*, 2007, **132**, 132.

180. J. Palmer, R. Lai, N. Thomas, J. Bullman, Z. Ali, A. Baines and J. Appleby, presented at the 6th Congress of the European Federation of IASP Chapters (EFIC), Portugal, 2009.

181. L. Alenmyr, L. Greiff, M. Andersson, O. Sterner, P. M. Zygmunt and E. D. Högestätt, *Basic Clin. Pharmacol. Toxicol.*, 2012, **110**, 264.

182. M. H. Norman in *Accounts in Drug Discovery: Case Studies in Medicinal Chemistry*, ed. J. C. Barrish, P. H. Carter, P. T. W. Cheng and R. Zahler, Royal Society of Chemistry, 2011, p. 287.

183. N. R. Gavva, A. W. Bannon, D. N. Hovland Jr, S. G. Lehto, L. Klionsky, S. Surapaneni, D. C. Immke, C. Henley, L. Arik, A. Bak, J. Davis, N. Ernst, G. Hever, R. Kuang, L. Shi, R. Tamir, J. Wang, W. Wang, G. Zajic, D. Zhu, M. H. Norman, J. C. Louis, E. Magal and J. J. Treanor, *J. Pharmacol. Exp. Ther.*, 2007, **323**, 128.

184. N. R. Gavva, J. J. S. Treanor, A. Garami, L. Fang, S. Surapaneni, A. Akrami, F. Alvarez, A. Bak, M. Darling, A. Gore, G. R. Jang, J. P. Kesslak, L. Ni, M. H. Norman, G. Palluconi, M. J. Rose, M. Salfi, E. Tan, A. A. Romanovsky, C. Banfield and G. Davar, *Pain*, 2008, **136**, 202.

185. N. R. Gavva, *Trends Pharmacol. Sci.*, 2008, **29**, 550.

186. A. A. Romanovsky, M. C. Almeida, A. Garami, A. A. Steiner, M. H. Norman, S. F. Morrison, K. Nakamura, J. J. Burmeister and T. B. Nucci, *Pharmacol. Rev.*, 2009, **61**, 228.

187. R. Bakthavatchalam, Preparation of diarylpiperazines as capsaicin receptor ligands, WO 2002/008221-A2, 2002.

188. D. J. Kyle and Q. Sun, Preparation of pyridin-2-ylpiperazine derivatives for treating pain, WO 2003/066595, 2003.

189. N. I. Carruthers, C. R. Shah and D. M. Swanson, Preparation of pyridyl piperazinyl ureas with VR1 antagonistic activity, WO 2005/014580-A1, 2005.

190. X. Zheng, K. J. Hodgetts, H. Brielmann, A. Hutchison, F. Burkamp, A. B. Jones, P. Blurton, R. Clarkson, J. Chandrasekar, R. Bakthavatchalam, S. De Lombaert, M. Crandall, D. Cotright and C. A. Blum, *Bioorg. Med. Chem. Lett.*, 2006, **16**, 5217.

191. Q. Sun, L. Tafesse, K. Islam, X. Zhou, S. F. Victory, C. Zhang, M. Hachicha, L. A. Schmid, A. Patel, Y. Rotshteyn, K. J. Valenzano and D. J. Kyle, *Bioorg. Med. Chem. Lett.*, 2003, **13**, 3611.

192. K. J. Valenzano, E. R. Grant, G. Wu, M. Hachicha, L. Schmid, L. Tafesse, Q. Sun, Y. Rotshteyn, J. Francis, J. Limberis, S. Malik, E. R. Whittemore and D. Hodges, *J. Pharm. Exp. Ther.*, 2003, **306**, 377.

193. D. M. Swanson, A. E. Dublin, C. Shah, N. Nasser, L. Chang, S. L. Dax, M. Jetter, J. G. Breitenbucher, C. Liu, C. Mazur, B. Lord, L. Gonzales, K. Hoey, M. Rizzolio, M. Bogenstetter, E. E. Codd, D. H. Lee, S. P. Zhang, S. R. Chaplan and N. I. Carruthers, *J. Med. Chem.*, 2005, **48**, 1857.

194. M. Crutchlow, presented at the American Society Clinical Pharmacology Therapeutics (ASCPT) Annual Meeting, 2009.
195. A. R. Moritz and F. C. Henriques Jr, *Arch. Pathol.*, 1947, **43**, 466.
196. W. S. Denney, Y. Hang, M. Dockendorf, C.-C. Li, S. R. Eid, R. Valesky, T. Laethem, P. Van Hoydonck, I. De Lepeliere, J. N. J. M. De Hoon, M. Cruthlow and R. Blanchard, presented at the Annual Meeting of the Population Approach Group in Europe, 2009.
197. A. Gomtsyan, E. K. Bayburt, R. G. Schmidt, G. Z. Zheng, R. J. Perner, S. Didomenicio, J. R. Koenig, S. Turner, T. Jinkerson, I. Drizin, S. M. Hannick, B. S. Macri, H. A. McDonald, P. Honore, C. T. Wismer, K. C. Marsh, J. Wetter, K. D. Stewart, T. Oie, M. F. Jarvis, C. S. Surowy, C. R. Faltynek and C.-H. Lee, *J. Med. Chem.*, 2005, **48**, 744.
198. A. Gomtsyan, E. K. Bayburt, R. Keddy, S. C. Turner, T. K. Jinkerson, S. Didomenico, R. J. Perner, J. R. Koenig, I. Drizin, H. A. McDonald, C. S. Surowy, P. Honore, J. Mikusa, K. C. Marsch, J. M. Wetter, C. R. Faltynek and C.-H. Lee, *Bioorg. Med. Chem. Lett.*, 2007, **17**, 3894.
199. C. S. Surowy, T. R. Neelands, B. R. Bianchi, S. McGaraughty, R. El Kouhen, P. Han, K. L. Chu, H. A. McDonald, M. Vos, W. Niforatos, E. K. Bayburt, A. Gomtsyan, C. H. Lee, P. Honore, J. P. Sullivan, M. F. Jarvis and C. R. Faltynek, *J. Pharmacol. Exp. Ther.*, 2008, **326**, 879.
200. K. J. Frank, U. Westedt, K. M. Rosenblatt, P. Hölig, J. Rosenberg, M. Mägerlein, M. Brandl and G. Fricker, *Eur J Pharm Sci.*, 2012, **47**, 16.
201. A. A. Othman, W. Nothaft, W. M. Awni and S. Dutta, *J. Clin. Pharmacol.*, 2012, **52**, 1028.
202. M. C. Rowbotham, W. Nothaft, W. R. Duan, Y. Wang, C. Faltynek, S. McGaraughty, K. L. Chu and P. Svensson, *Pain*, 2011, **152**, 1192.
203. K. Schaffler, P. Reeh, W. R. Duan, A. E. Best, A. A. Othman, C. R. Faltynek, C. Locke and W. Nothaft, *Br. J. Clin. Pharmacol.*, 2013, **75**, 404.
204. J. M. Laird, G. Martino, S. Visser, S. Oerther, X. H. Yu and M. N. Perkins, *13th World Congress on Pain*, IASP Press, 2010, PH 379.
205. R. Karlsten, B. Jonzon, H. Quiding, M. A. Carlsson, M. Segerdahl, R. Malamut, M. Björnsson and F. Miller, *13th World Congress on Pain*, IASP Press, 2010, PM 380.
206. H. Quiding, B. Jonzon, O. Svensson, L. Webster, A. Reimfelt, A. Karin, R. Karlsten and M. Segerdahl, *Pain*, 2013, **154**, 808.
207. O. Svensson, C. Thorne, F. Miller, M. Bjornsson, A. Reimfelt and R. Karlstén, *13th World Congress on Pain*, IASP Press, 2010, PM 379.
208. A. L. Krarup, L. Ny, M. Astrand, A. Bajor, F. Hvid-Jensen, M. B. Hansen, M. Simrén, P. Funch-Jensen and A. M. Drewes, *Aliment. Pharmacol. Ther.*, 2011, **33**, 1113.
209. E. Szoke, K. Boelcskei, K. Kvell, B. Bender, Z. Bosze, J. Szolcsanyi and Z. Sandor, *Cell. Mol. Life Sci.*, 2011, **68**, 2589.
210. A. A. Steiner, V. F. Turek, M. C. Almeida, J. J. Burmeister, D. L. Oliveira, J. L. Roberts, A. W. Bannon, M. H. Norman, J.-C. Louis, J. J. S. Treanor, N. R. Gavva and A. A. Romanovsky, *J. Neurosci.*, 2007, **27**, 7459.

211. S. McGaraughty, J. A. Segreti, R. M. Fryer, B. S. Brown, C. R. Faltynek and P. R. Kym, *Brain Res.*, 2009, **1268**, 58.
212. A. R. Gomtsyan, E. A. Voight, E. K. Bayburt, J. Chen, J. F. Daanen, S. DiDomenico, M. E. Kort, P. R. Kym, H. McDonald, R. J. Perner and R. G. Schmidt, Preparation of chromene urea compounds as therapeutic TRPV1 antagonists, WO 2010/045401, 2010.
213. (a) A. R. Gomtsyan, E. A. Voight, E. K. Bayburt, J. Chen, J. F. Daanen, S. DiDomenico, M. E. Kort, P. R. Kym, H. McDonald, R. J. Perner and R. G. Schmidt, Preparation of chromene urea compounds as therapeutic TRPV1 antagonists, WO 2010/045402, 2010; (b) M. E. Kort and P. R. Kym, *Prog. Med. Chem.*, 2012, **51**, 57.
214. R. M. Reilly, H. A. McDonald, P. S. Puttfarcken, S. K. Joshi, L. Lewis, M. Pai, P. H. Franklin, J. A. Segreti, T. R. Neelands, P. Han, J. Chen, P. W. Mantyh, J. R. Ghilardi, T. M. Turner, E. A. Voight, J. F. Daanen, R. G. Schmidt, A. Gomtsyan, M. E. Kort, C. R. Faltynek and P. R. Kym, *J. Pharmacol. Exp. Ther.*, 2012, **342**, 416.
215. A. Garami, Y. P. Shimansky, E. Pakai, D. L. Oliveira, N. R. Gavva and A. A. Romanovsky, *J. Neurosci.*, 2010, **30**, 1435.
216. R. Frank, G. Bahrenberg, T. Christoph, K. Schiene, J. DeVry, D. Saunders, M. Przewosny, B. Saundermann and J. Lee, Preparation of N-benzyl-2-phenylpropanamides as vanilloid receptor antagonists, WO 2008/125337, 2008.
217. R. Frank, G. Bahrenberg, T. Christoph, K. Schiene, J. DeVry, D. Saunders, B. Saundermann and J. Lee, Preparation of N-benzyl-2-phenylpropanamides as vanilloid receptor antagonists, WO 2008/125342, 2008.
218. R. Frank, G. Bahrenberg, T. Christoph, K. Schiene, J. DeVry, N. Damann, S. Frormann, B. Lesch, J. Lee, Y.-S. Kim and M.-S. Kim, Preparation of substituted aromatic carboxamide and urea derivatives as vanilloid receptor ligands for the treatment of pain, WO 2010/127855, 2010.
219. R. Frank, G. Bahrenberg, T. Christoph, K. Schiene, J. DeVry, N. Damann, S. Frormann, B. Lesch, J. Lee, Y.-S. Kim, M.-S. Kim, Preparation of substituted phenylamine derivatives for use as vanilloid receptor ligands, WO2010/127856, 2010.
220. P. G. Baraldi, P. A. Borea, P. Geppetti, M. G. Pavani, F. Fruttarolo and M. Trevisani, Preparation of N-aryl biarylcarboxamides as vanilloid-1 receptor modulators, WO 2008/006480, 2008.
221. M. Napoletano, M. G. Pavani, F. Fruttarolo and M. Trevisani, Preparation of nicotinamide N-oxide derivatives as vanilloid-1 receptor modulators, WO 2009/043582, 2009.
222. L. J. Wu, T. B. Sweet and D. E. Clapham, *Pharmacol. Rev.*, 2010, **62**, 381.
223. A. M. Peier, A. J. Reeve, D. A. Andersson, A. Moqrich, T. J. Earley, A. C. Hergarden, G. M. Story, S. Colley, J. B. Hogenesch, P. McIntyre, S. Bevan and A. Patapoutian, *Science*, 2002, **296**, 2046.
224. G. D. Smith, M. J. Gunthorpe, R. E. Kelsell, P. D. Hayes, P. Reilly, P. Facer, J. E. Wright, J. C. Jerman, J. P. Walhin, L. Ooi, J. Egerton,

K. J. Charles, D. Smart, A. D. Randall, P. Anand and J. B. Davis, *Nature*, 2002, **418**, 186.

225. H. Xu, I. S. Ramsey, S. A. Kotecha, M. M. Moran, J. A. Chong, D. Lawson, P. Ge, J. Lilly, I. Silos-Santiago, Y. Xie, P. S. DiStefano, R. Curtis and D. E. Clapham, *Nature*, 2002, **418**, 181.
226. A. Patapoutian, A. M. Peier, G. M. Story and V. Viswanath, *Nat. Rev. Neurosci.*, 2003, **4**, 529.
227. R. Vennekens, G. Owsianik and B. Nilius, *Curr. Pharma. Design*, 2008, **14**, 18.
228. U. Park, N. Vastani, Y. Guan, S. N. Raja, M. Koltzenburg and M. J. Caterina, *J. Neurosci.*, 2011, **31**, 11425.
229. M. J. Caterina, T. A. Rosen, M. Tominaga, A. J. Brake and D. Julius, *Nature*, 1999, **398**, 436.
230. M. P. Neeper, Y. Liu, T. L. Hutchinson, Y. Wang, C. M. Flores and N. Qin, *J. Biol. Chem.*, 2007, **282**, 15894.
231. A. Moqrich, S. W. Hwang, T. J. Earley, M. J. Petrus, A. N. Murray, K. S. Spencer, M. Andahazy, G. M. Story and A. Patapoutian, *Science*, 2005, **307**, 1468.
232. S. M. Huang, X. Li, Y. Yu, J. Wang and M. J. Caterina, *Mol. Pain*, 2011, **7**, 37.
233. S. Bang, S. Yoo, T. J. Yang, H. Cho and S. W. Hwang, *Br. J. Pharmacol.*, 2012, **165**, 683.
234. R. M. Reilly and P. R. Kym, *Curr. Topics Med. Chem.*, 2011, **11**, 2210.
235. X. Cheng, J. Jin, L. Hu, D. Shen, X. P. Dong, M. A. Samie, J. Knoff, B. Eisinger, M. L. Liu, S. M. Huang, M. J. Caterina, P. Dempsey, L. E. Michael, A. A. Dlugosz, N. C. Andrews, D. E. Clapham and H. Xu, *Cell*, 2010, **141**, 331.
236. W. Liedtke, Y. Choe, M. A. Marti-Renom, A. M. Bell, C. S. Denis, A. Sali, A. J. Hudspeth, J. M. Friedman and S. Heller, *Cell*, 2000, **103**, 525.
237. R. Strotmann, C. Harteneck, K. Nunnenmacher, G. Schultz and T. D. Plant, *Nat. Cell Biol.*, 2000, **2**, 695.
238. W. Liedtke and J. M. Friedman, *Proc. Natl. Acad. Sci. U. S. A.*, 2003, **100**, 13698.
239. A. Mizuno, N. Matsumoto, M. Imai and M. Suzuki, *Am. J. Physiol.*, 2003, **285**, C96.
240. S. G. Lechner, S. Markworth, K. Poole, E. S. Smith, L. Lapatsina, S. Frahm, M. May, S. Pischke, M. Suzuki, I. Ibanez-Tallon, F. C. Luft, J. Jordan and G. R. Lewin, *Neuron*, 2011, **69**, 332.
241. J. Taniguchi, S. Tsuruoka, A. Mizuno, J. Sato, A. Fujimura and M. Suzuki, *Am. J. Physiol.*, 2007, **292**, F667.
242. T. Gevaert, J. Vriens, A. Segal, W. Everaerts, T. Roskams, K. Talavera, G. Owsianik, W. Liedtke, D. Daelemans, I. Dewachter, F. Van Leuven, T. Voets, D. De Ridder and B. Nilius, *J. Clin. Invest.*, 2007, **117**, 3453.
243. R. N. Willette, W. Bao, S. Nerurkar, T. L. Yue, C. P. Doe, G. Stankus, G. H. Turner, H. Ju, H. Thomas, C. E. Fishman, A. Sulpizio, D. J. Behm,

S. Hoffman, Z. Lin, I. Lozinskaya, L. N. Casillas, M. Lin, R. E. Trout, B. J. Votta, K. Thorneloe, E. S. Lashinger, D. J. Figueroa, R. Marquis and X. Xu, *J. Pharmacol. Exp. Ther.*, 2008, **326**, 443.

244. K. Tabuchi, M. Suzuki, A. Mizuno and A. Hara, *Neurosci. Lett.*, 2005, **382**, 304.

245. K. Hamanaka, M. Y. Jian, D. S. Weber, D. F. Alvarez, M. I. Townsley, A. B. Al-Mehdi, J. A. King, W. Liedtke and J. C. Parker, *Am. J. Physiol.*, 2007, **293**, L923.

246. K. Hamanaka, M. Y. Jian, M. I. Townsley, J. A. King, W. Liedtke, D. S. Weber, F. G. Eyal, M. M. Clapp and J. C. Parker, *Am. J. Physiol.*, 2010, **299**, L353.

247. M. Suzuki, A. Mizuno, K. Kodaira and M. Imai, *J. Biol. Chem.*, 2003, **278**, 22664.

248. N. Alessandri-Haber, O. A. Dina, J. J. Yeh, C. A. Parada, D. B. Reichling and J. D. Levine, *J. Neurosci.*, 2004, **24**, 4444.

249. N. Alessandri-Haber, O. A. Dina, E. K. Joseph, D. Reichling and J. D. Levine, *J. Neurosci.*, 2006, **26**, 3864.

250. N. Alessandri-Haber, E. Joseph, O. A. Dina, W. Liedtke and J. D. Levine, *Pain*, 2005, **118**, 70.

251. H. Todaka, J. Taniguchi, J. Satoh, A. Mizuno and M. Suzuki, *J. Biol. Chem.*, 2004, **279**, 35133.

252. H. Lee, T. Iida, A. Mizuno, M. Suzuki and M. J. Caterina, *J. Neurosci.*, 2005, **25**, 1304.

253. F. Vincent and M. A. Duncton, *Curr. Topics Med. Chem.*, 2011, **11**, 2216.

254. F. Vincent, A. Acevedo, M. T. Nguyen, M. Dourado, J. DeFalco, A. Gustafson, P. Spiro, D. E. Emerling, M. G. Kelly and M. A. Duncton, *Biochem. Biophy. Res. Commun.*, 2009, **389**, 490.

255. Y. Chen, C. Yang and Z. J. Wang, *Neuroscience*, 2011, **193**, 440.

256. W. Everaerts, X. Zhen, D. Ghosh, J. Vriens, T. Gevaert, J. P. Gilbert, N. J. Hayward, C. R. McNamara, F. Xue, M. M. Moran, T. Strassmaier, E. Uykal, G. Owsianik, R. Vennekens, D. De Ridder, B. Nilius, C. M. Fanger and T. Voets, *Proc. Natl. Acad. Sci. U. S. A.*, 2010, **107**, 19084.

257. S. Materazzi, C. Fusi, S. Benemei, P. Pedretti, R. Patacchini, B. Nilius, J. Prenen, C. Creminon, P. Geppetti and R. Nassini, *Pflugers Arch., EJP*, 2012, **463**, 561.

258. H. X. Deng, C. J. Klein, J. Yan, Y. Shi, Y. Wu, F. Fecto, H. J. Yau, Y. Yang, H. Zhai, N. Siddique, E. T. Hedley-Whyte, R. Delong, M. Martina, P. J. Dyck and T. Siddique, *Nat. Genet.*, 2010, **42**, 165.

259. G. Landoure, A. A. Zdebik, T. L. Martinez, B. G. Burnett, H. C. Stanescu, H. Inada, Y. Shi, A. A. Taye, L. Kong, C. H. Munns, S. S. Choo, C. B. Phelps, R. Paudel, H. Houlden, C. L. Ludlow, M. J. Caterina, R. Gaudet, R. Kleta, K. H. Fischbeck and C. J. Sumner, *Nat. Genet.*, 2010, **42**, 170.

260. L. M. Duncan, J. Deeds, J. Hunter, J. Shao, L. M. Holmgren, E. A. Woolf, R. I. Tepper and A. W. Shyjan, *Cancer Res.*, 1998, **58**, 1515.

261. T. F. Wagner, S. Loch, S. Lambert, I. Straub, S. Mannebach, I. Mathar, M. Dufer, A. Lis, V. Flockerzi, S. E. Philipp and J. Oberwinkler, *Nat. Cell Biol.*, 2008, **10**, 1421.

262. C. Grimm, R. Kraft, G. Schultz and C. Harteneck, *Mol. Pharmacol.*, 2005, **67**, 798.

263. C. Grimm, R. Kraft, S. Sauerbruch, G. Schultz and C. Harteneck, *J. Biol. Chem.*, 2003, **278**, 21493.

264. N. Lee, J. Chen, L. Sun, S. Wu, K. R. Gray, A. Rich, M. Huang, J. H. Lin, J. N. Feder, E. B. Janovitz, P. C. Levesque and M. A. Blanar, *J. Biol. Chem.*, 2003, **278**, 20890.

265. J. Vriens, G. Owsianik, T. Hofmann, S. E. Philipp, J. Stab, X. Chen, M. Benoit, F. Xue, A. Janssens, S. Kerselaers, J. Oberwinkler, R. Vennekens, T. Gudermann, B. Nilius and T. Voets, *Neuron*, 2011, **70**, 482.

266. J. Naylor, C. J. Milligan, F. Zeng, C. Jones and D. J. Beech, *Br. J. Pharmacol.*, 2008, **155**, 567.

267. J. Naylor, J. Li, C. J. Milligan, F. Zeng, P. Sukumar, B. Hou, A. Sedo, N. Yuldasheva, Y. Majeed, D. Beri, S. Jiang, V. A. Seymour, L. McKeown, B. Kumar, C. Harteneck, D. O'Regan, S. B. Wheatcroft, M. T. Kearney, C. Jones, K. E. Porter and D. J. Beech, *Circ. Res.*, 2010, **106**, 1507.

268. C. Klose, I. Straub, M. Riehle, F. Ranta, D. Krautwurst, S. Ullrich, W. Meyerhof and C. Harteneck, *Br. J. Pharmacol.*, 2011, **162**, 1757.

269. L. Tsavaler, M. H. Shapero, S. Morkowski and R. Laus, *Cancer Res.*, 2001, **61**, 3760.

270. L. Zhang and G. J. Barritt, *Cancer Res.*, 2004, **64**, 8365.

271. D. D. McKemy, W. M. Neuhausser and D. Julius, *Nature*, 2002, **416**, 52.

272. A. M. Peier, A. Moqrich, A. C. Hergarden, A. J. Reeve, D. A. Andersson, G. M. Story, T. J. Earley, I. Dragoni, P. McIntyre, S. Bevan and A. Patapoutian, *Cell*, 2002, **108**, 705.

273. D. M. Bautista, J. Siemens, J. M. Glazer, P. R. Tsuruda, A. I. Basbaum, C. L. Stucky, S. E. Jordt and D. Julius, *Nature*, 2007, **448**, 204.

274. R. W. Colburn, M. L. Lubin, D. J. Stone Jr, Y. Wang, D. Lawrence, M. R. D'Andrea, M. R. Brandt, Y. Liu, C. M. Flores and N. Qin, *Neuron*, 2007, **54**, 379.

275. A. Dhaka, A. N. Murray, J. Mathur, T. J. Earley, M. J. Petrus and A. Patapoutian, *Neuron*, 2007, **54**, 371.

276. J. DeFalco, M. A. Duncton and D. Emerling, *Curr. Topics Med. Chem.*, 2011, **11**, 2237.

277. D. Preti, A. Szallasi and R. Patacchini, *Expert Opin. Ther. Pat.*, 2012, **22**, 663.

278. W. M. Knowlton, R. L. Daniels, R. Palkar, D. D. McCoy and D. D. McKemy, *PloS one*, 2011, **6**, e25894.

279. N. R. Gavva, C. Davis, S. G. Lehto, S. Rao, W. Wang and D. X. Zhu, *Mol. Pain*, 2012, **8**, 36.

280. D. N. Ruskin, R. Anand and G. J. LaHoste, *Eur. J. Pharmacol.*, 2007, **559**, 161.

281. Z. Ding, T. Gomez, J. L. Werkheiser, A. Cowan and S. M. Rawls, *Eur. J. Pharmacol.*, 2008, **578**, 201.
282. D. Jaquemar, T. Schenker and B. Trueb, *J. Biol. Chem.*, 1999, **274**, 7325.
283. G. M. Story, A. M. Peier, A. J. Reeve, S. R. Eid, J. Mosbacher, T. R. Hricik, T. J. Earley, A. C. Hergarden, D. A. Andersson, S. W. Hwang, P. McIntyre, T. Jegla, S. Bevan and A. Patapoutian, *Cell*, 2003, **112**, 819.
284. M. Bandell, G. M. Story, S. W. Hwang, V. Viswanath, S. R. Eid, M. J. Petrus, T. J. Earley and A. Patapoutian, *Neuron*, 2004, **41**, 849.
285. S. E. Jordt, D. M. Bautista, H. H. Chuang, D. D. McKemy, P. M. Zygmunt, E. D. Hogestatt, I. D. Meng and D. Julius, *Nature*, 2004, **427**, 260.
286. D. M. Bautista, S. E. Jordt, T. Nikai, P. R. Tsuruda, A. J. Read, J. Poblete, E. N. Yamoah, A. I. Basbaum and D. Julius, *Cell*, 2006, **124**, 1269.
287. K. Y. Kwan, A. J. Allchorne, M. A. Vollrath, A. P. Christensen, D. S. Zhang, C. J. Woolf and D. P. Corey, *Neuron*, 2006, **50**, 277.
288. Y. Karashima, K. Talavera, W. Everaerts, A. Janssens, K. Y. Kwan, R. Vennekens, B. Nilius and T. Voets, *Proc. Natl. Acad. Sci. U. S. A.*, 2009, **106**, 1273.
289. R. Madrid, E. de la Pena, T. Donovan-Rodriguez, C. Belmonte and F. Viana, *J. Neurosci.*, 2009, **29**, 3120.
290. C. Gentry, N. Stoakley, D. A. Andersson and S. Bevan, *Mol. Pain*, 2010, **6**, 4.
291. W. M. Knowlton, A. Bifolck-Fisher, D. M. Bautista and D. D. McKemy, *Pain*, 2010, **150**, 340.
292. R. Sharif-Naeini, A. Dedman, J. H. Folgering, F. Duprat, A. Patel, B. Nilius and E. Honore, *Pflugers Arch., EJP*, 2008, **456**, 529.
293. D. Vilceanu and C. L. Stucky, *PloS One*, 2010, **5**, e12177.
294. L. J. Macpherson, B. H. Geierstanger, V. Viswanath, M. Bandell, S. R. Eid, S. Hwang and A. Patapoutian, *Curr. Biol.*, 2005, **15**, 929.
295. L. J. Macpherson, B. Xiao, K. Y. Kwan, M. J. Petrus, A. E. Dubin, S. Hwang, B. Cravatt, D. P. Corey and A. Patapoutian, *J. Neurosci.*, 2007, **27**, 11412.
296. C. R. McNamara, J. Mandel-Brehm, D. M. Bautista, J. Siemens, K. L. Deranian, M. Zhao, N. J. Hayward, J. A. Chong, D. Julius, M. M. Moran and C. M. Fanger, *Proc. Natl. Acad. Sci. U. S. A.*, 2007, **104**, 13525.
297. J. A. Matta, P. M. Cornett, R. L. Miyares, K. Abe, N. Sahibzada and G. P. Ahern, *Proc. Natl. Acad. Sci. U. S. A.*, 2008, **105**, 8784.
298. A. Hinman, H. H. Chuang, D. M. Bautista and D. Julius, *Proc. Natl. Acad. Sci. U. S. A.*, 2006, **103**, 19564.
299. L. J. Macpherson, A. E. Dubin, M. J. Evans, F. Marr, P. G. Schultz, B. F. Cravatt and A. Patapoutian, *Nature*, 2007, **445**, 541.
300. T. E. Taylor-Clark, B. J. Undem, D. W. Macglashan Jr., S. Ghatta, M. J. Carr and M. A. McAlexander, *Mol. Pharmacol.*, 2008, **73**, 274.
301. D. A. Andersson, C. Gentry, S. Moss and S. Bevan, *J. Neurosci.*, 2008, **28**, 2485.

302. T. E. Taylor-Clark, M. A. McAlexander, C. Nassenstein, S. A. Sheardown, S. Wilson, J. Thornton, M. J. Carr and B. J. Undem, *J. Physiol.*, 2008, **586**, 3447.
303. Y. Dai, S. Wang, M. Tominaga, S. Yamamoto, T. Fukuoka, T. Higashi, K. Kobayashi, K. Obata, H. Yamanaka and K. Noguchi, *J. Clin. Invest.*, 2007, **117**, 1979.
304. S. R. Wilson, K. A. Gerhold, A. Bifolck-Fisher, Q. Liu, K. N. Patel, X. Dong and D. M. Bautista, *Nat. Neurosci.*, 2011, **14**, 595.
305. S. M. Brierley, P. A. Hughes, A. J. Page, K. Y. Kwan, C. M. Martin, T. A. O'Donnell, N. J. Cooper, A. M. Harrington, B. Adam, T. Liebregts, G. Holtmann, D. P. Corey, G. Y. Rychkov and L. A. Blackshaw, *Gastroenterology*, 2009, **137**, 2084.
306. F. Cattaruzza, I. Spreadbury, M. Miranda-Morales, E. F. Grady, S. Vanner and N. W. Bunnett, *Am. J. Physiol.*, 2010, **298**, G81.
307. E. Ceppa, F. Cattaruzza, V. Lyo, S. Amadesi, J. C. Pelayo, D. P. Poole, N. Vaksman, W. Liedtke, D. M. Cohen, E. F. Grady, N. W. Bunnett and K. S. Kirkwood, *Am. J. Physiol.*, 2010, **299**, G556.
308. E. S. Schwartz, J. A. Christianson, X. Chen, J. H. La, B. M. Davis, K. M. Albers and G. F. Gebhart, *Gastroenterology*, 2011, **140**, 1283.
309. A. I. Caceres, M. Brackmann, M. D. Elia, B. F. Bessac, D. del Camino, M. D'Amours, J. S. Witek, C. M. Fanger, J. A. Chong, N. J. Hayward, R. J. Homer, L. Cohn, X. Huang, M. M. Moran and S. E. Jordt, *Proc. Natl. Acad. Sci. U. S. A.*, 2009, **106**, 9099.
310. K. Raemdonck, J. de Alba, M. A. Birrell, M. Grace, S. A. Maher, C. G. Irvin, J. R. Fozard, P. M. O'Byrne and M. G. Belvisi, *Thorax*, 2012, **67**, 19.
311. J. Chen, S. K. Joshi, S. DiDomenico, R. J. Perner, J. P. Mikusa, D. M. Gauvin, J. A. Segreti, P. Han, X. F. Zhang, W. Niforatos, B. R. Bianchi, S. J. Baker, C. Zhong, G. H. Simler, H. A. McDonald, R. G. Schmidt, S. P. McGaraughty, K. L. Chu, C. R. Faltynek, M. E. Kort, R. M. Reilly and P. R. Kym, *Pain*, 2011, **152**, 1165.
312. C. E. Deering-Rice, E. G. Romero, D. Shapiro, R. W. Hughen, A. R. Light, G. S. Yost, J. M. Veranth and C. A. Reilly, *Chem. Res. Toxicol.*, 2011, **24**, 950.
313. M. S. Hazari, N. Haykal-Coates, D. W. Winsett, Q. T. Krantz, C. King, D. L. Costa and A. K. Farraj, *Environ. Health Perspect.*, 2011, **119**, 951.
314. S. R. Eid, E. D. Crown, E. L. Moore, H. A. Liang, K. C. Choong, S. Dima, D. A. Henze, S. A. Kane and M. O. Urban, *Mol. Pain*, 2008, **4**, 48.
315. H. Wei, M. M. Hamalainen, M. Saarnilehto, A. Koivisto and A. Pertovaara, *Anesthesiology*, 2009, **111**, 147.
316. L. Klionsky, R. Tamir, B. Gao, W. Wang, D. C. Immke, N. Nishimura and N. R. Gavva, *Mol. Pain*, 2007, **3**, 39.
317. M. Petrus, A. M. Peier, M. Bandell, S. W. Hwang, T. Huynh, N. Olney, T. Jegla and A. Patapoutian, *Mol. Pain*, 2007, **3**, 40.
318. S. McGaraughty, K. L. Chu, R. J. Perner, S. Didomenico, M. E. Kort and P. R. Kym, *Mol. Pain*, 2010, **6**, 14.

CHAPTER 10

Open Access to the KCNQ Channel: Retigabine and Second Generation M-current Openers

JOHANNES KRUPP,[a] ANTHONY M. RUSH,[a]
BRITT-MARIE SWAHN[a] AND MARTIN MAIN*[b]

[a] AstraZeneca R&D Södertälje, Sweden; [b] Discovery Sciences,
Alderley Park, UK
*Email: Martin.Main@astrazeneca.com

10.1 Introduction

Potassium channels form the largest group of ion channels in the human genome, with approximately 78 pore-forming 'alpha' subunits and 13 auxiliary 'beta' subunits identified to date.[1] Heteromerization between different subunits and the existence of alternative splicing leads to an additional level of diversity. Within the potassium channel classification, voltage-gated potassium channels (Kv) comprise the largest sub-family. Structurally, they are characterized by six trans-membrane (TM) domains and one pore-forming loop and they assemble into tetramers to form functional ion channels. Voltage-gated K^+ channels are expressed in all excitable cells, where they have an essential role in action potential repolarisation, but they are also found in non-excitable cells such as the

RSC Drug Discovery Series No. 39
Ion Channel Drug Discovery
Edited by Brian Cox and Martin Gosling
© The Royal Society of Chemistry 2015
Published by the Royal Society of Chemistry, www.rsc.org

epithelia of the gastrointestinal tract and immune cells where they play a role in fluid secretion and cell proliferation, respectively.

A large number of voltage-gated potassium channel blockers have developed, many of which have been used clinically to treat cardiac arrhythmia and hypertension. This chapter focuses on a more recent advance, namely the development of potassium channel openers for the treatment of epilepsy, pain and other disorders. Of the classical 6 TM voltage-gated K^+ channels, to date, only one voltage-gated potassium channel family has proved amenable to identification of channels openers—the KCNQ or Kv7 family.

10.2   KCNQ (Kv7) Potassium Channel Family

The KCNQ (Kv7) potassium channel comprises five members, named KCNQ1-5 (Kv7.1-7.5). KCNQ1 (Kv7.1, KVLQT1), the founding member of the family was identified through positional cloning at a locus for hereditary long QT syndrome, a cardiac disorder.[2] Expression of KCNQ1 in Xenopus oocytes or HEK293 cells revealed a voltage-gated potassium channel that activated at relatively negative membrane potentials, exhibited slow activation and a complete absence of inactivation. The biophysical signature and function of KCNQ1 varies across different tissues. Thus, in cardiac muscle KCNQ1 assembles with a single trans-membrane auxiliary sub-unit, KCNE1, to reconstitute the cardiac slow delayed rectifier current (I_{KS}). In contrast, in gut epithelial cells, KCNQ1 assembles with KCNE3 to form a constitutively active potassium channel that mediates recycling of potassium ions at the basolateral membrane.

Subsequently, two additional family members, KCNQ2 and KCNQ3, were identified through positional cloning in studies of an inherited form of epilepsy, Benign Familial Neonatal Convulsions (BFNC). Two loci were mapped for BFNC by linkage analysis—EBN1 on chromosome 20q13 maps to KCNQ2[3,4] and EBN2 on chromosome 8q24 maps to KCNQ3.[5] In both, KCNQ2 and KCNQ3, the BFNC mutation lies in the pore region of the channel. Functional analysis of mutant channels (KCNQ2 Y284C; KCNQ3 G310V), recombinantly expressed in Xenopus oocytes, revealed that the BFNC mutations decrease whole-cell current by approximately 20%, without affecting selectivity or gating kinetics.[6] Northern blot and *in situ* hybridisation experiments revealed co-expression of KCNQ2 and KCNQ3 in many brain regions,[6] suggesting the formation of KCNQ2/KCNQ3 heteromers. This is supported by functional experiments in Xenopus oocytes and CHO cells, where co-expression of KCNQ2 and KCNQ3[6–8] leads to a large increase in KCNQ currents. Likewise, expression of dominant negative mutations of either subunit reduced currents from KCNQ2/KCNQ3 heteromeric channels.[6] Modulation of surface expression is thought to underlie the increase in functional expression observed following KCNQ2/KCNQ3 co-expression.[9] Currents of KCNQ2/KCNQ3 heteromers resemble M-currents (see below) in their biophysics and sensitivity to pharmacological inhibitors such as linopirdine and XE991.[7]

Two further KCNQ channel subtypes have been identified: KCNQ4 was cloned following a screen of a human retina cDNA library using a KCNQ3 probe.[10] Similarly, KCNQ5 was identified by screening a thalamus cDNA library.[11] Functional analysis of these channels revealed a similar biophysical signature—slow activation and lack of time-dependent inactivation. Both KCNQ4 and KCNQ5 can form heteromeric channels with KCNQ3.

10.3   KCNQ Channels Underlie M-current

With the exception of KCNQ1, all of the KCNQ family members are expressed in neuronal tissue, where they reconstitute the low-threshold, non-inactivating 'M-current (I_M)' that was first described by Brown and Adams[12] in bullfrog sympathetic neurons. Stimulation of muscarinic receptors in this neuronal preparation causes a dramatic change in membrane excitability, such that quiescent neurons begin to tonically fire action potentials. It was demonstrated that the mechanism underlying this change in excitability is suppression of a time and voltage-dependent potassium current, the 'M' current where 'M' refers to muscarinic. M-current was subsequently measured in a wide range of cell-types, where it is dominant in regulating membrane excitability.

In most neurons, M-channels are composed of KCNQ2 plus KCNQ3 subunits,[7] although a contribution from KCNQ2 homomeric channels[13,14] and KCNQ5 channels[15] has also been demonstrated. KCNQ4 is predominantly expressed in auditory and vestibular pathways, but also probably contributes to M-current in central dopaminergic neurons.

The unique biophysical properties of M-current mean that its presence has a profound effect upon cell physiology. Thus, M-channels activate at membrane potentials below the threshold for action potential firing and do not exhibit inactivation, giving rise to a steady-state outward current at membrane potentials positive to −60mV.

In central neurons, M-current plays a key role in controlling action potential threshold and in suppression of repetitive firing. Immuno-histochemistry experiments in hippocampal neurons have revealed that the highest density of KCNQ2 and KCNQ3 channels lies in the initial axon segments, where the action potential is generated.[16] Indeed, these channels have been shown to co-localize with voltage-gated sodium channels through binding to the cytoskeletal protein ankyrin.[17] M-current suppresses action potential firing in two ways—by hyperpolarizing the resting membrane potential away from the threshold for action potential firing, and by decreasing the input resistance of the cell. The consequence of these changes is that a larger depolarizing stimulus is required to activate a neuronal action potential. These mechanisms are illustrated by experiments using the small molecule inhibitor XE991: inhibition of M-current leads to depolarization of the resting membrane potential and spontaneous action potential firing.[18] Similar results were reported using a peptide that inhibits KCNQ/ankyrin binding, thereby demonstrating the importance of sub-cellular localization.[19] M-current also plays a key role in regulating repetitive action potential firing: the activation kinetics of M-current are too slow for this current to play a role in action potential repolarization. However, during a train of action potentials cumulative activation of M-current takes place and suppresses subsequent action potential discharge.

Similar mechanisms have been reported in sympathetic and sensory neurons. Thus, rat sympathetic neurons exhibit phasic behavior in response to a depolarizing stimulus, firing a single action potential. Inhibition of M-current with XE991 leads to a shift to phasic firing whereby depolarization triggers a train of action potentials. Rat DRG neurons express KCNQ2, 3 and 5 sub-units and exhibit clear M-currents.[18] Modulation of M-current alters transmission of Aδ and C-fiber responses into the dorsal horn of the spinal cord.

A key feature of the M-current, indeed a feature that led to its naming, is inhibition following muscarinic receptor activation. A wide range of neurotransmitter receptors including muscarinic, 5-HT, mGlu and opioid receptors, have been shown to be coupled to M-current. Most, but not all, receptors that inhibit M-current are coupled to G_q/G_{11} G-proteins and hydrolysis of PIP_2. Intriguingly, evidence suggests that it is depletion of PIP2 in the membrane, rather than release of the products of hydrolysis—IP_3 and DAG—that inhibits the M-current.[20–22] Thus, PIP2 has been shown to bind directly to the C-terminus of KCNQ channels and muscarinic receptor induced inhibition of M-current is reduced or prevented when PIP2 levels are elevated by over-expressing PI5 kinase. As outlined above, suppression of M-current leads to an increase in neuronal excitability and synaptic suppression of M-current underlies the slow EPSP in frog and rat sympathetic neurons and in hippocampal CA3 neurons.

Given that M-current plays a key role in regulation of neuronal excitability, and that KCNQ2/3 channels have been genetically linked to epilepsy, the development of KCNQ2/3 channel activators is an attractive mechanism for developing therapeutic agents for treatment of disorders of neuronal excitability. As described below, significant progress has been made in this area, starting with the development of retigabine.

10.4 Retigabine: Discovery, Molecular Target and Mechanism of Action

Retigabine (D-20443), a structural analogue of the non-opioid analgesic and muscle relaxant flupirtine, was discovered by researchers at ASTA Medica AG and was first reported as a novel anti-convulsant in 1993.[23,24] Using sharp electrode recordings in rat hippocampal slices, the authors reported that retigabine abolished the stimulatory effects of 4-AP on action potential bursting and EPSP duration.[23–25]

The first indication of retigabine's mechanism of action was provided by patch-clamp analysis in NG108-15 neuronal cells, primary mouse cortical neurons and differentiated hNT cells.[26] In each of these preparations, retigabine application led to a dose-dependent increase in potassium conductance. This effect was not seen with the established anti-convulsants phenytoin, carbamazepine and sodium valproate. A more detailed analysis in NGF-treated PC12 cells revealed that retigabine activated a Ba^{2+} and TEA-sensitive/4-AP insensitive potassium current.[27]

The molecular target of retigabine was finally revealed as the M-current when the compound was tested against recombinant KCNQ2/3 channels expressed in CHO cells and Xenopus oocytes.[8,28,29] Retigabine activates KCNQ2/3 channels and native M-current by activating a hyperpolarizing shift in the voltage dependence of activation.[8,29,30] The importance of the M-current in setting resting membrane potential is illustrated by these experiments in recombinant systems—the resting membrane potential of *Xenopus* oocytes is significantly more negative following expression of KCNQ2/3 channels and application of retigabine leads to a further hyperpolarization.[29] In addition to the shift in voltage dependence, an increase in activation kinetics and decrease in deactivation kinetics was also reported, consistent with a stabilizing influence of the open channel state.[8,29] Subsequent single channel electrophysiology experiments revealed that retigabine shifts the open–closed time distribution such that channels spend more time in the open state.[31] Pharmacological studies utilizing recombinantly expressed KCNQ channels, have demonstrated that retigabine activates all of the neuronal KCNQ channel subtypes: Tatulian *et al.*[30] calculated IC50 values in transiently transfected CHO cells by measuring the shift in $V_{1/2}$ activation and reported the following rank order of potency: KCNQ3 $(0.6 \pm 0.3\mu M) >$ KCNQ2/3 $>$ KCNQ2 $>$ KCNQ4 $(5.2 \pm 0.9 \ \mu M)$. Retigabine had no activity as a KCNQ1 channel opener. Similar data were generated for KCNQ2-4 expressed in HEK293 cells.[32] Finally, Wickenden *et al.*[33] demonstrated that retigabine is active against KCNQ3/5 heteromeric channels over a similar potency range.

10.5   Retigabine in Preclinical Models

Retigabine has been tested in a wide range of preclinical models of epilepsy. A detailed analysis of this work is provided in a recent review by Large *et al.*[34] The first report of *in vivo* activity was provided by Dailey *et al.*[35] who reported anticonvulsant effects against sound-induced seizures in two strains of genetically epilepsy prone rats. Subsequently, efficacy has been demonstrated in a wide range of *in vivo* models, including kindling models of partial epilepsy[36,37] and generalized seizure models that involve administration of an electrical or chemical stimulus.[38] There is some *in vivo* evidence that the efficacy of retigabine is mediated through KCNQ channels, as the anticonvulsant action of retigabine is reduced in Szt1 transgenic mice, which have a C-terminal deletion of the KCNQ2 gene.[39]

Straub *et al.*[40] provided evidence that retigabine's anticonvulsant activity will translate to man: Using cortical slices prepared from the brain of patients with intractable, drug resistant seizures, these authors reported that retigabine is effective in inhibiting 'sharp waves' of electrical activity.

In summary, pre-clinical studies have revealed that retigabine is a broad-spectrum anticonvulsant, with efficacy in a wide range of models including those that are known to be resistant to other anticonvulsants.

10.6 Retigabine in Clinical Trials

Encouraged by the positive outcome of preclinical studies, the usefulness of retigabine in treating epilepsy and neuropathic pain has been tested in several clinical trials, some of which are still ongoing.

Trials testing the efficacy of retigabine in different epilepsy conditions have typically delivered a positive outcome. Thus, in a 2005 study sponsored by Valeant Pharmaceuticals,[41] as well as a 2007 study sponsored by Wyeth,[42] positive outcomes of phase II trials studying the effects of retigabine in combination with other antiepileptic drugs (AEDs), such as valproic acid, topiramate, phenytoin or carbamazepine, or as monotherapy were reported. In both studies retigabine was well tolerated up to a daily dose of 1200 mg/day when administered as a *b.i.d.* or *t.i.d.* regimen and caused a reduction in seizure frequency. In the 2005 study, in which 60 patients were studied, the median seizure rate per 28 days was 8.1 at the outset of the study and 6.4 during treatment. 44% of patients were found to be responders, defined as a more than 50% reduction in seizure frequency. The most common adverse side-effects observed in this study were dizziness, asthenia, somnolence, nausea, speech disorder and tremor, with dizziness and somnolence being dose limiting. Eight of the 60 patients discontinued from the study due to side-effects and two additional patients discontinued from the study for other reasons. In the 2007 study, a randomized, multicenter, dose-ranging study of retigabine for partial-onset seizures, 399 patients were enlisted, of whom 279 (69.9%) completed the double-blind treatment period of 16 weeks. There was a significant, dose-dependent median percent change in seizure frequency from baseline (-23% for 600 mg/day, -29% for 900 mg/day, and -35% for 1200 mg/day, *versus* -13% for placebo). Likewise, there also was a dose-dependent increase in responder rates, again defined as more than 50% reduction in seizure frequency (23% for 600 mg/day, 32% for 900 mg/day (p = 0.021), and 33% for 1200 mg/day (p = 0.016), *versus* 16% for placebo). The most common treatment-emergent adverse events were somnolence, dizziness, confusion, speech disorder, vertigo, tremor, amnesia, abnormal thinking, abnormal gait, paresthesia, and diplopia.

In 2005, encouraged by the positive outcome of the phase II study of the same year, Valeant Pharmaceuticals started two phase III studies, the RE-STORE (Retigabine Efficacy and Safety Trial for partial Onset Epilepsy) trials. Both of these trials were concluded in 2008 and their outcome has been presented at meetings and in full papers.[43–46] The RESTORE 1 trial evaluated a 1200 mg daily dose of retigabine *versus* placebo. 305 patients were enlisted of which 256 entered a 12 week maintenance phase of the trial.

The RESTORE 2 trial evaluated a 600 or 900 mg daily dose of retigabine *versus* placebo. 538 patients were enlisted in this trial of which 471 entered the 12 week maintenance phase. Both trials validated the therapeutic usefulness of retigabine, with positive outcome in the primary endpoints relevant for approval by the U.S. Food and Drug Administration (FDA) as well as European Medicines Agency (EMEA). Thus, in the RESTORE 1 trial, the median reduction in 28 day total partial seizure frequency from baseline to the end of the double-blind period (the FDA primary efficacy endpoint) in the intent-to-treat population was 44.3% (n = 151) and 17.5% (n = 150) in the retigabine 1200 mg arm and placebo arm of the trial, respectively. In the RESTORE 2 trial, the median percentage seizure reduction also was significantly greater in retigabine-treated patients (600 mg = 27.9%; 900 mg = 39.9%) compared with placebo (15.9%). The median percentage of seizure-free days during the double-blind treatment period (titration and maintenance) was significantly greater in each retigabine treatment group than placebo (RESTORE 1: placebo = 77.3% *versus* 1200 mg/day = 84.1%; RESTORE 2: placebo = 77.8% *versus* 600 mg/day = 79.5%, 900 mg/day = 82.1%). Likewise, the responder rate, defined as a more than 50% reduction in 28 day total partial seizure frequency during maintenance was 55.5% in the retigabine 1200 mg group and 22.6% in the placebo group of the RESTORE 1 trial. In the RESTORE 2 trial, 38.6% (600 mg) and 47% (900 mg) of the retigabine-treated patients were responders as compared to 18.9% for placebo.

Treatment discontinuations due to adverse events were more common with retigabine than with placebo, but in most cases adverse events were mild-to-moderate in intensity. The most common side effects associated with retigabine in the RESTORE trials included dizziness, somnolence, fatigue, confusion, dysarthria (slurring of speech), ataxia (loss of muscle coordination), blurred vision, tremor, and nausea. In addition, urinary adverse events were more frequent in patients receiving retigabine as compared to placebo. There is an increased risk of urinary retention with possible secondary renal effects caused by an inability to empty the bladder. This particular aspect of the safety profile of retigabine has recently been critically reviewed.[47]

On October 30th 2009, Valeant Pharmaceuticals, along with its collaboration partner GSK, submitted a New Drug Application to the FDA, as well as a Marketing Authorization Application to the EMEA, requesting marketing approval for the investigational drug retigabine, a neuronal potassium channel opener for the adjunctive treatment for adult epilepsy patients with partial-onset seizures. In 2011 both applications received a green light from the authorities; in March 2011 marketing authorization as an adjunctive treatment for partial-onset seizures in adult patients was granted by the EMEA,[48] and three months later retigabine was approved for the same indication by the FD.[49] Retigabine will be sold under the name ezogabine in the US.

In a 2009 statement to investors, Valeant Pharmaceuticals reported on a randomized, double-blind, placebo-controlled phase IIa study in which the efficacy of retigabine in reducing the pain associated with post-herpetic

neuralgia had been tested (http://phx.corporate-ir.net/phoenix.zhtml?
c = 119269andp = irol-newsArticleandID = 1323443andhighlight). 187 patients of both sexes and a wide variety of ages were followed in this 10 week
long study. The study was conducted in approximately 50 trial locations,
with patients being titrated to their individually determined maximum tolerated dose within the range of 300 mg to 900 mg retigabine per day. In
general, retigabine was well tolerated. However, the primary efficacy endpoint of the outcome measure—the comparison of the average pain intensity
over the last seven days of maintenance therapy with retigabine *versus*
placebo—was not met.

10.7 Molecular Pharmacology of Retigabine: Options for Improvement

Although retigabine has been progressed into clinical trials, many companies in the pharmaceutical sector have realized that there is scope for
improvement on the compound, as also highlighted by some of the adverse
events seen with retigabine.

One major issue with retigabine is that it does not have good selectivity
between subtypes KCNQ2 through to KCNQ5. The main targets for anti-
epileptic and anti-nociceptive agents are thought to be KCNQ2/KCNQ3
heteromers and avoiding activity at the other channel subtypes may be important in limiting side effects. The primary channel to avoid is KCNQ1, as
this subunit is expressed in the heart and has been associated with prolonged QT interval.[50,51] Yet, retigabine demonstrates little activity at
KCNQ1,[30] and while KCNQ1 is thus not a concern for retigabine, second
generation KCNQ channel openers should avoid hitting KCNQ1.

Retigabine does act on KCNQ4.[32] This channel is expressed in cochlear
hair cells and interaction with this channel may cause high frequency
hearing problems.[52–54] KCNQ4 channel activity is thus a feature that needs
to be closely monitored in any second generation KCNQ channel opener.

The potency of retigabine is also relatively low for a small molecule, with an
EC50 of around 1 μM on KCNQ2/3 channels,[55,56] which may be improved upon
with today's screening technologies. In addition, the compound has effects on
targets unrelated to the Kv7 family, for instance, potentiation of inhibitory
post-synaptic currents *via* enhancement of $GABA_A$[57] and in pre-clinical tests it
can impair motor function,[58,59] making the likely therapeutic window narrow.

There is thus a clear opportunity to improve on several aspects of the
pharmacodynamic properties of a second generation KCNQ channel opener.

10.8 Chemistry and Preclinical Drug Discovery of KCNQ Openers

Retigabine (**1**) (ethyl N-[2-amino-4-[(4-fluorophenyl)methylamino]phenyl]-
carbamate) is a 1,2,4-triaminobenzene derivative, where the 1-aminogroup is

protected as an ethylcarbamate and the 4-amino moiety with a p-F-benzyl group. Retigabine has a molecular weight of 303 and it is a weak base with a reported pKa of 3.7. The logD is estimated to be 2.7 (at pH 7) and the solubility is quite low, estimated to be 0.07g/L in a neutral aqueous medium at 20 °C.[60]

Several preclinical drug discovery programs have emerged after the discovery of retigabine[23,24] as being an opener of KCNQ channels. Retigabine (**1**) itself has been the starting point for the development of new compounds described in several patent applications by Lundbeck.[61–63] Generally the 4-amino-1-carbonylaminophenyl functionality was preserved (**2**) but the 2-amino group could be replaced by other small groups like methyl, halogen or cyano (**3**, Figure 10.1). The amino group in the 4-position has been substituted with a variety of groups and also condensed into a ring as in the indoline (**4**) described by Lundbeck and Valeant.[64,65] Other 4-amino substituted anilides have been disclosed by Valeant Pharmaceuticals where the flexible benzylamine part has been replaced by the more rigid 1,2,3,4-tetra-hydroisoquinoline moiety as in **5**[66] or by the corresponding naphthyridine analogues.[67]

Neurosearch[68] recently disclosed close analogues to flupirtine (the pyridine analogue of retigabine) where the benzylamine part was replaced by non-aromatic counterparts such as in the tetrahydropyranylmethyl derivative **6**. The 1,4-diamino pattern has also been preserved in pyridine **7** and pyrimidine **8** derivatives disclosed by Lundbeck[69,70] the preferred amino group seemed to be morpholine and both carbamates as well as amides were described.

A different series of compounds consisting of pyridine-3-yl benzamides were disclosed by Icagen.[71,72] The SAR revealed that the 6-chloro substituent on the pyridine moiety was important for potency. The derivative ICA-27243 (**9**, Figure 10.2) has been extensively studied and shown to alleviate pain and

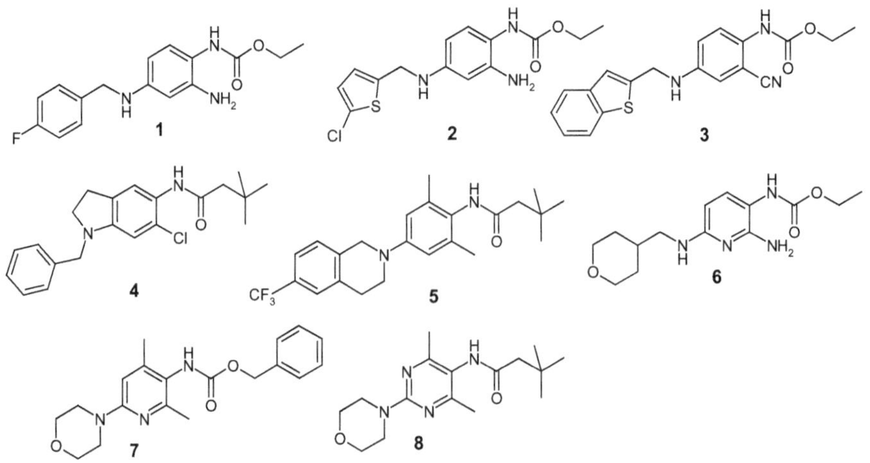

Figure 10.1

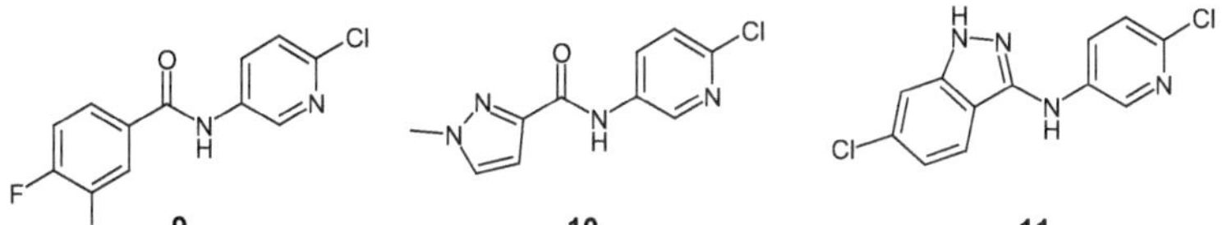

Figure 10.2

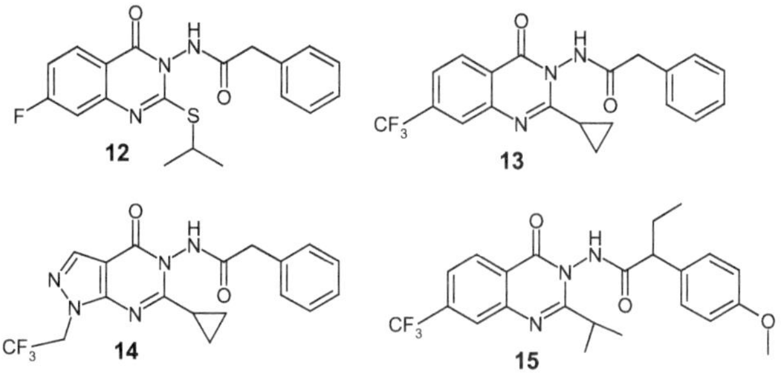

Figure 10.3

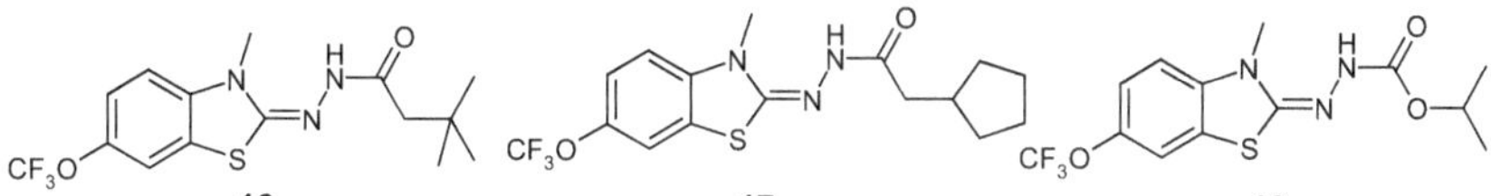

Figure 10.4

epileptic seizures.[73,74] Compound **9** was reported to be selective for opening the KCNQ2/3 ion channel and not the KCNQ4 or KCNQ1 subtypes. The left-hand side phenyl could be replaced by different heterocycles (**10**), and amide bond isosteres such as indazole (**11**) or benzisoxazoles were also disclosed.[75]

Several quinazolinone derivatives have been shown to be KCNQ openers, the first one being discovered by Icagen,[76] a representative example is illustrated by compound **12** (Figure 10.3). The thioether group was successfully replaced by a variety of other groups such as alkyl, cycloalkyl and aryls (**13**),[77] and the quinazolinone core could be modified into the more polar pyrazolo-pyrimidinone core as in **14**.[78] The linker between the amide and the phenyl group has been extensively explored by Neurosearch;[79,80] one of these variations is shown in derivative **15**.

Another class of compounds, which also contains the hydrazide moiety, was recently published by Icagen.[81] The development and SAR of these benzothiazole derivatives were described and the most potent compounds **16**, **17** and **18** (Figure 10.4) were shown to have an effect in animal models for epilepsy and pain.

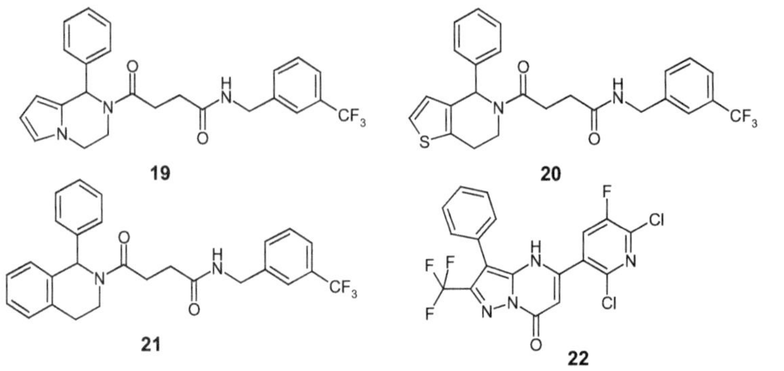

Figure 10.5

A different series of compounds described as being KCNQ2/3 openers have quite recently been disclosed by Grunenthal[82–84] and representative examples are shown (Figure 10.5), these include different bicyclic moieties such as tetrahydropyrrolopyrazin (**19**), tetrahydrothienopyridine (**20**) or tetrahydroisoquinoline (**21**). Compounds **20** and **21** were examined in an animal model for pain and were shown to alleviate pain in the rat formalin model. Other heterocyclic scaffolds such as pyrazolopyrimidinones[85] have been reported to be useful as KCNQ2/3 openers. The pyrazolopyrimidinones were evaluated in an atomic absorption Rb^+ efflux assay and compound **22** (Figure 10.5) was estimated to be the most potent compound in this series.

10.9 Mapping the Site of Molecular Interaction of Retigabine and other KCNQ Channel Openers

As detailed earlier, retigabine induces its effect on the M-current by shifting the voltage dependence of activation to more hyperpolarized potentials, largely due to an increase in the channel open probability and greatly reduced closed time.[31] Importantly, and unlike the KCNQ blockers linopirdine and XE991, retigabine is inactive against the KCNQ1 cardiac channel.[30] This selectivity has been used to map the retigabine binding site by generating chimeric channel constructs, and the site of action has been narrowed down to a hydrophobic pocket between the S5 and S6 transmembrane domains. When the tryptophan residue (W236 in Kv7.2) is replaced with less hydrophobic amino acids, retigabine is no longer active.[86,87] The site of action explains why retigabine does not act on the KCNQ1 channel, as this site is not conserved in the heart channel. Other openers have also been developed that try to exploit this, but also attempt to get selectivity between KCNQ2-KCNQ5. Both the acrylamide compound (S)-1 and BMS-204352 act through the same site on KCNQ4 channels, although (S)-1 has some blocking activity at KCNQ1.[32,58,88] Adding to the complexity, compound (S)-2 also has activity at KCNQ2-KCNQ5, but the effect on KCNQ4 was not abolished by the relevant site mutation W242L and (S)-2 actually became an inhibitor of KCNQ2 channels with the analogous mutation W236L.[89] Interestingly, the R-enantiomer of BMS-204352 is an inhibitor of KCNQ4 channels[58] and thus, it appears that small changes in compound structure can markedly alter channel interaction.

Openers known as phenomates, such as meclofenamic acid and diclofenac, are thought to act at a separate site on KCNQ2/KCNQ3 channels as they showed additive effects to retigabine, although these compounds are not particularly potent (25 µM and 2.6 µM, respectively).[90] These compounds are COX inhibitors and this led to the hypothesis that activity at the two distinct targets could be separated by looking at structurally related compounds. This resulted in a series of diphenyl carboxylate derivatives being found that were KCNQ channel openers.[91] Using modeling studies, one of these (NH29) was used to show docking with an external groove to interact with the voltage sensor.[92] However, the voltage sensor domain is also common to TRPV1 channels, where NH29 is also a potent blocker. This may mean that targeting this site on Kv7 channels may lead to unintended promiscuity of the compound activity. Zinc pyrithione has also been identified as an opener, producing both a gating shift and an increase in maximal conductance. It also acts *via* the S5/S6 domains but at a site separate from the retigabine site. Zinc pyrithione is not very selective among the KCNQ channels.[93]

More recently, another opener has been discovered, known as ICA-27243. This compound has made some progress towards greater selectivity for KCNQ2/KCNQ3 over general activity at KCNQ2-KCNQ5. ICA-27243 is around 25 times more potent at KCNQ2/KCNQ3 *versus* KCNQ4 and has little activity at KCNQ3/KCNQ5 heteromers,[94] which may assist in improving the side-effect profile. The site of action of the compound has now been mapped to a site distant from the S5-S6 domain, explaining why the KCNQ2W236L/KCNQ3W265L double mutant disrupts retigabine action, but not ICA-27243 activity.[94] These authors, using chimera studies, found that the site resides in the S1–S4 voltage sensor domain of KCNQ2 and is likely to involve the non-conserved regions at the C-terminus end of S2 and the N-terminus end of S3. Recently, further screening work from a small library of compounds has confirmed the potentiating activity of ICA-27243 and a number of related compounds, also acting away from the retigabine site.[95] Present and future rationales for subtype selective openers may be best to focus on targeting these more heterologous regions of the channels, as is the case for the ICA-27243, rather than the more conserved regions where retigabine binds. Zhang *et al.*[96] noted that their data suggested that selectivity between KCNQ2/3 and KCNQ2-5 homotetramers was challenging, in particular *versus* KCNQ2, followed by KCNQ4. Identifying which counterscreens are essential through translational efforts would help to streamline an effective screening cascade.

10.10   Future KCNQ Channel Openers: Lead Generation Approaches

The action of flupirtine and retigabine as M-current activators was only discovered after their effects on nerve firing were known and subsequently the molecular identity of the M-current was revealed. As we now know that the likely mechanism of action of the anti-epileptic and anti-nociceptive properties of these compounds is *via* opening of KCNQ2/KCNQ3 channel complexes, it is possible to directly develop compounds that have this characteristic. However, for many years, progress in identification of ion channel targets was held back by the lack of any high or even medium throughput screening technologies and relying on the gold standard technique of manual patch-clamp, an information rich technique but with a low throughput of a few compounds per day. In recent times, this is no longer the case with the advent of ion channel platforms such as IonWorks™ (Molecular Devices), where thousands of potential compounds can be rapidly screened directly on the target of interest. However, the pharmaceutical industry often wishes to screen extremely large compound libraries, perhaps consisting of up to around 1 million substances. This level of throughput is still beyond the present ion channel platforms due to time and cost restraints and thus, some other techniques have been used. For instance, ICA-27243 was initially discovered using a fluorescence based assay[73] and Abbott have used thallium influx fluorescence to compare selectivity across the subtypes[75] in a high throughput manner. Icagen utilised SH-SY5Y cells as a primary screen, as these cells were found to express KCNQ2, KCNQ3 and KCNQ5 channels. Activity was detected *via* a membrane potential sensitive probe reacting to the hyperpolarizing response produced by M-current activators. Confirmation of hits and counter-screening were done using a second technique—[86]Rb efflux was measured from a variety of cell lines overexpressing the KCNQ channel of interest. Similar flux-based screening assays have been developed using a non-radiometric method, namely atomic absorption spectroscopy.[74] The problem with these methods is that they are indirect measurements of the background activity of the target of interest. It is possible that there are false positives but also, perhaps more worrying, is that there may be false negatives, which can never be followed up. To eliminate any false positives and, in addition, get a more accurate measurement of the potency of the compounds, the next step in a screening cascade is most likely to be an electrophysiological assay. One such platform is IonWorks™, which is able to screen compounds in a 384 well format[97] (Figure 10.6). Although the seal resistances in the recording plate are not as high as would be normal in manual patch-clamp recording, the stability of the recordings and leak subtraction are of good enough quality for reproducible pharmacological screening. This has been successfully used as a KCNQ2/KCNQ3 assay.[98] One major breakthrough with IonWorks was the development of population patch plates, where the current from a potential 64 cells is averaged in each well. In practice, this means that the chance of

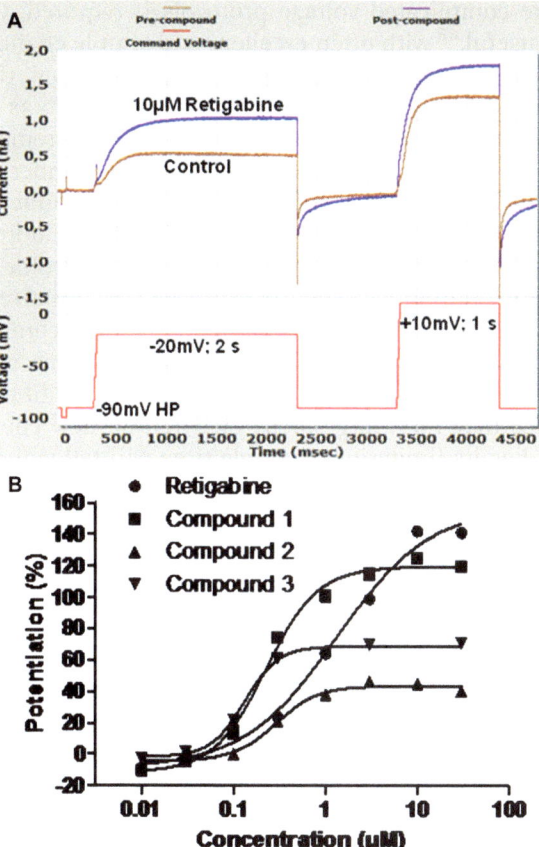

Figure 10.6 Use of IonWorks to investigate pharmacology on the KCNQ2/3 channel. Population patch technology on IonWorks was used to record
from CHO cells expressing KCNQ2/3 channels as shown in **A** with the
current traces in the top panel in response to the voltage protocol
shown in the bottom panel (annotated screen shot from the software).
Cells were held at −90 mV and depolarized to −20 mV for 2 s (Pulse 1)
to activate the current before (brown) and after (blue) application of
Retigabine at 10 μM producing a clear potentiation of the current. In
this protocol, cells were also depolarized to +10 mV (Pulse 2), after a
recovery period, to provide additional information on the compounds.
Concentration response curves are shown in **B** for retigabine and 3
test compounds using data from Pulse 1. This highlights one of the
additional challenges of developing potentiators of this current where
some compounds produce a large potentiation and some have a less
pronounced effect.

losing each data point is drastically reduced, which allows for an extremely
high success rate and requirement for few re-runs. This helps enormously
with delivery of data and a reduced cycle time for feedback to chemistry.
Other electrophysiology platforms include Qpatch and the PatchXpress, both
of which have a lower throughput but can arguably produce higher quality

data, with more complicated voltage protocols if required. Oocyte preparations are also useful,[68] with often excellent and simple channel expression, good voltage control and stability but the proviso that the immediate channel environment may not translate as well to human as a mammalian cellular background. One way of using the direct measurement techniques as the first level of a screening cascade is to limit the number of compounds but screen these smaller libraries on medium throughput devices. Compound selection can be done using multiple methods, including eliminating those compounds that have an extremely unattractive profile, such as high predicted protein binding or likely toxicity issue. Alternatively, the library can be based on already known pharmacology of similar compounds. It may also be possible to limit the numbers of compounds that are near neighbours, so that potential clusters are 'represented' in the library. Of course, selected libraries may miss large areas of potential new chemistry and so each approach has its limitation and only time will tell which method will lead to better, solid clinical candidates.

References

1. D. H. Jenkinson, *Br J Pharmacol.*, 2006, **147**(Suppl. 1), S63.
2. Q. Wang, M. E. Curran, I. Splawski, T. C. Burn, J. M. Millholland, T. J. VanRaay, J. Shen, K. W. Timothy, G. M. Vincent, T. de Jager, P. J. Schwartz, J. A. Toubin, A. J. Moss, D. L. Atkinson, G. M. Landes, T. D. Connors and M. Y. Keating, *Nat. Genet.*, 1996, **12**, 17.
3. C. Biervert, B. C. Schroeder, C. Kubisch, S. F. Berkovic, P. Propping, T. J. Jentsch and O. K. Steinlein, *Science*, 1998, **279**, 403.
4. N. A. Singh, C. Charlier, D. Stauffer, B. R. DuPont, R. J. Leach, R. Melis, G. M. Ronen, I. Bjerre, T. Quattlebaum, J. V. Murphy, M. L. McHarg, D. Gagnon, T. O. Rosales, A. Peiffer, V. E. Anderson and M. Leppert, *Nat. Genet.*, 1998, **18**, 25.
5. C. Charlier, N. A. Singh, S. G. Ryan, T. B. Lewis, B. E. Reus, R. J. Leach and M. Leppert, *Nat. Genet.*, 1998, **18**, 53.
6. B. C. Schroeder, C. Kubisch, V. Stein and T. J. Jentsch, *Nature*, 1998, **396**, 687.
7. H. S. Wang, Z. Pan, W. Shi, B. S. Brown, R. S. Wymore, I. S. Cohen, J. E. Dixon and D. McKinnon, *Science*, 1998, **282**, 1890.
8. A. D. Wickenden, W. Yu, A. Zou, T. Jegla and P. K. Wagoner, *Mol. Pharmacol.*, 2000, **58**, 591.
9. M. Schwake, M. Pusch, T. Kharkovets and T. J. Jentsch, *J. Biol. Chem.*, 2000, **275**, 13343.
10. C. Kubisch, B. C. Schroeder, T. Friedrich, B. Lütjohann, A. El-Amraoui, S. Marlin, C. Petit and T. J. Jentsch, *Cell*, 1999, **96**, 437.
11. B. C. Schroeder, M. Hechenberger, F. Weinreich, C. Kubisch and T. J. Jentsch, *J. Biol. Chem.*, 2000, **275**, 24089.
12. D. A. Brown and P. R. Adams, *Nature*, 1980, **283**, 673.

13. J. K. Hadley, G. M. Passmore, L. Tatulian, M. Al-Qatari, F. Ye, A. D. Wickenden and D. A. Brown, *J. Neurosci.*, 2003, **23**, 5012.
14. J. R. Schwarz, G. Glassmeier, E. C. Cooper, T. C. Kao, H. Nodera, D. Tabuena, R. Kaji and H. Bostock, *J. Physiol.*, 2006, **573**, 17.
15. M. M. Shah, M. Mistry, S. J. Marsh, D. A. Brown and P. Delmas, *J. Physiol.*, 2002, **544**, 29.
16. H. J. Chung, Y. N. Jan and L. Y. Jan, *Proc. Natl. Acad. Sci. U. S. A.*, 2006, **103**, 8870.
17. Z. Pan, T. Kao, Z. Horvath, J. Lemos, J. Y. Sul, S. D. Cranstoun, V. Bennett, S. S. Scherer and E. C. Cooper, *J. Neurosci.*, 2006, **26**, 2599.
18. G. M. Passmore, A. A. Selyanko, M. Mistry, M. Al-Qatari, S. J. Marsh, E. A. Matthews, A. H. Dickenson, T. A. Brown, S. A. Burbidge, M. Main and D. A. Brown, *J. Neurosci.*, 2003, **23**, 7227.
19. M. M. Shah, M. Migliore, I. Valencia, E. C. Cooper and D. A. Brown, *Proc. Natl. Acad. Sci. U. S. A.*, 2008, **105**, 7869.
20. D. A. Brown, S. A. Hughes, S. J. Marsh and A. Tinker, *J. Physiol.*, 2007, **582**, 917.
21. B. C. Suh and B. Hille, *J. Physiol.*, 2007, **582**, 911.
22. C. C. Hernandez, O. Zaika, G. P. Tolstykh and M. S. Shapiro, *J. Physiol.*, 2008, **586**, 1811.
23. B. Nickel, A. Shandra, L. Godlevsky, A. Mazarati, H. Kupferberg and I. Szelenyi, *Epilepsia*, 1993, **34**(Suppl. 2): Abst 95.
24. B. Nickel, A. Shandra, L. Godlevsky, A. Mazarati, H. Kupferberg and I. Szelenyi, *Arch. Pharmacol.*, 1993, **347**(Suppl.): Abst R-142.
25. W. D. Yonekawa, I. M. Kapetanovic and H. J. Kupferberg, *Epilepsy Res.*, 1995, **20**, 137.
26. C. Rundfeldt, *Eur. J. Pharmacol.*, 1997, **336**, 243.
27. C. Rundfeldt, *Epilepsy Res.*, 1999, **35**, 99.
28. C. Rundfeldt and R. Netzer, *Neurosci. Lett.*, 2000, **282**, 73.
29. M. J. Main, J. E. Cryan, J. R. Dupere, B. Cox, J. J. Clare and S. A. Burbidge, *Mol. Pharmacol.*, 2000, **58**, 253.
30. L. Tatulian, P. Delmas, F. C. Abogadie and D. A. Brown, *J. Neurosci.*, 2001, **21**, 5535.
31. L. Tatulian and D. A. Brown, *J. Physiol.*, 2003, **549**, 57.
32. R. L. Schroder, T. Jespersen, P. Christophersen, D. Strobaek, B. S. Jensen and S. P. Olesen, *Neuropharmacology*, 2001, **40**, 888.
33. A. D. Wickenden, A. Zou, P. K. Wagoner and T. Jegla, *Br. J. Pharmacol.*, 2001, **132**, 381.
34. C. H. Large, D. M. Sokal, A. Nehlig, M. J. Gunthorpe, M. Sankar, C. S. Crean, K. E. VanLandingham and H. S. White, *Epilepsia*, 2012, **53**, 425.
35. J. W. Dailey, J. H. Cheong, K. H. Ko, L. E. Adams-Curtis and P. C. Jobe, *Neurosci. Lett.*, 1995, **195**, 77.
36. C. Tober, A. Rostock, C. Rundfeldt and R. Bartsch, *Eur. J. Pharmacol.*, 1996, **303**, 163.
37. A. Mazarati, J. Wu, D. Shin, Y. S. Kwon and R. Sankar, *Epilepsia*, 2008, **49**, 1777.

38. A. Rostock, C. Tober, C. Rundfeldt, R. Bartsch, J. Engel, E. E. Polymeropoulos, B. Kutscher, W. Löscher, D. Hönack, H. S. White and H. H. Wolf, *Epilepsy Res.*, 1996, **23**, 211.
39. J. F. Otto, Y. Yang, W. N. Frankel, K. S. Wilcox and H. S. White, *Epilepsia*, 2004, **45**, 1009.
40. H. Straub, R. Kohling, J. Hohling, C. Rundfeldt, I. Tuxhorn, A. Ebner, P. Wolf, H. Pannek and E. Speckmann, *Epilepsy*, 2001, **44**, 155.
41. R. Sachdeo, R. Porter, V. Biton, W. Rosenfeld, W. Alves and V. Nohria, *Epilepsia*, 2005, **46**(Suppl.8), 185.
42. R. J. Porter, A. Partiot, R. Sachdeo, V. Nohria and W. M. Alves, *Neurology*, 2007, **68**, 1197.
43. M. J. Brodie, H. Lerche, A. Gil-Nagel, C. Elger, S. Hall, P. Shin, V. Nohria and H. Mansbach on behalf of RESTORE 2 Study group, *Neurology*, 2010, **75**, 1817.
44. D. Burdette, C. Elger, H. Mansbach, P. Shin and S. Hall, *Epilepsia*, 2009, **50**(Suppl.11), 116.
45. J. A. French, B. W. Abou-Khalil, R. F. Leroy, E. M. T. Yacubian, P. Shin, S. Hall, H. Mansbach and V. Nohria on behalf of the RESTORE 1/Study 301 Investigators, *Neurology*, 2011, **76**, 1555.
46. R. J. Porter, J. A. French, M. J. Brodie, K. E. VanLandingham, K. P. Nanry and S. T. Hall, *62nd Annual American Academy of Neurology Meeting, Toronto, Canada*, 2010, **Poster 5.189**. http://www.abstracts2view.com/aan/view.php?nu = AAN10L_P05.189.
47. N. Brickel, P. Gandhi, K. VanLandingham, J. Hammond and S. DeRossett, *Epilepsia*, 2012, **53**, 606.
48. European Medicines Agency (EMA). European Public Assessment Report—Trobalt. EMA website http://www.ema.europa.eu/docs/en_GB/document_library/EPAR_-_Product_Information/human/001245/WC500104835.pdf (2011).
49. US Food and Drug Administration. FDA labeling information—Potiga. FDA website http://www.accessdata.fda.gov/drugsatfda_docs/label/2011/022345s000lbl.pdf (2011).
50. D. Peroz, N. Rodriguez, F. Choveau, I. Baro, J. Merot and G. Loussouarn, *J. Physiol.*, 2008, **586**, 1785.
51. A. Lundby, L. S. Ravn, J. H. Svendsen, S. P. Olesen and N. Schmitt, *Heart Rhythm*, 2007, **4**, 1532.
52. K. W. Beisel, S. M. Rocha-Sanchez, K. A. Morris, L. Nie, F. Feng, B. Kachar, E. N. Yamoah and B. Fritzsch, *J. Neurosci.*, 2005, **25**, 9285.
53. P. J. Coucke, H. P. Van, P. M. Kelley, H. Kunst, I. Schatteman, V. D. Van, J. Meyers, R. J. Ensink, M. Verstreken, F. Declau, H. Marres, K. Kastury, S. Bhasin, W. T. McGuirt, R. J. Smith, C. W. Cremers, P. Van de Heyning, P. J. Willems, S. D. Smith and C. G. Van, *Hum. Mol. Genet.*, 1999, **8**, 1321.
54. T. Kharkovets, K. Dedek, H. Maier, M. Schweizer, D. Khimich, R. Nouvian, V. Vardanyan, R. Leuwer, T. Moser and T. J. Jentsch, *EMBO J.*, 2006, **25**, 642.

55. Y. J. Wu, C. G. Boissard, J. Chen, W. Fitzpatrick, Q. Gao, V. K. Gribkoff, D. G. Harden, H. He, R. J. Knox, J. Natale, R. L. Pieschl, J. E. Jr. Starrett, L. Q. Sun, M. Thompson, D. Weaver, D. Wu and S. I. Dworetzky, *Bioorg. Med. Chem. Lett.*, 2004, **14**, 1991.
56. Y. J. Wu, L. Q. Sun, H. He, J. Chen, J. E. Starrett Jr., P. Dextraze, J. P. Daris, C. G. Boissard, R. L. Pieschl, V. K. Gribkoff, J. Natale, R. J. Knox, D. G. Harden, M. W. Thompson, W. Fitzpatrick, D. Weaver, D Wu, Q Gao and S. I. Dworetzky, *Bioorg. Med. Chem. Lett.*, 2004, **14**, 4533.
57. J. F. Otto, M. M. Kimball and K. S. Wilcox, *Mol. Pharmacol.*, 2002, **61**, 921.
58. M. P. Korsgaard, B. P. Hartz, W. D. Brown, P. K. Ahring, D. Strobaek and N. R. Mirza, *J. Pharmacol. Exp. Ther.*, 2005, **314**, 282.
59. G. Blackburn-Munro, W. Dalby-Brown, N. R. Mirza, J. D. Mikkelsen and R. E. Blackburn-Munro, *CNS Drug Rev.*, 2005, **11**, 1.
60. W. Thiel, *Pharmazie*, 1998, **53**, 865.
61. M. Rottlaender, A. Ritzen, M. B. Norgaard, N. Khanzhin and C. W. Tornoe, H Lundbeck A/S. 1,2,4-Triaminobenzene derivatives useful for the treating disorders of the central nervous system, *United States Patent Application*, US20060014822A1, 2006.
62. D. R. Greve, M. Rottlaender and W. P. Watson, Substituted aniline derivatives, *United States Patent Application*, US20060167087A1, 2006.
63. N. Khanzhin, M. Rottlaender, A. Ritzen and W. P. Watson, Substituted p-diaminobenzene derivatives, *United States Patent Application*, US20060183791A1, 2006.
64. N. Khanzhin, M. Rottlaender and W. P. Watson, H Lundbeck A/S, Substituted indoline and indole derivatives, *United States Patent Application*, US20060264496A, 2006.
65. J.-M. Vernier, H. Chen and J. Song, Derivatives of 5-amino-4,6-disubstituted indole and 5-amino-4,6-disubstituted indoline as potassium channel modulators, WO2009/023667, 2009.
66. J.-M. Vernier, M. A. De La Rosa, H. Chen, J. Z. Wu, G. L. Larson and I. W. Cheney, Valeant pharmaceuticals, Derivatives of 4-(Nazacycloalkyl) anilides as potassium channel modulators. *United States Patent Application*, US20080139610A1, 2008.
67. J.-M. Vernier, Valeant pharmaceuticals, Naphthyridine derivatives as potassium channel modulators, WO2009/018466, 2009.
68. C. Jessen, W. D. Brown and D. Stroebeck, Neurosearch A/S, 4-Tetrahydropyran- aminopyridine derivatives and their medical use, WO2010/026104, 2010.
69. N. Khanzhin, D. R. Greve and M. Rottlaender, Lundbeck Research USA, Substituted pyrimidine derivatives, *United States Patent Application*, US20070066612A1, 2007.
70. C. W. Tornroe, N. Khanzhin, M. Rottlaender, W. P. Watson and D. R. Greve, H. Lundbeck A/S Substituted pyridine derivatives, WO2006/092143, 2006.
71. G. A. McNaughton-Smith, M. F. Gross and A. D. Wickenden, Icagen Inc., Benzanilides as potassium channel openers, WO2001010380, 2001.

72. G. A. McNaughton-Smith, M. F. Gross, G. C. Rigdon and A. D. Wickenden, Icagen Inc. Methods for treating or preventing pain and anxiety, WO2001010381, 2001.

73. A. D. Wickenden, J. L. Krajewski, B. London, P. K. Wagoner, W. A. Wilson, S. Clark, R. Roeloffs, G. McNaughton-Smith and G. C. Rigdon, *Mol. Pharmacol.*, 2008, **73**, 977.

74. R. Roeloffs, A. D. Wickenden, C. Crean, S. Werness, G. McNaughton-Smith and J. Stables, *J. Pharmacol. Exp. Ther.*, 2008, **326**, 818.

75. G. A. McNaughton-Smith and G. S. Amato, Icagen Inc. Bisarylamines as potassium channel openers, *United States Patent*, US20026593349, 2002.

76. G. A. McNaughton-Smith, J. B. Thomas and G. Amato, Icagen, Inc.Quinazolinones as potassium channel modulators, WO2004/058704, 2004.

77. G. A. McNaughton-Smith, G. S. Amato and J. B. Thomas, Icagen, Inc., Fused ring heterocycles as potassium channel modulators, *United States Patent*, US7223768, 2007.

78. G. A. McNaughton-Smith, G. S. Amato and J. B. Thomas, Icagen, Inc. Fused ring heterocycles as potassium channel modulators, *United States Patent*, US2008/0058319, 2008.

79. W. D. Brown, C. Jessen, J. Demnitz, T. Dyhring and D. Neurosearch A/S, Quniazolinones and their use as potassium channel activators, WO2007/104717, 2007.

80. W. D. Brown, L. Teuber, T. Dyhring, D. Strøbæk and C. Jessen, Neurosearch A/S, Novel quinazoline derivatives and their medical use, WO2007/057447, 2007.

81. P. C. Fritch, G. A. McNaughton-Smith, G. S. Amato, J. F. Burns, W. Eargle, R. Roeloffs, W. Harrison, L. Jones and A. D. Wickenden, *J. Med. Chem.*, 2010, **53**, 887.

82. B. Merla, T. Christoph, S. Oberbörsch, K. Schiene, G. Bahrenberg, R. Frank, S. Kuhnert and W. Schröder, Grunenthal GMBH, Substituted tetrahydropyrrolopyrazine compounds having affinity with the KCNQ2/3 K+ channel and use thereof in medicaments, WO2008/046582, 2008.

83. S. Kuhnert, G. Bahrenberg, A. Kless, B. Merla, K. Schiene and W. Schröder, Grunenthal GMBH, Substituted 4,5,6,7-tetra-hydrothienopyridines as KCNQ2/3 modulators for treating pain, epilepsy and urinary incontinence, WO2010/04618, 2010.

84. S. Kuhnert, G. Bahrenberg, B. Merla, K. Schiene and W. Schröder, Grunenthal GMBH, Substituted 4-(1,2,34-tetrahydroisoquinolin-2-yl)-4-oxobutyric acid amide as KCNQ2/3 modulators, US2010/0152234(A1), 2010.

85. J. Qi, F. Zhang, Y. Mi, Y. Fu, W. Xu, D. Zhang, Y. Wu, X. Du, Q. Jia, K. Wang and H. Zhang, *Eur. J. Med. Chem.*, 2011, **46**, 934.

86. T. V. Wuttke, G. Seebohm, S. Bail, S. Maljevic and H. Lerche, *Mol. Pharmacol.*, 2005, **67**, 1009.

87. A. Schenzer, T. Friedrich, M. Pusch, P. Saftig, T. J. Jentsch, J. Grotzinger and M. Schwake, *J. Neurosci.*, 2005, **25**, 5051.

88. B. H. Bentzen, N. Schmitt, K. Calloe, B. W. Dalby, M. Grunnet and S. P. Olesen, *Neuropharmacology*, 2006, **51**, 1068.

89. S. M. Blom, N. Schmitt and H. S. Jensen, *PLoS One*, 2009, **4**, e8251.
90. A. Peretz, N. Degani, R. Nachman, Y. Uziyel, G. Gibor, D. Shabat and B. Attali, *Mol. Pharmacol.*, 2005, **67**, 1053.
91. A. Peretz, N. Degani-Katzav, M. Talmon, E. Danieli, A. Gopin, E. Malka, R. Nachman, A. Raz, D. Shabat and B. Attali, *PLoS One*, 2007, **2**, e1332.
92. A. Peretz, L. Pell, Y. Gofman, Y. Haitin, L. Shamgar, E. Patrich, P. Kornilov, O. Gourgy-Hacohen, N. Ben-Tal and B. Attali, *Proc. Natl. Acad. Sci. U. S. A.*, 2010, **107**, 15637.
93. Q. Xiong, H. Sun and M. Li, *Nat. Chem. Biol.*, 2007, **3**, 287.
94. K. Padilla, A. D. Wickenden, A. C. Gerlach and K. McCormack, *Neurosci. Lett.*, 2009, **465**, 138.
95. Z. Gao, T. Zhang, M. Wu, Q. Xiong, H. Sun, Y. Zhang, L. Zu, W. Wang and M. Li, *J. Biol. Chem.*, 2010, **285**, 28322.
96. D. Zhang, R. Thimmapaya, X. F. Zhang, D. J. Anderson, J. L. Baranowski, M. Scanio, A. Perez-Medrano, S. Peddi, Z. Wang, J. R. Patel, D. A. DeGoey, M. Gopalakrishnan, P. Honore, B. B. Yao and C. S. Surowy, *J. Neurosci. Meth.*, 2011, **200**, 54.
97. K. Schroeder, B. Neagle, D. J. Trezise and J. Worley, *J. Biomol. Screen.*, 2003, **8**, 50.
98. F. Jow, R. Shen, P. Chanda, E. Tseng, H. Zhang, J. Kennedy, J. Dunlop and M. R. Bowlby, *J. Biomol. Screen.*, 2007, **12**, 1059.

CHAPTER 11

The Therapeutic Potential of hERG1 K^+ Channels for Treating Cancer and Cardiac Arrhythmias

JOHN MITCHESON*[a] AND ANNAROSA ARCANGELI*[b]

[a] University of Leicester, Department of Cell Physiology and Pharmacology, Medical Sciences Building, University Road, Leicester, LE1 9HN, UK; [b] Department of Experimental Pathology and Oncology, University of Florence, Viale GB Morgagni, 50, 50134 Firenze, Italy
*Email: jm109@le.ac.uk; annarosa.arcangeli@unifi.it

11.1 Introduction

hERG was first identified by Drosophila geneticists, Warmke and Ganetsky.[1] They had identified mutant flies that when anaesthetised with ether exhibited an unusual twitching behaviour that resembled the action of go-go dancers. The fly mutation was in a potassium (K^+) channel that they named *ether-á-go-go* (eag). When they subsequently screened a human hippocampal cDNA library they identified an eag-like gene that they named human ether-a-go-go related gene or *hERG*. Within a few months another group of geneticists working entirely independently had shown that mutations in *hERG* were present in patients with chromosome 7 linked long QT syndrome, a cardiac disease that causes arrhythmias and sudden death.[2] When *hERG* was subsequently cloned and characterised it was shown to mediate K^+

RSC Drug Discovery Series No. 39
Ion Channel Drug Discovery
Edited by Brian Cox and Martin Gosling
© The Royal Society of Chemistry 2015
Published by the Royal Society of Chemistry, www.rsc.org

currents with similar biophysical properties to the rapid delayed rectifier K$^+$ current (I_{Kr}), a vital current for cardiac action potential repolarisation.[3,4] It was at this point that the importance of hERG as a target for numerous different drugs first became apparent. It was already known that certain medications could lengthen the cardiac action potential by blocking I_{Kr}.[5] While this could be advantageous and was the basis for type III antiarrhythmic drugs, which increase refractoriness and help protect against re-entrant stimuli, in other circumstances it resulted in increased risk of arrhythmias and sudden cardiac death.[6] In recent years, hERG has received bad press as a target to avoid in drug development. hERG channel block has become synonymous with unacceptable levels of proarrhythmic risk and hERG is more often than not considered as an anti-target rather than a target. The reality is more nuanced and complicated than this. Furthermore, there is now substantial evidence that hERG is aberrantly expressed in cancers and that it is involved in distinct processes in tumour progression.[7] A variety of studies suggest that hERG inhibitors have therapeutic potential for treating certain types of cancer.[7,183] In this article we will describe the physiological functions of hERG, mechanisms of hERG channel gating and why so many compounds block the channel. Later sections will discuss the expression and role of hERG channels in human cancers, new insights from molecules that activate rather than reduce hERG channel activity and the pitfalls and therapeutic potential of targeting hERG channels for treating arrhythmias and cancer.

11.1.1 hERG Channel Family Members and Alternative Isoforms

hERG channels belong to an evolutionarily conserved family of voltage activated potassium channels that also includes hEAG and human eag like potassium (hELK) channels. In the IUPHAR nomenclature for ion channels, hERG, hEAG, and hELK channels correspond to Kv11, Kv10 and Kv12 channel families, respectively.[8] These channels consist of four individual subunits that each have transmembrane domain structures similar to other voltage gated K$^+$ channels—consisting of six α-helical segments (S1–S6). In addition, they have large intracellular amino- (N-) and carboxy- (C-) terminal domains (see Figure 11.1). The C-terminus contains a region with homology to cyclic nucleotide binding domains (cNBD) of cyclic nucleotide gated channels. The N-terminus contains the eag-domain, the unique and defining feature of the eag channel family, which folds to form a Per-Arnt-Sim (PAS) domain and is involved in regulating hERG channel activity and trafficking to the cell surface.

In humans (and mammals), the *hERG* gene family is composed by at least three members: *hERG1* (KCNH2), *hERG2* (KCNH6) and *hERG3* (KCNH7). The *hERG1* gene has sometimes been subsequently referred to as *hERG1a*, since an alternative transcript of this gene, named *hERG1b*, has been identified both in the heart and in tumours.[9–13] *hERG1b* encodes a protein, hERG1B,

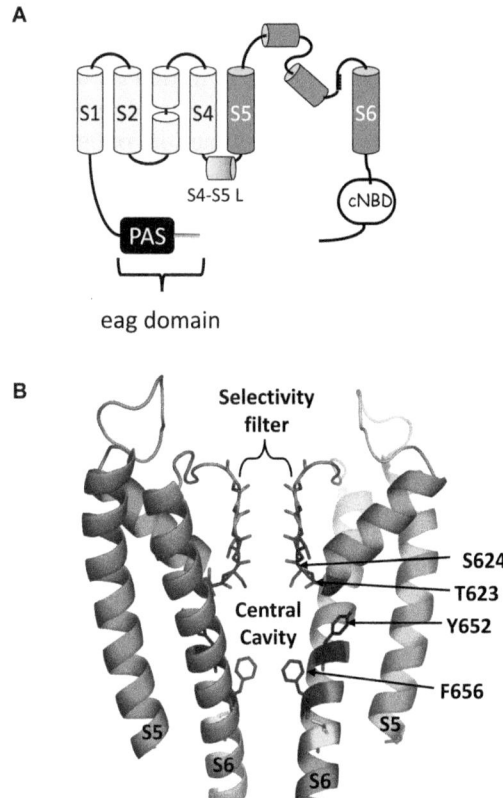

Figure 11.1   Structure of hERG1 potassium channels. (A) Secondary structure of
hERG1a K$^+$ channels. The transmembrane α-helices are labelled S1-
S6. The N- and C-termini contain large intracellular domains. The
N-terminus of hERG1a contains an eag domain (that is absent in
hERG1b isoforms) that consists of a Per-Arnt-Sim (PAS) domain
(amino acids 26–135) and a less highly structured region (amino acids
1–26) indicated by thick straight line. The C-terminus contains a
region with homology to the cyclic nucleotide binding domains
(cNBD) found in cyclic nucleotide gated channels, although hERG1
channels lack key residues for binding cyclic nucleotides and are not
regulated by physiological levels of cAMP. (B) Homology model of the
pore of hERG1 K$^+$ channels. The pore is formed by the S5–S6 domains
of four subunits co-assembling to form a tetrameric complex. For
clarity, only two of four subunits are shown. Residues that line the
inner cavity of the pore and are important for drug interactions are
labelled.

in which the entire N-terminus is substituted by a short chain of 34 amino
acids.[10] Another hERG1 isoform has been described, hERG1$_{USO}$. Indeed, the
hERG1$_{USO}$ transcript presents a unique exon at the 3′ end, the USO exon,
which substitutes most of the intracytoplasmic C-terminus, encoded by

exons 9 to 15, with a short stretch of 88 amino acids.[14] While *hERG1b* is an alternative transcript, with its own promoter region, *hERG1$_{USO}$* is a splice product of the *hERG1* gene.[14] A new hERG1 transcript was cloned and characterised, which contained the 1b exon at the 5′ end and the USO exon at the 3′ end, and hence was named *hERG1b$_{USO}$*.[15] The proteins encoded by both transcripts, hERG1$_{USO}$ and hERG1B$_{USO}$, do not give rise to any detectable ionic current, since they are not expressed on the plasma membrane, but are retained in the endoplasmic reticulum. hERG1B$_{USO}$ accomplishes a fundamental function in cancer cells, since it regulates hERG1 channel surface expression and hence the amount of hERG1 currents.[14] Recently, a primate-specific KCNH2 isoform (KCNH2-3.1) has been described, whose expression is roughly limited to the brain.[16] In rat smooth muscle cells, both erg1b and a tissue-specific erg1 isoform (erg1-sm) exist, lacking 101 amino acids in the C-terminus.[17]

Table 11.1 shows the expression pattern of the various transcripts of the *hERG1* genes, as well as the corresponding proteins and currents, both in humans and in rodents. It is evident that the two hERG1 principal isoforms are expressed in the heart, in selected neuronal cell populations of the central (CNS) and peripheral nervous system (PNS), in some neuroendocrine cells, such as anterior pituitary cells (both lactotropes and gonadotropes), chromaffin cells and pancreatic beta cells. hERG1 was also found to be expressed in gastrointestinal smooth muscle fibres, in the pacemaker cells of Cajal, as well as in non-excitable cells, like bone marrow CD34$^+$ hematopoietic precursors. The expression of hERG1 in various types of cancer cells is discussed in a later section.

The other two genes (*i.e. hERG2* and *hERG3*) have an expression pattern which is almost restricted to the nervous system. Detailed expression profiles of the three rodent *erg* gene transcripts and of the corresponding proteins (erg1, erg2 and erg3) in selected areas of the CNS have been reported elsewhere.[18–20] Since the different hERG proteins can co-assemble to form hetero-tetrameric functional channels,[21,22] neuronal cells can express functional ion channels with mixed biophysical properties. In some cases, just one of the four hERG subunits can modify biophysical properties of the tetrameric channel: this mechanism has been shown for erg3 in rat embryonic serotonergic neurons.[23]

The expression pattern of the *erg* genes has been also studied during embryo development in quails[24] and mice.[25] A broader *erg* expression pattern emerged early in development, with the three *erg* genes showing significant expression in areas and selected cell populations which then became progressively reduced later on. For example, the *erg* genes turned out to be highly expressed in neural crest cells, while their expression decreased in neural crest derivatives, like PNS ganglia. This progressive decrease in functional ERG channel expression has profound and interesting consequences for neuronal excitability (see below). Moreover, some cancer tissues apparently re-express hERG1, mimicking the embryonic pattern of expression.

Table 11.1 Summary of expression and functional roles of ERG channel family members in different tissues and mammalian species.

Tissue type	Species	Transcript	Protein	Current	Role	References
Heart	Human	herg1A herg1B herg1_USO	hERG1A hERG1B	–	Cardiac repolarisation (I_{KR})	2–4, 12, 14, 208
	Rat	r-erg1A r-erg1B	rERG1A rERG1B	Yes		12, 208, 209
	Mouse	m-erg1A m-erg1B	mERG1	Yes		9, 208, 210
Nervous System CNS	Human (brain)	herg1 herg 3.1	hERG1 hERG 3.1	–	Neuronal firing	16, 211
	Rat (brain, hippocampal astrocytes, substantia nigra)	r-erg1 r-erg2 r-erg3	rERG1 rERG2 rERG3	Yes	K^+ homeostasis, Neuronal firing	18, 19, 23, 211–213
	Mouse (brain, spinal cord, cerebellar Purkinje neurons)	m-erg1A m-erg1B m-erg2 m-erg3	mERG1A mERG1B mERG2 mERG3	Yes	Excitability, Firing Pattern, Complex spike modulation	20, 25, 68, 214
Nervous System PNS	Human	–	–	–	Unknown	
	Rat (celiac ganglia)	r-erg1 r-erg2 r-erg3	rERG1 rERG2 rERG3	–	Unknown	215
	Mouse (dorsal root ganglia)	m-erg1 m-erg2 m-erg3	mERG1 mERG2 mERG3	–	Unknown	25

	Cell type	Gene	Protein	Functional	Function	References
Endocrine System	Human (α and β cells islets)	herg1 herg2 herg3	hERG1 hERG2 hERG3	Yes	Secretion	73, 216
	Rat (pituitary cells, pancreas β cells, chromaffin cells)	r-erg1A r-erg1B r-erg2 r-erg3	rERG1A rERG1B rERG12 rERG13	Yes	AP repolarisation and firing frequency accommodation	74, 75, 217–220
	Mouse (gonadotropes, α and β cells islets,)	m-erg1	mERG1	Yes	Mediator of α and β cells repolarisation	216, 221
Muscle	Human (smooth muscle cells of the colon)	herg1A herg1B herg1$_{USO}$	hERG1A hERG1B	Yes	Repolarisation of action potential and of gastrointestinal waves	222, 223
	Rat (tibialis, stomach)	r-erg1A r-erg1B r-erg1-sm	rERG1A	–	Unknown	17, 211
	Mouse (interstitial cells of Cajal, smooth muscle cells of the jejunum)	m-erg1 m-erg2 m-erg3	mERG1 mERG2 mERG3	Yes	Contributes to pacemaker activity	224, 225
Blood cell and hemopoietic precursors	Human (PB hemopoietic progenitor cells)	herg1		Yes	Proliferation, adhesion	164, 165, 226
	Human (PB CD3 +)	herg1		–	Proliferation	227
	Human (BM CD10 +)		hERG1	–		166

11.1.2 hERG1 Channel Gating

Broadly speaking, the functional roles of hERG1 in the different tissues in which it is expressed are either to repolarise membrane potential during action potentials, suppress or dampen depolarising stimuli, maintain cells at a stable resting membrane potential or a combination of the above. These functions are intimately associated with the unique voltage and time dependent channel opening and closing (gating) properties of hERG1 channels (Figure 11.2). hERG1 channel gating is distinctive, in large part due to its unusual inactivation gating properties.[26–28] At depolarised potentials, currents are generally small in amplitude because the channels can rapidly enter a non-conducting inactivated state.[29] However, with repolarisation, hERG1 channels can rapidly (within a few milliseconds) recover from inactivation into the open state giving rise to a resurgent outward K^+ current that will drive membrane potential towards E_K, the equilibrium potential for K^+ ions, which is -80 to -90 mV in most cells (Figure 11.2). Once open, hERG channels close (deactivate) slowly, in a voltage dependent manner with time constants ranging from hundreds of milliseconds to seconds at physiologically relevant membrane potentials.[3,30,31] It is this combination of rapidly recovering from

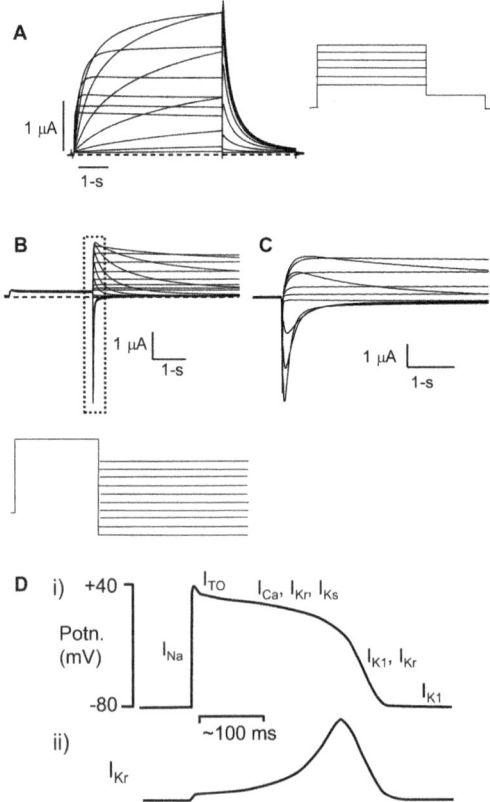

inactivation to pass large amplitude currents, together with slow channel closure that enables hERG channels to quickly repolarise membrane potential, rapidly respond to depolarising stimuli (such as unsynchronised depolarisations in the heart) and decrease cellular excitability.[32,33] Reduction of hERG currents invariably results in increases in cellular excitability.[34,35]

11.1.3 hERG1 Channel Structure

So far, there are no crystal structures for the pore and voltage sensing domains of hERG1. Nevertheless, structures from other K$^+$ channels coupled with results from structure-function studies have provided valuable insight into the likely structural basis of hERG1 channel gating. Within each hERG subunit, the S1–S4 helices form a voltage sensor domain (VSD) that senses transmembrane potential and is coupled to a central K$^+$-selective pore domain *via* the S4–S5 linker (*e.g.* [36–44]). Each pore domain is comprised of an outer helix (S5) and inner helix (S6) that together coordinate the pore helix and selectivity filter (Figure 11.1). The C-terminal end of the pore helix and

---

Figure 11.2 hERG1a K$^+$ currents elicited by voltage step protocols (A–C) and during a cardiac action potential (D). (A) hERG currents during a current-voltage protocol (inset) in which a series of test pulses are applied to potentials from −60 to +40 mV and tail currents measured upon repolarisation to −70 mV. (B and C) hERG currents during a fully activated current-voltage protocol. The first voltage step is to +40 mV and is sufficient to activate (open) all the functional channels, although the current amplitude is small because at this potential most channels also rapidly inactivate and are thus non-conducting. The second voltage step is to a series of potentials ranging from −140 to +30 mV. Upon repolarisation to more negative potentials during the second voltage step, channels rapidly recover from inactivation giving rise to a quick increase in current amplitudes. The subsequent exponential decay in the current is due to the channels closing (deactivating). These gating processes are highly voltage and time dependent. This can be seen more clearly in C, in which selected current traces from (B) at the time points indicated by the dotted rectangle, are shown with greater temporal resolution to illustrate the differences in current deactivation at different potentials. Currents in A–C were recorded from Xenopus oocytes expressing WT hERG1a. D) Schematic representation of a cardiac ventricular action potential (i) and the physiological time course of I$_{Kr}$ (ii) carried by hERG1 channels. The time course of hERG1 currents during an action potential is clearly very different from that during step-changes in membrane potential. Note that most current flows during the late stage of the action potential when this current has an important role in ending the plateau phase. The other major ionic currents involved in different stages of the action potential are indicated in (i). I$_{Na}$ and I$_{Ca}$ are voltage gated sodium and calcium currents respectively, I$_{TO}$ is the transient outward potassium current, I$_{Kr}$ and I$_{Ks}$ are the rapidly and slowly activating delayed rectifier K$^+$ currents respectively and I$_{K1}$ is the inwardly rectifying K$^+$ current.

selectivity filter contain the highly conserved K^+ channel signature sequence, which in hERG is Thr-Ser-Val-Gly-Phe-Gly. This sequence forms a narrow conduction pathway at the extracellular end of the pore in which K^+ ions are coordinated by the backbone carbonyl oxygen atoms of the signature sequence residues. Inactivation gating in hERG and other channels is not fully understood, but is likely to involve subtle conformational changes to the backbone of the selectivity filter (*e.g.* [45]) that impair K^+ ion coordination and block conduction. Inactivation can be perturbed by mutations at other sites, which probably reflect changes in the stability of the selectivity filter or altered coupling between the voltage sensor and pore.[46]

Below the selectivity filter the pore widens to form a water-filled central cavity that is lined by residues from the S6 helices. As with other K_v channels, the cytoplasmic ends of the inner (S6) helices of hERG are thought to act as a barrier (gate) to movement of ions across the pore.[47] Upon depolarisation, conformational changes in the voltage sensor are coupled to the pore so that the S6 helices splay apart to create a wide aperture at the interface with the cytoplasm, which ions and drugs can move through to enter the inner cavity. Opening of K_v channels requires the S6 helices to bend.[48] The precise conformation of the hERG channel pore in the open state is unknown. The S6 helices probably bend at or close to conserved glycine residues. These glycines are not required for their flexibility but are retained because in these positions their small side chains permit close packing with neighbouring helices when the channels close.[49] How wide the pore opens is not known, but it must be sufficient to accommodate relatively large blockers of the channel, which bind within the inner cavity (see later section and Figure 11.1B).

Slow channel deactivation is a trait of hERG1 currents and important for the channels physiological roles and pharmacological properties (Figure 11.2). Recent studies have significantly clarified the role of intracellular structural domains on the N- and C-termini (Figure 11.1A). The N-terminus of hERG1 is ~400 residues long, with residues 1–135 forming the eag domain.[1] hERG1 channel deactivation is greatly accelerated by deleting either the whole N-terminus or just the eag domain[26,28,50,51] and it is because hERG1b channels lack an eag domain that they deactivate so rapidly.[9,10] The crystal structure for residues 26–135 of the eag domain was solved by Rod MacKinnon and colleagues and was shown to fold into a *Per-Arnt-Sim* (PAS) domain.[52] PAS domains are structural folds that frequently impart a sensor capability in signalling proteins by binding cofactors such as heme, light sensitive chromophores and redox sensitive flavins.[53,54] The PAS domain of hERG1 was crystallised without a ligand and its role is likely to bind to structures in the channel to regulate deactivation gating. Unfortunately, the crystal structure did not resolve the first 26 amino acids (NT1-26), which is now known to be the functionally critical domain for slowing deactivation gating.[51,52] Recently, three groups have solved the structure of NT1-26 using NMR spectroscopy.[55–57] The NT1-26 region was shown to be structurally independent from the PAS domain and composed of two parts—a well-defined amphipathic α-helix (Gln11–Gly24) and a highly dynamic extended structure (Met1-Pro10).[57] Calculation of the electrostatic surface of the NT1-26 domain revealed that it is highly positively charged on one side and this charge is required for interactions with the C-terminus cNBD that slow deactivation.[57] hERG channels are not regulated directly by cyclic nucleotides since key residues necessary for binding are not conserved.[58,59] However, it has become clear that the cNBD (and perhaps also the neighbouring C-linker region that links with the pore of the channel), form the receptor for interactions with the eag domain that regulate deactivation. Both an intact cNBD + C-linker and the NT1-26 are required for normal slow deactivation gating.[57,60] Together, a consistent model is emerging in which slow deactivation of hERG1a results from stabilisation of the open state of the channel pore through constraints imposed by the cNBD + C-linker when it is bound to the NT1-26 on the eag domain. The precise role of the PAS domain remains uncertain but it may control access of the NT1-26 domain to its binding site and dynamically regulate deactivation gating through interactions with the pore, cNBD + C-helix, and voltage sensing domain.[61]

11.2   Physiological and Pathophysiological Roles of hERG1

11.2.1   Repolarisation of the Cardiac Action Potential

The physiological role of hERG channel subunits is most fully understood in the heart. hERG1 and hERG1b subunits coassemble to form heterologous channels that mediate the rapid delayed rectifier potassium current, I_{Kr}, which is necessary for normal action potential repolarisation (Figure 11.2D).[12,62] The ventricular cardiac action potential has five distinct phases. Phase 1 is the rapid upstroke due to activation of Na channels. Repolarisation then takes place during the next three phases. There is an initial fast repolarisation (phase 1) due to activation of transient outward K$^+$ current (I_{TO}), which lasts only a few milliseconds because the I_{TO} channels rapidly inactivate; this is followed by a very slow repolarisation (phase 2) that gives rise to the prolonged action potential plateau, and then the final repolarisation (phase 3) that terminates the action potential and returns membrane potential back to the resting membrane potential (phase 4). Phases 2 and 3 are so prolonged, in part because of the slow inactivation of L-type calcium current, but also because the K$^+$ currents responsible for repolarisation are small in amplitude, slow to activate or have a low conductance at depolarised potentials. The long duration of the cardiac action potential is important for allowing excitation-contraction coupling to occur, but also makes ventricular muscle refractory to premature or unsynchronised depolarisations. I_{Kr} is the most important of the K$^+$ currents for terminating the action potential. In the early part of Phase 2, I_{Kr} amplitude is small as the channels have not had sufficient time to activate and are also inactivated (Figure 11.2D). As other currents start to repolarise membrane potential, I_{Kr} channels rapidly recover from inactivation and the resulting resurgent current repolarises membrane potential and brings it into the range of potentials when I_{K1} channels are open—bringing about Phase 3 repolarisation. I_{Kr} also contributes to repolarisation of atrial cells and myocytes in the cardiac conduction system. Indeed, in pacemaker cells of the sinoatrial and atrioventricular nodes, I_{Kr} also helps to control the diastolic depolarisation; initially opposing depolarisation—an influence that decreases as the channels slowly deactivate.[63,64] MiRP1 (KCNE2) is an accessory β subunit that in heterologous expression systems decreases trafficking of channels to the cell surface and accelerates deactivation.[65] In the heart, MiRP1 expression is mostly limited to cells of the conduction system (particularly Purkinje cells) and thus it is unlikely to modify properties of I_{Kr} channels in the atria and ventricles.[66]

11.2.2 Neuronal and Smooth Muscle Cell Excitability

Data are emerging on the functional role of ERG K^+ currents ($I_{K(ERG)}$) in the non-cardiac cell populations where ERG expression has been detected.[67] In neurones, $I_{K(ERG)}$ contributes in shaping the neuronal discharge patterns, due to its unusual kinetic properties.[35,68] Indeed $I_{K(ERG)}$ can regulate cell firing in several types of neuronal cells.[35,68] For example, in cerebellar Purkinje cells it was found that the specific pharmacological blockade of the current increased the number of action potentials during climbing fibre stimulated activation. $I_{K(ERG)}$ also controls the excitability of medial vestibular nucleus neurones (MVNn), their discharge regularity and probably their resonance properties.[69] During early stages of spinal cord development, the excitability of GABAergic ventral interneurons, which is important for circuit maturation, depends, at least in part, on ERG currents.[70] In these cells, as in developing neuroblasts from quail neural crest cells,[71] $I_{K(ERG)}$ progressively disappears during maturation,[71,72] with profound implications for neuronal excitability. $I_{K(ERG)}$ also regulates hormonal secretion in neuroendocrine cells,[73–75] mainly through the regulation of Ca^{2+} concentration. In these cells, $I_{K(ERG)}$ has an inhibitory activity on secretion of the specific hormone, so that the blockade of $I_{K(ERG)}$ suddenly leads to increased hormone secretion. Finally, $I_{K(ERG)}$ controls smooth muscle cells contraction. In particular, $I_{K(ERG)}$ serves the function of a subthreshold current: hence, the pharmacological block of $I_{K(ERG)}$ in these cells depolarizes the membrane potential, with an increase in muscle contraction.[17]

Based on functional and expression data, we can conclude that channels of the ERG family can accomplish modulatory functions in the excitability of neurons, endocrine cells and smooth muscle cells.

11.3 hERG1 as an Antitarget

The importance of hERG1 channels in cardiac repolarisation came into sharp focus when it was discovered that inherited mutations in hERG1 were found in patients suffering from congenital long QT syndrome (LQTS).[2] LQTS is defined by a prolongation of the QT interval on the body surface electrocardiogram and indicates a lengthening of the time required for ventricular repolarisation. Afflicted patients are at a high risk of *torsade de pointes* (TdP), a characteristic arrhythmia in which the QRS complex on the ECG appears to oscillate around the isoelectric line. This polymorphic ventricular tachycardia causes a sudden loss of cardiac output, which if short-lived, causes syncope, but which can also spontaneously degenerate into ventricular fibrillation and lead to sudden cardiac death.[76] LQTS associated mutations in hERG1 reduce the amplitude of I_{Kr} (so called loss of function mutations) resulting in a slowing of ventricular repolarisation. The heart becomes more susceptible to arrhythmias because at the single cell level there is an increased risk of early after depolarisations—spontaneous depolarisations that may trigger arrhythmogenic events.[77] At the whole heart level there is also an exacerbation of differences in the time-course of action potential repolarisation across the myocardium (dispersion of repolarisation) providing an environment for propagation of re-entrant arrhythmias.[77,78] More recently congenital *short* QT syndrome has been described in which the QT interval is abnormally short (QTc less than 330–340 ms).[79,80] SQTS patients are also at a much greater risk of atrial and ventricular arrhythmias than the normal population, with disease having a high penetrance and lethality. As with LQTS, mutations in hERG1 are one of the most frequent causes of SQTS, but in this case the mutations cause a gain in function of hERG1 channels, increasing the amount of repolarising current.[81–84] What these congenital diseases demonstrate is the importance of normal hERG1 function for controlling action potential duration and protecting the heart against arrhythmic events.

11.3.1 Drug Induced Ventricular Arrhythmias

The incidence of congenital LQTS is relatively rare, with an estimated frequency of 1 in 3000 to 5000 people across the globe.[76] In the clinic, it is the drug induced form of LQTS that has been more prevalent.[6] It should be pointed out that QT prolongation *per se* is not necessarily proarrhythmic. There are drugs that cause QT prolongation, but are nonetheless not associated with TdP, and there are others that cause TdP without any detectable change in QT interval.[6,85] Over the years QT prolongation has become a surrogate for TdP, but this disguises the reality that multiple mechanisms are probably involved—including a loss of a clear plateau phase (action potential triangulation), reverse rate dependence and beat to beat instability of action potential duration.[86,87] It was in the early 1990s that it was widely recognized that relatively commonly used medications could induce TdP and sudden death. Large scale clinical investigations of class I antiarrhythmic agents that block sodium channels showed that drugs used to treat arrhythmias were proarrhythmic. At about the same time it also became clear that many non-cardiac medications including antibiotics, antipsychotics, antidepressants and antimalarials were also causing TdP.[78] Currently, there are about 30 medications that are widely accepted to cause TdP, but a much larger number than this have been implicated. Lists of drugs linked with TdP and estimates of risks are maintained on a variety of web sites including www.qtdrugs.org/. However, these drugs represent just the tip of the iceberg. Once the pharmaceutical industry and the drug regulatory agencies became aware of this issue, and started to investigate it more systematically, it became clear that very large numbers of compounds exhibited QT prolongation liabilities and that this represented a significant issue for drug development. The problem of drug induced TdP has been to some extent ameliorated by testing for it early in compound development.[88–92] However, this is not without drawbacks in terms of financial cost and because large numbers of potentially therapeutically valuable compounds are excluded from further development.

11.3.1.1 *Mechanisms of hERG1 Current Inhibition*

The mechanism by which compounds cause TdP is almost always by reducing hERG1 channel function. Indeed this is one of the most remarkable characteristics of this phenomenon since hERG1 is only one of many ion channels involved in regulating cardiac repolarisation and yet compounds from many different therapeutic classes and with diverse chemical structures interact with the channel to reduce its function.[93–95] In the vast majority of cases the drugs block the pore of the channel, but disruption of hERG1 channel trafficking so that fewer functional channels reach the cell surface is another mechanism that has been described for a small number of drugs.[96–102]

Most hERG1 channel blockers block from the intracellular side of the membrane and exhibit an open state dependent block. In voltage clamp studies the onset of block is seen once membrane potential has been stepped to a potential at which the channels open. Block is absent if the channels are maintained in the closed state. For most (but not all) high affinity blockers, recovery from block is slow and only occurs when the channels are open. This led to the idea that the activation gate acts a barrier to drugs gaining access to their binding site and also that once bound they become trapped. Support for this drug trapping hypothesis was provided by an unusual hERG1 channel mutant D540K, which not only opened upon depolarisation but also slowly opened upon hyperpolarisation, yielding an inward current reminiscent of hyperpolarisation activated cyclic nucleotide gated non-selective cation (HCN) current.[47,103] Drugs were found to block D540K at depolarised potentials, but recovery from block was relatively rapid and complete when the channels were opened by hyperpolarisation.[47,104–107] These studies strongly indicated that the inner cavity of the channel was the drug binding site and that when channels closed upon hyperpolarisation they were trapped and unable to escape. Mutagenesis studies have since provided valuable insight into the molecular nature of the inner cavity drug binding site and why so many drugs bind to hERG with a far higher affinity than other channels. The inner cavity is lined by residues from S6 and the pore helices (Figure 11.1B). The most important determinants of drug binding are the aromatic residues Tyr652 and Phe656 on S6 and the polar residues Thr623 and Ser624 on the pore helices.[94,104–106,108–120] The importance of the aromatic residues is particularly significant because most other K$^+$ channels have hydrophobic residues (usually Leu, Ile or Val) in the analogous positions.[114] Aromatic residues are capable of highly diverse interactions with other molecules because not only can they be considered hydrophobic, but also polar because of the quadrupolar charge distribution of their phenyl rings.[121] Electronegative faces formed by π-electron orbitals can make cation-π interactions with ligands and the electropositive hydrogen ring edge may also allow electrostatic interactions with anionic or electronegative molecules. This is in addition to π–π and T-stacking interactions with phenyl rings. A study in which the Tyr652 and Phe656 were systematically mutated to different residues suggested that a hydrophobic residue at position 656 is sufficient for high affinity binding of terfenadine, cisapride and MK-499, whereas an aromatic residue is required at position 652.[110] The fact that there are eight aromatic residues (two per hERG subunit) facing into the cavity of the channel probably accounts for why so many diverse molecules can bind to hERG—especially since many hERG blockers are charged amines with one or more phenyl rings.

11.3.1.2 State Dependent Differences in Drug Potency

The importance of channel gating for drug potency has been known about for a long time. Early studies clearly demonstrated the requirements for channels to open before drug block could occur. However, the situation appears to be more complex than just drugs gaining access to their receptor. EAG (Kv10) channels have Tyr and Phe residues in analogous positions to hERG and yet most blockers have a much lower potency for this closely related channel.[122–124] The explanation is likely to be related to gating dependent differences in the orientation of these residues relative to the central cavity. Mutation of EAG, so that the Tyr or Phe residues are repositioned towards the C-terminal ends of S6 by just one residue (which reorientates the side chain on the helix by $\sim110°$), increased the sensitivity to cisapride so that it was similar to hERG.[112] Repositioning the residues in hERG reduced cisapride sensitivity, suggesting that the aromatic residues are already optimally positioned for blocker interactions. EAG channels do not inactivate. One intriguing hypothesis is that inactivation gating of hERG1 channels includes conformational changes that position the aromatic residues for interactions with compounds in the central cavity.[112] This may explain the widely reported decrease of drug potency in inactivation-deficient hERG1 channel constructs (*e.g.* [122,123,125–127]).

A reduction of potency when channels close may also explain observations that a number of groups (including the Mitcheson lab) have made regarding recovery from channel block. As described earlier, many drugs show very slow recovery from block consistent with drug trapping (*e.g.* bepridil, domperidone, E-4031, MK499). On the other hand, other molecules with similar potencies dissociate more rapidly (*e.g.* amiodarone, cisapride, droperidol and haloperidol). Although it was thought that this was because they were too large to fit within the inner cavity when the channel was closed (so-called foot in the door type recovery from block), the latest evidence suggests that is not the case—with no clear relationship between molecule size and dissociation rate.[128] It appears that some molecules can form specific interactions with the closed state of the channel that retain binding affinity,[104] while others show very striking closed state dependent losses of binding affinity. A better understanding of the basis for these differences could help with developing drugs more suitable as antiarrhythmics. Many hERG1 blockers exhibit reverse use dependence, resulting in reduced block at faster heart rates and potentially beat to beat instability. Drugs with rapid kinetics may have more favourable properties than more slowly acting ones.

11.3.2 Medicinal Chemistry Strategies for Avoiding hERG Channel Block

In the past two decades a number of marketed drugs have been withdrawn because of TdP liabilities. Consequently, a great deal of effort has been made within the pharmaceutical industry to predict and detect compounds at risk of blocking hERG1 channels early in development and to design strategies to minimise hERG1 activity without losing efficacy at the primary target. The advent of medium to high throughput assays for measuring hERG activity—including the use of automated electrophysiology platforms—has provided a huge amount of information on drug block and the means to quickly determine if modifications have a favourable outcome for reducing hERG1 potency. The main strategies that have been utilised (reviewed in ref. 139) include: (1) specific structural modifications that utilise knowledge of hERG1 binding sites, hERG1 homology models and pharmacophore models; (2) the incorporation of a carboxylic group to compounds that contain charged amines to generate zwitterions; (3) reduction of logP; (4) modifications of pK_a. In practice, modifications rarely alter a single parameter in isolation so that identifying the means by which a change renders a positive outcome may be open to interpretation. Strategies published in the literature generally appear to be empirical rather than to reflect systematic application of medicinal chemistry approaches. A useful starting point may be to first assess the lipophilicity of compounds, since drugs need to cross the plasma membrane to access the inner cavity of hERG1 and hydrophobic interactions with inner cavity residues make important contributions to binding.[110] If the lipophilicity of a compound is high then reducing it by introducing polar substituents may be a useful strategy to reduce hERG1 binding and has been successfully used by a number of groups.[129–132] If compound lipophilicity is already low then other strategies need to be investigated.

The case for using strategies that modify the pK_a of compounds with basic amines is less straightforward. Arguments exist in favour of both increasing and reducing pK_a. On the one hand, mutagenesis data suggests that the aromatic inner cavity residues may form cation-π interactions with charged amine groups. Thus reducing pK_a, so that the proportion of molecules in the protonated charged form is reduced, would be expected to reduce these types of electrostatic interactions in the inner cavity. On the other hand, increasing the polarity of compounds by increasing pK_a, would be favourable through a similar mechanism to reducing lipophilicity. The literature suggests that both approaches have been utilised successfully and that there is a poor correlation between measures of basicity and hERG1 activity.[133–135]

The zwitterion approach was one of the first strategies to be successfully utilised when it was realised that the principal metabolite of terfenadine was the carboxylated form fexofenadine. Fexofenadine retains its activity for its primary target, H1 histamine receptors, but has a markedly reduced affinity for hERG1 channels.[136] Several different explanations have been provided for the reduced block of hERG1.[137] We have found that even if fexofenadine is applied intracellularly, block of hERG1 channels is dramatically attenuated compared to terfenadine (Wernhoener and Mitcheson, unpublished observations). This is consistent with the proposal that the carboxylate enables a salt bridge to be formed with the charged amine so that the conformation of fexofenadine is more folded—reducing favourable interactions with the inner cavity.[138] In general though, the rationale behind the zwitterionic approach is that compound modifications reduce membrane permeability. While this can be favourable it may also cause a concomitant loss of accessibility of compounds to their primary targets, reductions in absorption and failure of CNS targeted drugs to cross the blood–brain barrier. Therefore, the zwitterion approach has substantial limitations in practise and is not suitable as a general resolution that can be applied across compound series.[139]

The most successful strategy for reducing hERG1 block has been to make discrete structural changes. These have frequently been made in an apparently empirical manner. However, the availability of homology models of hERG1 and descriptive pharmacophore models of TdP associated compounds increases the prospects for more rational approaches to reducing hERG1 block.[45,90,94,95,138,140–146] Removing or modifying phenyl groups, particularly at the ends of molecules, to minimise interactions with inner cavity aromatic residues can be highly beneficial.[109,147,148] Changing para-phenyl substituents has also been used—for example a methyl sulfonamide to an ethyl sulfonamide,[149] a nitro group to an amine or amide[109] and re-positioning a para-substituent to a meta position[150] have all been shown to reduce hERG potency while having minimal effects on efficacy for primary targets. These changes may well reduce interactions with residues deep within the inner cavity such as Ser624, Thr623 and Tyr652.[109] These types of discrete structural modification should become easier to utilise as our knowledge of the structure of the inner cavity improves.

11.4   hERG1 as a Target

Concerns over the Torsadogenic potential of hERG1 blockers have meant that hERG is considered by many as a pariah protein or antitarget, which should be avoided in drug development. This ignores the evidence that there are several examples of compounds that block hERG1 and would have been discarded under current regulatory guidelines, but that have excellent safety records. Thus, hERG1 block *per se* is not arrhythmogenic any more than QT prolongation is. In addition, a few molecules have been identified that actually increase hERG1 channel function (reviewed in [151]). Studies on these hERG activators have provided valuable insight into hERG1 gating and offer the potential of enhancing repolarising currents in patients with congenital LQTS. The other reason why hERG1 should be considered as a target is that there is now compelling evidence that hERG1 is aberrantly expressed in cancer cells. *In vivo* and *in vitro* studies suggest that hERG1 blockers are worth considering for oncological therapy. The following paragraphs will focus on hERG1 expression, its role in human cancers, its diagnostic and therapeutic potential in oncology, as well as on strategies aimed at overcoming potential adverse cardiac side effects of hERG1 blocker. A subsequent section will look at the properties and therapeutic potential of hERG1 activators.

11.4.1   hERG1 Channels as Potential Drug Targets in Oncology

Cancer is a multistep process, characterized by the progressive accumulation of genetic defects (somatic mutations, over expression of cancer related genes, inactivation or deletion of tumour suppressor genes) that ultimately leads to the acquisition of an uncontrolled malignancy.[152] In the last two decades, increasing evidence has clearly indicated that ion channels and transporters are involved in distinct steps of tumour progression (reviewed in [7,153,154]). Indeed, ion channels can control cell proliferation, escape from apoptotic death, migration and invasiveness, both fundamental for the establishment of metastases.[155] Ample evidence indicates that hERG1 is one of the K^+ channels aberrantly expressed in several human cancers (see Table 11.2). hERG1 expression/activity is implicated in neoplastic progression, due to the pleiotropic effects that hERG1 channels exert on different aspects of cancer cell behaviour. Such differences often depend on the specific cancer cell types where hERG1 channels are expressed (Table 11.2).[7,156]

11.4.1.1   *hERG1 Channels are Expressed in Tumour Cells.*

Both the *hERG1* gene and the corresponding protein and current have been detected in several types of tumour cell lines and primary human cancers,

Table 11.2 hERG1A, hERG1B and hERG2/3 expression in cell lines and primary tumours. The first column (Cell type) shows the cell lines or primary tumours where the hERG channels are expressed. In the second and third columns are reported ion channel types found to be expressed in the different tumour cells and tissues. The fourth column shows the cellular mechanisms (Role) affected by hERG ion channels. The fifth column (Biomarker) indicates if the ion channel represents a biomarker or not. (AML: acute myeloid leukaemia; ALL: acute lymphoblastic leukaemia; CLL: chronic lymphoblastic leukaemia)

Cell type (cell line and primary tumour)	hERG1A	hERG1B	Role in tumour	Biomarker	References
Blood/bone marrow					
Myeloid leukaemia (FLG 29.1, U937, HL60, K562, UT-7)	medium	high	Control of cell proliferation and invasion Correlation with shorter survival Control of cell differentiation	yes	158 163 164 11 165 226
Lymphoid leukemia (CEM, REH,RS,697)	medium	high	Contribution to chemoresistance		227 166
Lymphoma (BL2, RAJI)	low	medium			227
Primary					
AML	medium	high	Control of cell proliferation		164 165 228
B-CLL	medium	high	Control of cell proliferation		227
B-ALL	low	high	Contribution to chemoresistance		166
Brain					
Neuroblastoma (SH-SY5Y,SH-NBE)	high	high	Control of cell differentiation Control of cell proliferation	–	158 171 11
Glia					
Glioblastoma (U138)	medium		Control of cell proliferation		160
Muscle					
Rhabdomyosarcoma (TE671)	medium				223
Colon					
Colorectal carcinoma (Colo 205, C26, HCT8, HCT116, HT29, T84, DLD1, H630)	medium	negative	Control of cell invasion	yes	161 229 230

Table 11.2 (Continued)

Cell type (cell line and primary tumour)	hERG1A	hERG1B	Role in tumour	Biomarker	References
Primary Colorectal carcinoma	high	negative	Marker of advanced stages		161 184
Stomach Gastric cancer (AGS, MKN45, MGC803, SGC7901)			Control of VEGF-A secretion Control of cell cycle	yes	169 231
Primary Gastric cancer	medium	medium			
Kidney *Primary* Clear cell carcinoma	medium	low			167
Lung adenocarcinoma (PG, A549) microcytoma (NCI-N592) small cell lung cancer (H69, GLC8)	medium				158
Breast Breast cancer (MCF-7, SKBr-3[a])	low		Control of cell proliferation		158 232
Skin Melanoma (MDA-MB-435S)	medium				168
Eye Retinoblastoma (Y79[b])	medium				11
Primary Retinoblastoma	high	medium			233
Prostate Prostatic adenocarcinoma (LNCaP)					234

[a] = SkBr3 cell line express hERG3.
[b] = Y79 cell line express hERG2.

whereas they are absent in their disease-free counterparts (Table 11.2). Evidence was first provided for neuroblastoma and myeloid leukaemia cell lines[157,158] and for primary endometrial adenocarcinomas.[159] Subsequent work showed hERG1 expression in several types of human cancers (malignant astrocytomas,[160] neuroblastomas,[11,157,158] colorectal and gastric cancers,[161,162] acute leukaemias,[158,163–166] renal carcinomas,[167] melanomas[168]) and in both cell cultures and primary samples. hERG1 also turned out to be expressed in precancerous lesions of the gastrointestinal (GI) tract: the Barrett's oesophagus[169] and gastric dysplasia.[162] RNAase Protection Assay experiments allowed the comparison of the *hERG1* transcript levels in the heart and in some tumour cells: it emerged that tumour cells indeed express the transcript at high levels, comparable to those found in the heart.[11] Moreover, tumour cells, mainly neuroblastomas and leukaemias, preferentially express the *hERG1b* transcript, at levels much higher than in the heart and, sometimes, even higher than the full length *hERG1* transcript.[11,237] Consequently, the "tumour hERG1 channel" was preferentially composed, totally or at least partially, by the hERG1B subunit, which often dictates the biophysical features of the ensuing current (Wanke E. *et al.*, personal communication). The opposite occurs in the heart, where the hERG1A subunit prevails on hERG1B, both quantitatively and functionally, to form active channels.[15,170] Also hERG1B$_{USO}$ (see above) was preferentially expressed in cancer cells.[15] As described above, the USO-containing isoforms (*e.g.* hERG1$_{USO}$ and hERG1B$_{USO}$) regulate hERG1 channel surface expression through a peculiar post-translational mechanism. This process results in decreased hERG1 current density, a fact that is relevant in cancer cells. In fact the "USO dependent effect" is apparently a check point in the regulation of different biological functions of cancers cells, from cell differentiation to apoptosis.[15]

11.4.1.2   What are hERG1 Channels Doing in Cancer Cells?

hERG1 channels can regulate diverse aspects of cancer cell behaviour. Differences are mainly related to the specific tumour cell type in which the channel is expressed. In fact, hERG1 can modulate: a) cell proliferation in neuroblastomas[11] and acute myeloid leukemias (AML);[164] b) apoptosis, and in turn chemoresistance, in acute lymphoblastic leukemias (ALL)[166] and colorectal cancers (Fortunato A., unpublished); c) invasiveness in AML[165] and colorectal cancers;[161] d) the secretion of the Vascular Endothelial Growth Factor-A (VEGF-A) in tumours of the brain[160] and AML.[162,165,235,236]

HERG1 can be considered a proliferation related gene in AML and neuroblastoma cells. In these cell types, the addition of drugs that block hERG1 channels impairs cell proliferation, albeit at relatively high concentrations. In leukaemia cells, such impairment is due to the block of the cells in the G1 phase of the cell cycle.[164] The pro-proliferative role of hERG1 channels in myeloid leukaemia cells was also supported by the demonstration that, in AML cell lines and primary AML blasts, the amount of hERG1 expressed correlates with the capacity of leukaemia cells to engraft the bone marrow, when injected into immunodeficient mice.[165] This fact can be related to a hERG1-dependent autocrine regulation of VEGF-A secretion (see below) and hence of leukaemia cell survival and proliferation. In AML, the role of hERG1 is far more complex, since it also affects the capacity of leukaemia cells to migrate outside the bone marrow into the peripheral blood and extra-medullary organs (the so-called transendothelial migration, TEM). This effect occurs through the modulation of adhesion receptors of the integrin family (see below). On the whole, hERG1 expression in AML represents a crucial step in the malignant phenotype progression.[165] In ALL hERG1 has a different, but still relevant, function since it mediates resistance to chemotherapy-induced apoptosis.[166] The role of hERG1 in apoptosis is controversial: some reports indicated hERG1 as a mediator of TNF-[171] or H_2O_2-induced[172] apoptosis. On the other hand, the inhibition of hERG1 by cisapride was shown to mediate the induction of apoptosis in gastric cancer cells.[173] The same effect was shown for doxazosin, an antihypertensive drug which can block hERG1 and induce apoptosis of hERG1 expressing cells.[174] We showed that hERG1 mediates the protective effect of bone marrow mesenchymal cells (MSCs) against chemotherapeutic drugs, thus behaving as an anti-apoptotic device.[166] Similarly to the hERG1 effect on TEM of AML, the hERG1-dependent chemoresistance in ALL relies on a complex interplay between the channel and integrin receptors (see below). Moreover, specific hERG1 blockers can overcome the MSCs-induced chemoresistance in leukaemia cells (see below).

In cancers of the gastrointestinal (GI) tract (colorectal and gastric cancers) hERG1 has a dual function: in colorectal cancer cells, it is an important determinant for the acquisition of an invasive phenotype.[161] This fact can explain the high incidence of hERG1 expression in invasive and metastatic colorectal adenocarcinomas, whilst no expression of the channel is detected in proliferating adenomas.[161] In both types of cancer, hERG1 regulates tumour angiogenesis, since it mediates the intracellular signalling pathway which ultimately triggers the secretion of the pro-angiogenic factor VEGF-A.[235,236] hERG1 drives VEGF-A secretion also in high grade gliomas[160] and AML cells.[162,165]

In summary, a progressively more complex and interconnected picture is emerging in which hERG1 exerts different effects on tumour cell behaviour, often with tissue specificity. How can such specificity be explained? hERG1 could affect different biological aspects in tumours, since it regulates the dynamics of membrane potential (V_m) in tumour cells, and this regulation could exert diverse effects depending on the total biophysical profile of the plasma membrane of the cell. It is worth noting here that because of the crossover of activation and inactivation curves, hERG1 produces steady state currents at potentials positive to -60 mV, which are maximal at around -40 mV.[33] Therefore, it can determine tonic modulation of both excitable (firing threshold) and non-excitable (resting potential) cells. The range of V_m between -30 and -50 mV is particularly relevant for cycling cells.[175]

Moreover, when hERG1b subunits are the main component of the hERG1 channels, as occurs in many tumour cell types, they display a positive shift in the voltage dependence of activation and inactivation, and hence steady-state currents may sustain more depolarized values (-20 mV).[11,157] This is indeed the V_m value of many tumour cells,[175] including leukaemias.[164] In addition, at least in certain neuroblastoma cells, hERG1B is significantly up regulated during the S phase of the cell cycle, and this could account for the depolarized V_m value detected in cycling cells.[11] These concepts imply that the functioning of hERG1 is different in tumour cells compared to the heart. In fact, in the heart hERG1 channels cycle relatively rapidly through different gating states in response to Vm changes during cardiac action potentials. In cancer cells on the other hand, hERG1 channels exhibit a significant steady-state conductance that contributes to the resting membrane potential. In other words, while in the heart hERG1 spends much of the time in the closed and inactivated states, but in tumour cells it is more frequently in the acti-vated state. These differences might have implications for pharmaceutical interventions.

Finally, it may be possible that hERG1 exerts such diverse effects by switching on or modulating cell-specific signalling pathways. This effect may or may not depend on K^+ fluxes and V_m alterations. Indeed, we showed that hERG1 can trigger and modulate intracellular signalling cascades.[7] This role depends on the formation of macromolecular complexes on the plasma membrane of tumour cells with adhesion (in particular integrin) receptors, alone or in association with other plasma membrane proteins (growth factor receptors, chemokine receptors) (reviewed in ref. 238). In other words, we can consider hERG1 channels as bifunctional proteins, in that they can both regulate V_m and modulate different membrane or cytosolic proteins in both a voltage-dependent and a voltage-independent way. As anticipated above, the main partner that is physically and functionally linked to hERG1 is represented by adhesion receptors belonging to the integrin family. In general, the functional association between ion channels and integrins can rely on two different mechanisms: an effect of integrins on ion channel function ("outside-to-in") and, *vice versa*, a modulatory effect of ion channel activity on integrin function and signalling ("inside-to-out"). In different experimental systems either of the two mechanisms can occur (reviewed in [176–178]). When hERG1 channels are involved in the cross talk with integ-rins, a complex interplay between the two proteins takes place and both mechanisms are operative. Integrins, once engaged by the proper ligand, can activate hERG1 channels which, in turn, modulate integrin function. This dual mechanism relies on the assembly of a plasma membrane com-plex between hERG1 and the β1 subunit of integrins. The β1/hERG1 complex has been detected in tumour cells and tissues, while it is absent in normal tissues, and in particular in the heart (Pillozzi S., personal com-munication). In epithelial cancers (gastric and colorectal cancer cells), the β1/hERG1 complex localizes at the adhesion sites and recruits FAK and Rac1. Both FAK phosphorylation and Rac1 activity turned out to depend on hERG1 currents.[179] hERG1 activity also regulates VEGF-A secretion in these types of cancers, through a complex modulation of HIF-1α transcriptional activity.[235,236] In AML, the β1/hERG1 complex includes the VEGF receptor 1 (Flt-1). The complex regulates signalling downstream to Flt-1 (MAP kinases and PI3K) and, in this way, AML proliferation and migration.[165] In ALL, the complex is triggered by adhesion onto MSCs and comprises the chemokine receptor CXCR4. When adherent to MSCs, ALL cells undergo activation of multiple pro-survival signalling pathways, centred on the activation of the Integrin Linked Kinas (ILK). The activity of hERG1 channels is relevant to these processes, while the blockade of hERG1 channels induces the apoptotic cell death of leukaemia cells.[166] Recent results, obtained employing the Föster Resonance Energy Transfer (FRET) technique have clearly demonstrated that the β1/hERG1 complex relies on a direct interaction between the two proteins.[180]

Finally, it is important to note that hERG1 channel effects on intracellular signalling pathways could rely on mechanisms that are independent from ion fluxes. Some voltage-dependent ion channels have been shown to behave as bifunctional proteins that, besides gating ion fluxes as usual, exert an ion conduction-independent control of several intracellular responses.[181] Evidence is particularly extensive for K^+ channels of the EAG family.[182] Based on our current evidence, we favour the hypothesis that the activity of hERG1 channels inside the complex with the β1 integrin modulates the function of the partner protein(s) mainly through conformational coupling, instead of alterations of ion flow (Arcangeli and Mitcheson, personal communication). Interestingly, long term application (for 2–3 weeks) of therapeutically relevant concentrations of dofetilide can alter the migration properties and morphology of transformed hERG expressing cell lines,[183] again providing supporting evidence that therapeutic interventions that target hERG1 channels may be worthwhile oncology therapies.

11.4.1.3   hERG1 as a Biomarker

The expression of hERG1 in various types of human cancers, and its absence in their normal counterparts, led us to consider the channel as a potential tumour biomarker. Indeed, hERG1 expression correlated with a more aggressive AML phenotype both *in vitro* and *in vivo*. In a cohort of patients affected by AML, hERG1 expression was associated with a higher probability of relapse and a shorter overall survival.[165] This was one of the first clinical and prognostic applications of an expression screen for a Kv channel. In ALL, the amount of hERG1 expressed on the plasma membrane of leukemic cells can be used to predict chemoresistance.[166] Finally, hERG1 immunoreactivity, measured by immunohistochemistry, turned out to be a reliable biomarker in solid cancers, such as colorectal,[161,162,184] gastric[173,185] and renal[167] cancers. In particular, in colorectal cancer, it was shown that the positivity for hERG1 expression in conjunction with the negativity for the expression of the Glucose Transporter 1(Glut-1) are in-dependent prognostic factors, in conjunction with the tumour staging ac-cording to the "Tumour-Node-Metastasis" (TNM) classification. Risk scores calculated from the final multivariate model allowed stratification of patients into different risk groups, where hERG1 positivity with Glut-1 negativity identified a poor prognosis for patient groups with stage I–II colorectal cancer.[162]

In line with this negative prognostic role of hERG1 in colorectal cancer, hERG1 was found to be expressed in pre-cancerous lesions of both the stomach and the lower tract of the oesophagus, and its expression could predict later progression towards a true cancer.[169]

Besides using hERG1 expression as a prognostic and, hopefully, predictive marker, another possibility is to apply the above concepts to design a new therapeutic and diagnostic (theranostic) approach employing hERG1 chan-nels as tumour targets. In particular, there is the intriguing possibility of using nanoparticles targeted with anti-hERG1 antibodies that are able to kill targeted cells by the delivery of a specific antineoplastic drug, previously loaded inside the nanoparticle. Such a strategy, already under development employing cancer-specific/cancer-associated proteins, could take advantage of the easy accessibility of ion channels on the plasma membrane of tumour cells.

11.4.1.4 Therapeutic Potential of hERG1 as an Anticancer Target

Blocking hERG1 activity tends to arrest cell proliferation in a variety of cultured neoplastic cells,[11,156,164] blocks the invasiveness of colorectal cancer cells[161] and the VEGF-A secretion from gliomas,[160] AML[165] and GI cancer cells.[235,236] More recently, we studied the effects of hERG1 blockers on chemosensitivity in ALL. It is known that ALL cells, when cultured in the presence of MSC, show a significant resistance to the apoptotic death in-duced by chemotherapeutic drugs. As described above, this effect relies on the fact that the co-culture of ALL with MSC induces the expression of a plasma membrane signalling complex constituted by hERG1, the β1 integrin subunit and the chemokine receptor CXCR4 (see above). When ALL cells were cultured with MSC and treated with the classical chemotherapeutic drugs used to treat ALL (*e.g.* doxorubicin, prednisone and methothrexate), they were protected by MSC from the drug-induced apoptosis. The addition of classical hERG1 blockers, the antiarrhythmics E4031 or Way 123,398, al-most completely abrogated the MSC-induced chemoresistance. The same effect was also observed with sertindole and erythromycin, two drugs which block hERG1 channels but are still used clinically. We conclude that the activity of hERG1 channels inside the β1 integrin/hERG1/CXCR4 complex, triggered by culture on MSC, mediates the MSC- induced drug resistance to apoptosis, and that different hERG1 blockers can overcome such resistance.[166]

These results were corroborated by studies in a murine model of ALL. In a first set of experiments, NOD-SCID mice were inoculated with 697 cells and treated daily with E4031 (20 mg/kg) for two weeks, starting one week after the inoculation. At the end of treatment, some of the mice were sacrificed and the degree of bone marrow, peripheral blood and extramedullary organs invasion by ALL cells was quantified. E4031 treatment significantly reduced the leukaemia burden and amount of liver and spleen infiltration by leukemic cells, with significant prolongation of mice survival. In a second set of experiments, we tested the effects *in vivo* of the combined treatment with E4031 and dexamethasone on ALL cells (the REH cells), which have been reported to be resistant to corticosteroids *in vivo*. Mice were treated for two weeks with dexamethasone, E4031, or both. E4031 reduced bone marrow engraftment of REH cells. This effect was related to an increased apoptosis of ALL cells, and was higher than that produced by dexamethasone. Treatment with dexamethasone and E4031 together nearly abolished bone marrow engraftment while producing substantial apoptosis. A marked reduction in leukemic cell infiltration of the spleen in mice treated with dexamethasone plus E4031 was also observed. On the whole, these data clearly indicate that hERG1 blockers can treat the leukaemia disease *in vivo*, both alone and in combination with classical chemotherapeutic drugs and are capable of reverting drug resistance both *in vitro* and *in vivo*.[166]

Notwithstanding this, the adverse risk of TdP by hERG1 blockers has to be considered.[6,88,93,142,186] As described earlier, hERG1 is blocked by many compounds. What is interesting from our perspective, and with regard to hERG1 as a possible antineoplastic target, is that not all hERG blockers produce arrhythmogenicity.[187] Sertindole, for example, inhibits hERG1 and prolongs QT, without inducing Torsadogenic effects.[188] The Torsadogenic potential of channel blockers is generally unknown because of the limited knowledge we have about the precise mechanistic relationship between QT prolongation and arrhythmia.[6] Another example that points to the feasibility of this approach is offered by R-roscovitine.[189] A supplementary problem presented by many of the hERG1-targeting inhibitors is that they often bind to the intracellular side of the channel protein. To develop more convenient drugs, a possible starting point is the structure of the several peptide toxins recently found to bind different extracellular channel portions, with high specificity.[190] To be proposed for antineoplastic purposes in human beings, it is hence necessary to test and produce less harmful hERG1 blocking compounds, devoid of cardiotoxicity and more specific for the hERG1 channels expressed in cancer cells.[7,191] Besides the molecular and biophysical features which characterize the "tumour" hERG1 channel, it is also possible to exploit the ion channel/integrin complex, which could be targeted and disassembled through different strategies, including inhibitory peptides or bifunctional antibodies directed against the channel and the integrin proteins.[7,191]

11.4.2 Therapeutic Potential of hERG1 Activators as Antiarrhythmic Compounds

Several small molecule activators of hERG channels that increase current amplitudes have recently been identified.[192–197] Many of these act by attenuating or slowing inactivation gating.[193–196,198,199] hERG channel activators can be separated into two distinct classes based on their mode of action. Type 1 activators, such as RPR260243, enhance current by attenuating inactivation and severely slowing the rate of channel deactivation.[192,198] Although slowed deactivation has been observed with other hERG channel activators,[194–196] this effect is much more profound with RPR260243.[198] Type 2 activators are the more frequent form and include PD-118057,[193] its analogue PD307243,[199] NS1643,[194] A935142[195] and ICA-105574.[196] They act primarily to attenuate inactivation through a dual mechanism involving both a shift in the voltage dependence of inactivation to more depolarized membrane potentials and a slowing of the onset rate.[195,199,200]

hERG activators are providing useful insights into the molecular and structural basis for the unusual gating properties of hERG. Type 1 activators bind to a site at the interface between the pore and the voltage sensor and hamper conformational changes to S6 as the pore closes.[192,198] In addition, they shift the voltage dependence of inactivation by an allosteric mechanism that has yet to be determined. By contrast, type II activators (such as ICA-105574 and PD-118057) bind closer to the selectivity filter, forming inter-subunit interactions between the pore helix and S6 that shifts the voltage dependence of inactivation to more depolarised potentials.[200,201] Impaired inactivation appears to be a common mechanism for most hERG channel activators. Two notable exceptions to this are mallotoxin[202] and the amiodarone derivative KB130015.[197] Neither of these molecules affects inactivation, but instead enhance hERG channel current by accelerating the rate of activation and shifting its voltage dependence to more negative potentials. The location of the binding residues for either of these molecules remains to be elucidated.

In principle, hERG activators have significant therapeutic potential. They could be used to enhance hERG currents in patients with LQTS. They have also been proposed as a novel antiarrhythmic strategy (reviewed in [203], although their potential has yet to be demonstrated clinically. Antiarrhythmic effects have been demonstrated by several groups using a number of single cell preparations, isolated cardiac tissue, Langendorff perfused hearts and *in vivo* models.[204,205] The actions of these compounds are not necessarily limited to hERG channels. For example, NS1643 inhibits a variety of other cardiac currents, including $I_{Ca,L}$, I_{Ks} and I_{TO} at similar (or lower) concentrations to those effective for activation of I_{Kr}.[205] Proarrhythmic actions of hERG activators have also been reported.[206,207] There are also concerns about the management of these compounds and correct dosing. There is the risk that these compounds could cause excessive shortening or triangulation of action potentials (drug induced short QT syndrome)[195] and increase dispersion of repolarisation. Despite this, hERG activators are an attractive proposition since they increase repolarising K^+ current, stabilise the repolarisation phase of the cardiac action potential and may protect against unsynchronised depolarisations triggered by delayed or early after-depolarisations.[203] The characterisation of the binding sites of activators may facilitate development of more potent and specific agents that could be developed as a novel alternative to current anti-arrhythmic treatments.

11.5 Conclusions

Since hERG1 channels were first discovered they have attracted a lot of interest because disrupting their function, either due to inherited mutations or drug block, carries a high risk of potentially lethal arrhythmias. Considerable progress has been made in understanding the reasons why many compounds have the ability to block this channel and in the early detection of compounds with QT liability. The ability to rationally modify compounds to reduce unwanted hERG channel block is also improving and is facilitated by advances in ligand docking, quantitative structure–activity relationships and *in silico* modelling of hERG. One of the unexpected consequences of screening compounds for inhibition of hERG has been the identification of novel compounds that do the opposite and activate the channel. These compounds offer the potential of combating arrhythmias by increasing the amplitude of I_{Kr}, but the discovery of short QT syndrome has demonstrated that increasing hERG channel function can also be arrhythmogenic. hERG inhibitors and activators interact with the channel in multiple ways. More research is needed to understand how to target these channels in a safe manner. This has become even more urgent now that the role of hERG in cancer is more fully understood. Initial studies indicate that hERG has great potential as a biomarker for the early identification of cancerous and pre-cancerous tissue. Further research is needed to assess the potential of hERG channel block for treating different types of cancer. Up until now industry has been reluctant to consider this because of the long QT issue. However, not all hERG channel blockers are unsafe. Alternative ways of treating cancer are urgently needed and it may be time to consider hERG as a tractable target rather than an anti-target.

References

1. J. W. Warmke and B. Ganetzky, *Proc. Natl. Acad. Sci. U. S. A.*, 1994, **91**, 3438–3442.
2. M. E. Curran, I. Splawski, K. W. Timothy, G. M. Vincent, E. D. Green and M. T. Keating, *Cell*, 1995, **80**, 795–803.
3. M. C. Sanguinetti, C. Jiang, M. E. Curran and M. T. Keating, *Cell*, 1995, **81**, 299–307.

4. M. C. Trudeau, J. W. Warmke, B. Ganetzky and G. A. Robertson, *Science*, 1995, **269**, 92–95.

5. D. M. Roden, *Am. J. Cardiol.*, 1993, **72**, 44B–49B.

6. P. Kannankeril, D. M. Roden and D. Darbar, *Pharmacol. Rev.*, 2010, **62**, 760–781.

7. A. Arcangeli, O. Crociani, E. Lastraioli, A. Masi, S. Pillozzi and A. Becchetti, *Curr. Med. Chem.*, 2009, **16**, 66–93.

8. G. A. Gutman, K. G. Chandy, S. Grissmer, M. Lazdunski, D. Mckinnon, L. A. Pardo, G. A. Robertson, B. Rudy, M. C. Sanguinetti, W. Stühmer and X. Wang, *Voltage-Gated Potassium Channels. IUPHAR database*, http://www.iuphar-db.org/DATABASE/FamilyMenuForward? familyId = 81.

9. J. P. Lees-Miller, C. Kondo, L. Wang and H. J. Duff, *Circ. Res.*, 1997, **81**, 719–726.

10. B. London, M. C. Trudeau, K. P. Newton, A. K. Beyer, N. G. Copeland, D. J. Gilbert, N. A. Jenkins, C. A. Satler and G. A. Robertson, *Circ. Res.*, 1997, **81**, 870–878.

11. O. Crociani, L. Guasti, M. Balzi, A. Becchetti, E. Wanke, M. Olivotto, R. S. Wymore and A. Arcangeli, *J. Biol. Chem.*, 2003, **278**, 2947–2955.

12. E. M. Jones, E. C. Roti Roti, J. Wang, S. A. Delfosse and G. A. Robertson, *J. Biol. Chem.*, 2004, **279**, 44690–44694.

13. G. A. Robertson, E. M. Jones and J. Wang, *Novartis Found. Symp.*, 2005, **266**, 4–15; discussion 15–18, 44–15.

14. S. Kupershmidt, D. J. Snyders, A. Raes and D. M. Roden, *J. Biol. Chem.*, 1998, **273**, 27231–27235.

15. L. Guasti, O. Crociani, E. Redaelli, S. Pillozzi, S. Polvani, M. Masselli, T. Mello, A. Galli, A. Amedei, R. S. Wymore, E. Wanke and A. Arcangeli, *Mol. Cell. Biol.*, 2008, **28**, 5043–5060.

16. S. J. Huffaker, J. Chen, K. K. Nicodemus, F. Sambataro, F. Yang, V. Mattay, B. K. Lipska, T. M. Hyde, J. Song, D. Rujescu, I. Giegling, K. Mayilyan, M. J. Proust, A. Soghoyan, G. Caforio, J. H. Callicott, A. Bertolino, A. Meyer-Lindenberg, J. Chang, Y. Ji, M. F. Egan, T. E. Goldberg, J. E. Kleinman, B. Lu and D. R. Weinberger, *Nat. Med.*, 2009, **15**, 509–518.

17. S. Ohya, K. Asakura, K. Muraki, M. Watanabe and Y. Imaizumi, *Am. J. Physiol.: Gastrointest. Liver Physiol.*, 2002, **282**, G277–287.

18. M. J. Saganich, E. Machado and B. Rudy, *J. Neurosci.*, 2001, **21**, 4609–4624.

19. M. Papa, F. Boscia, A. Canitano, P. Castaldo, S. Sellitti, L. Annunziato and M. Taglialatela, *J. Comp. Neurol.*, 2003, **466**, 119–135.

20. L. Guasti, E. Cilia, O. Crociani, G. Hofmann, S. Polvani, A. Becchetti, E. Wanke, F. Tempia and A. Arcangeli, *J. Comp. Neurol.*, 2005, **491**, 157–174.

21. S. Wimmers, C. K. Bauer and J. R. Schwarz, *Pflugers Arch., EJP*, 2002, **445**, 423–430.

22. S. Wimmers, I. Wulfsen, C. K. Bauer and J. R. Schwarz, *Pflugers Arch., EJP*, 2001, **441**, 450–455.

23. W. Hirdes, M. Schweizer, K. S. Schuricht, S. S. Guddat, I. Wulfsen, C. K. Bauer and J. R. Schwarz, *J. Physiol.*, 2005, **564**, 33–49.

24. O. Crociani, A. Cherubini, E. Piccini, S. Polvani, L. Costa, L. Fontana, G. Hofmann, B. Rosati, E. Wanke, M. Olivotto and A. Arcangeli, *Mech. Dev.*, 2000, **95**, 239–243.

25. S. Polvani, A. Masi, S. Pillozzi, L. Gragnani, O. Crociani, M. Olivotto, A. Becchetti, E. Wanke and A. Arcangeli, *Gene Expr. Patterns*, 2003, **3**, 767–776.

26. P. S. Spector, M. E. Curran, A. Zou, M. T. Keating and M. C. Sanguinetti, *J. Gen. Physiol.*, 1996, **107**, 611–619.

27. P. L. Smith, T. Baukrowitz and G. Yellen, *Nature*, 1996, **379**, 833–836.

28. R. Schonherr and S. H. Heinemann, *J. Physiol.*, 1996, **493**, 635–642.

29. S. Wang, S. Liu, M. J. Morales, H. C. Strauss and R. L. Rasmusson, *J. Physiol.*, 1997, **502**, 45–60.

30. Z. F. Zhou, Q. M. Gong, B. Ye, Z. Fan, J. C. Makielski, G. A. Robertson and C. T. January, *Biophys. J.*, 1998, **74**, 230–241.

31. J. Chen, A. Zou, I. Splawski, M. T. Keating and M. C. Sanguinetti, *J. Biol. Chem.*, 1999, **274**, 10113–10118.

32. J. C. Hancox, A. J. Levi and H. J. Witchel, *Pflugers Arch., EJP*, 1998, **436**, 843–853.

33. Y. Lu, M. P. Mahaut-Smith, A. Varghese, C. L. H. Huang, P. R. Kemp and J. I. Vandenberg, *J. Physiol.*, 2001, **537**, 843–851.

34. L. Faravelli, A. Arcangeli, M. Olivotto and E. Wanke, *J. Physiol.*, 1996, **496**(Pt 1), 13–23.

35. N. Chiesa, B. Rosati, A. Arcangeli, M. Olivotto and E. Wanke, *J. Physiol.*, 1997, **501**, 313–318.

36. J. Liu, M. Zhang, M. Jiang and G. N. Tseng, *J. Gen. Physiol.*, 2003, **121**, 599–614.

37. D. R. Piper, M. C. Sanguinetti and M. Tristani-Firouzi, *Novartis Found. Symp.*, 2005, **266**, 46–52; discussion 52–46, 95–49.

38. D. R. Piper, A. Varghese, M. C. Sanguinetti and M. Tristani-Firouzi, *Proc. Natl. Acad. Sci. U. S. A.*, 2003, **100**, 10534–10539.

39. P. L. Smith and G. Yellen, *J. Gen. Physiol.*, 2002, **119**, 275–293.

40. R. N. Subbiah, C. E. Clarke, D. J. Smith, J. Zhao, T. J. Campbell and J. I. Vandenberg, *J. Physiol.*, 2004, **558**, 417–431.

41. R. N. Subbiah, M. Kondo, T. J. Campbell and J. I. Vandenberg, *J. Physiol.*, 2005, **569**, 367–379.

42. M. Zhang, J. Liu and G. N. Tseng, *J. Gen. Physiol.*, 2004, **124**, 703–718.

43. T. Ferrer, J. Rupp, D. R. Piper and M. Tristani-Firouzi, *J. Biol. Chem.*, 2006, **281**, 12858–12864.

44. A. C. Van Slyke, S. Rezazadeh, M. Snopkowski, P. Shi, C. R. Allard and T. W. Claydon, *Biophys. J.*, 2010, **99**, 2841–2852.

45. P. J. Stansfeld, A. Grottesi, Z. A. Sands, M. S. Sansom, P. Gedeck, M. Gosling, B. Cox, P. R. Stanfield, J. S. Mitcheson and M. J. Sutcliffe, *Biochemistry*, 2008, **47**, 7414–7422.

46. D. R. Piper, W. A. Hinz, C. K. Tallurri, M. C. Sanguinetti and M. Tristani-Firouzi, *J. Biol. Chem.*, 2005, **280**, 7206–7217.
47. J. S. Mitcheson, J. Chen and M. C. Sanguinetti, *J. Gen. Physiol.*, 2000, **115**, 229–240.
48. O. Yifrach and R. MacKinnon, *Cell*, 2002, **111**, 231–239.
49. R. M. Hardman, P. J. Stansfeld, S. Dalibalta, M. J. Sutcliffe and J. S. Mitcheson, *J. Biol. Chem.*, 2007, **282**, 31972–31981.
50. J. Wang, M. C. Trudeau, A. M. Zappia and G. A. Robertson, *J. Gen. Physiol.*, 1998, **112**, 637–647.
51. J. Wang, C. D. Myers and G. A. Robertson, *J. Gen. Physiol.*, 2000, **115**, 749–758.
52. J. H. Morais Cabral, A. Lee, S. L. Cohen, B. T. Chait, M. Li and R. Mackinnon, *Cell*, 1998, **95**, 649–655.
53. M. A. Gilles-Gonzalez and G. Gonzalez, *J. Appl. Physiol.*, 2004, **96**, 774–783.
54. B. L. Taylor and I. B. Zhulin, *Microbiol. Mol. Biol. Rev.*, 1999, **63**, 479–506.
55. C. A. Ng, M. J. Hunter, M. D. Perry, M. Mobli, Y. Ke, P. W. Kuchel, G. F. King, D. Stock and J. I. Vandenberg, *PLoS One*, 2011, **6**, e16191.
56. Q. Li, S. Gayen, A. S. Chen, Q. Huang, M. Raida and C. Kang, *Biochem. Biophys. Res. Commun.*, 2010, **403**, 126–132.
57. F. W. Muskett, S. Thouta, S. J. Thomson, A. Bowen, P. J. Stansfeld and J. S. Mitcheson, *J. Biol. Chem.*, 2011, **286**, 6184–6191.
58. W. N. Zagotta, N. B. Olivier, K. D. Black, E. C. Young, R. Olson and E. Gouaux, *Nature*, 2003, **425**, 200–205.
59. T. I. Brelidze, A. E. Carlson and W. N. Zagotta, *J. Biol. Chem.*, 2009, **284**, 27989–27997.
60. A. S. Gustina and M. C. Trudeau, *J. Gen. Physiol.*, 2011, **137**, 315–325.
61. A. S. Gustina and M. C. Trudeau, *Proc. Natl. Acad. Sci. U. S. A.*, 2009, **106**, 13082–13087.
62. H. Sale, J. Wang, T. J. O'Hara, D. J. Tester, P. Phartiyal, J. Q. He, Y. Rudy, M. J. Ackerman and G. A. Robertson, *Circ. Res.*, 2008, **103**, e81–95.
63. J. S. Mitcheson and J. C. Hancox, *Pflugers Arch., EJP*, 1999, **438**, 843–850.
64. K. Ono and H. Ito, *Am J. Physiol.*, 1995, **269**, H453–462.
65. G. W. Abbott, F. Sesti, I. Splawski, M. E. Buck, M. H. Lehmann, K. W. Timothy, M. T. Keating and S. A. Goldstein, *Cell*, 1999, **97**, 175–187.
66. M. Pourrier, S. Zicha, J. Ehrlich, W. Han and S. Nattel, *Circ. Res.*, 2003, **93**, 189–191.
67. J. R. Schwarz and C. K. Bauer, *J. Cell. Mol. Med.*, 2004, **8**, 22–30.
68. T. Sacco, A. Bruno, E. Wanke and F. Tempia, *J. Neurophysiol.*, 2003, **90**, 1817–1828.
69. M. Pessia, I. Servettini, R. Panichi, L. Guasti, S. Grassi, A. Arcangeli, E. Wanke and V. E. Pettorossi, *J. Physiol.*, 2008, **586**, 4877–4890.
70. F. Furlan, G. Taccola, M. Grandolfo, L. Guasti, A. Arcangeli, A. Nistri and L. Ballerini, *J. Neurosci.*, 2007, **27**, 919–928.
71. A. Arcangeli, B. Rosati, A. Cherubini, O. Crociani, L. Fontana, C. Ziller, E. Wanke and M. Olivotto, *Eur. J. Neurosci.*, 1997, **9**, 2596–2604.

72. A. Arcangeli, B. Rosati, A. Cherubini, O. Crociani, L. Fontana, B. Passani, E. Wanke and M. Olivotto, *Biochem. Biophys. Res. Commun.*, 1998, **244**, 706–711.

73. B. Rosati, P. Marchetti, O. Crociani, M. Lecchi, R. Lupi, A. Arcangeli, M. Olivotto and E. Wanke, *FASEB J.*, 2000, **14**, 2601–2610.

74. M. Lecchi, E. Redaelli, B. Rosati, G. Gurrola, T. Florio, O. Crociani, G. Curia, R. R. Cassulini, A. Masi, A. Arcangeli, M. Olivotto, G. Schettini, L. D. Possani and E. Wanke, *J. Neurosci.*, 2002, **22**, 3414–3425.

75. F. Gullo, E. Ales, B. Rosati, M. Lecchi, A. Masi, L. Guasti, M. F. Cano-Abad, A. Arcangeli, M. G. Lopez and E. Wanke, *FASEB J.*, 2003, **17**, 330–332.

76. I. Goldenberg, W. Zareba and A. J. Moss, *Curr. Probl. Cardiol.*, 2008, **33**, 629–694.

77. S. Poelzing and D. S. Rosenbaum, *Novartis Found. Symp.*, 2005, **266**, 204–217; discussion 217–224.

78. Y. G. Yap and A. J. Camm, *Heart*, 2003, **89**, 1363–1372.

79. I. Gussak, P. Brugada, J. Brugada, R. S. Wright, S. L. Kopecky, B. R. Chaitman and P. Bjerregaard, *Cardiology*, 2000, **94**, 99–102.

80. P. Bjerregaard and I. Gussak, *Nat. Clin. Pract. Cardiovasc. Med.*, 2005, **2**, 84–87.

81. R. Brugada, K. Hong, R. Dumaine, J. Cordeiro, F. Gaita, M. Borggrefe, T. M. Menendez, J. Brugada, G. D. Pollevick, C. Wolpert, E. Burashnikov, K. Matsuo, Y. S. Wu, A. Guerchicoff, F. Bianchi, C. Giustetto, R. Schimpf, P. Brugada and C. Antzelevitch, *Circulation*, 2004, **109**, 30–35.

82. J. M. Cordeiro, R. Brugada, Y. S. Wu, K. Hong and R. Dumaine, *Cardiovasc. Res.*, 2005, **67**, 498–509.

83. M. J. McPate, R. S. Duncan, J. T. Milnes, H. J. Witchel and J. C. Hancox, *Biochem. Biophys. Res. Commun.*, 2005, **334**, 441–449.

84. Y. Sun, X. Q. Quan, S. Fromme, R. H. Cox, P. Zhang, L. Zhang, D. Guo, J. Guo, C. Patel, P. R. Kowey and G. X. Yan, *J. Mol. Cell. Cardiol.*, 2011, **50**, 433–441.

85. R. R. Shah and L. M. Hondeghem, *Heart Rhythm*, 2005, **2**, 758–772.

86. L. M. Hondeghem, *Novartis Found. Symp.*, 2005, **266**, 235–244; discussion 244–250.

87. L. M. Hondeghem, *Heart Rhythm*, 2008, **5**, 1210–1212.

88. J. C. Hancox, M. J. McPate, A. El Harchi and Y. H. Zhang, *Pharmacol. Ther.*, 2008, **119**, 118–132.

89. H. J. Witchel, J. T. Milnes, J. S. Mitcheson and J. C. Hancox, *J. Pharmacol. Toxicol. Methods*, 2002, **48**, 65–80.

90. M. Recanatini, E. Poluzzi, M. Masetti, A. Cavalli and F. De Ponti, *Med. Res. Rev.*, 2005, **25**, 133–166.

91. M. H. Bridgland-Taylor, A. C. Hargreaves, A. Easter, A. Orme, D. C. Henthorn, M. Ding, A. M. Davis, B. G. Small, C. G. Heapy, N. Abi-Gerges, F. Persson, I. Jacobson, M. Sullivan, N. Albertson, T. G. Hammond, E. Sullivan, J. P. Valentin and C. E. Pollard, *J. Pharmacol. Toxicol. Methods*, 2006, **54**, 189–199.

92. C. E. Pollard, N. Abi Gerges, M. H. Bridgland-Taylor, A. Easter, T. G. Hammond and J. P. Valentin, *Br. J. Pharmacol.*, 2010, **159**, 12–21.
93. M. C. Sanguinetti and J. S. Mitcheson, *Trends Pharmacol. Sci.*, 2005, **26**, 119–124.
94. P. J. Stansfeld, M. J. Sutcliffe and J. S. Mitcheson, *Expert Opin. Drug Metab. Toxicol.*, 2006, **2**, 81–94.
95. P. J. Stansfeld, P. Gedeck, M. Gosling, B. Cox, J. S. Mitcheson and M. J. Sutcliffe, *Proteins*, 2007, **68**, 568–580.
96. A. T. Dennis, D. Nassal, I. Deschenes, D. Thomas and E. Ficker, *J. Biol. Chem.*, 2011.
97. L. Wang, B. A. Wible, X. Wan and E. Ficker, *J. Pharmacol. Exp. Ther.*, 2007, **320**, 525–534.
98. B. A. Wible, P. Hawryluk, E. Ficker, Y. A. Kuryshev, G. Kirsch and A. M. Brown, *J. Pharmacol. Toxicol. Methods*, 2005, **52**, 136–145.
99. Y. A. Kuryshev, E. Ficker, L. Wang, P. Hawryluk, A. T. Dennis, B. A. Wible, A. M. Brown, J. Kang, X. L. Chen, K. Sawamura, W. Reynolds and D. Rampe, *J. Pharmacol. Exp. Ther.*, 2005, **312**, 316–323.
100. E. Ficker, Y. A. Kuryshev, A. T. Dennis, C. Obejero-Paz, L. Wang, P. Hawryluk, B. A. Wible and A. M. Brown, *Mol. Pharmacol.*, 2004, **66**, 33–44.
101. H. Takemasa, T. Nagatomo, H. Abe, K. Kawakami, T. Igarashi, T. Tsurugi, N. Kabashima, M. Tamura, M. Okazaki, B. P. Delisle, C. T. January and Y. Otsuji, *Br. J. Pharmacol.*, 2008, **153**, 439–447.
102. S. Rajamani, L. L. Eckhardt, C. R. Valdivia, C. A. Klemens, B. M. Gillman, C. L. Anderson, K. M. Holzem, B. P. Delisle, B. D. Anson, J. C. Makielski and C. T. January, *Br. J. Pharmacol.*, 2006, **149**, 481–489.
103. M. C. Sanguinetti and Q. P. Xu, *J. Physiol.*, 1999, **514**, 667–675.
104. M. Perry, M. J. de Groot, R. Helliwell, D. Leishman, M. Tristani-Firouzi, M. C. Sanguinetti and J. Mitcheson, *Mol. Pharmacol.*, 2004, **66**, 240–249.
105. H. J. Witchel, C. E. Dempsey, R. B. Sessions, M. Perry, J. T. Milnes, J. C. Hancox and J. S. Mitcheson, *Mol. Pharmacol.*, 2004.
106. K. Kamiya, R. Niwa, J. S. Mitcheson and M. C. Sanguinetti, *Mol. Pharmacol.*, 2006, **69**, 1709–1716.
107. D. Stork, E. N. Timin, S. Berjukow, C. Huber, A. Hohaus, M. Auer and S. Hering, *Br. J. Pharmacol.*, 2007, **151**, 1368–1376.
108. K. Kamiya, R. Niwa, M. Morishima, H. Honjo and M. C. Sanguinetti, *J. Pharmacol. Sci.*, 2008, **108**, 301–307.
109. M. Perry, P. J. Stansfeld, J. Leaney, C. Wood, M. J. de Groot, D. Leishman, M. J. Sutcliffe and J. S. Mitcheson, *Mol. Pharmacol.*, 2006, **69**, 509–519.
110. D. Fernandez, A. Ghanta, G. W. Kauffman and M. C. Sanguinetti, *J. Biol. Chem.*, 2004, **279**, 10120–10127.
111. J. A. Sanchez-Chapula, R. A. Navarro-Polanco, C. Culberson, J. Chen and M. C. Sanguinetti, *J. Biol. Chem.*, 2002, **277**, 23587–23595.
112. J. Chen, G. Seebohm and M. C. Sanguinetti, *Proc. Natl. Acad. Sci. U. S. A.*, 2002, **99**, 12461–12466.

113. K. Kamiya, J. S. Mitcheson, K. Yasui, I. Kodama and M. C. Sanguinetti, *Mol. Pharmacol.*, 2001, **60**, 244–253.

114. J. S. Mitcheson, J. Chen, M. Lin, C. Culberson and M. C. Sanguinetti, *Proc. Natl. Acad. Sci. U. S. A.*, 2000, **97**, 12329–12333.

115. J. P. Lees-Miller, Y. Duan, G. Q. Teng and H. J. Duff, *Mol. Pharmacol.*, 2000, **57**, 367–374.

116. D. Thomas, B. C. Hammerling, K. Wu, A. B. Wimmer, E. K. Ficker, G. E. Kirsch, M. C. Kochan, B. A. Wible, E. P. Scholz, E. Zitron, S. Kathofer, V. A. Kreye, H. A. Katus, W. Schoels, C. A. Karle and J. Kiehn, *Br. J. Pharmacol.*, 2004, **142**, 485–494.

117. D. Thomas, A. B. Wimmer, K. Wu, B. C. Hammerling, E. K. Ficker, Y. A. Kuryshev, J. Kiehn, H. A. Katus, W. Schoels and C. A. Karle, *Naunyn-Schmiedeberg's Arch. Pharmacol.*, 2004, **369**, 462–472.

118. J. M. Ridley, P. C. Dooley, J. T. Milnes, H. J. Witchel and J. C. Hancox, *J. Mol. Cell. Cardiol.*, 2004, **36**, 701–705.

119. J. T. Milnes, H. J. Witchel, J. L. Leaney, D. J. Leishman and J. C. Hancox, *Biochem. Biophys. Res. Commun.*, 2006, **351**, 273–280.

120. Y. Hosaka, M. Iwata, N. Kamiya, M. Yamada, K. Kinoshita, Y. Fukunishi, K. Tsujimae, H. Hibino, Y. Aizawa, A. Inanobe, H. Nakamura and Y. Kurachi, *Channels*, 2007, **1**, 198–208.

121. D. A. Dougherty, *Science*, 1996, **271**, 163–168.

122. E. Ficker, W. Jarolimek, J. Kiehn, A. Baumann and A. M. Brown, *Circ. Res.*, 1998, **82**, 386–395.

123. E. Ficker, W. Jarolimek and A. M. Brown, *Mol. Pharmacol.*, 2001, **60**, 1343–1348.

124. I. M. Herzberg, M. C. Trudeau and G. A. Robertson, *J. Physiol.*, 1998, **511**, 3–14.

125. M. J. Perrin, P. W. Kuchel, T. J. Campbell and J. I. Vandenberg, *Mol. Pharmacol.*, 2008, **74**, 1443–1452.

126. J. T. Milnes, O. Crociani, A. Arcangeli, J. C. Hancox and H. J. Witchel, *Br. J. Pharmacol.*, 2003, **139**, 887–898.

127. M. J. McPate, R. S. Duncan, J. C. Hancox and H. J. Witchel, *Br. J. Pharmacol.*, 2008, **155**, 957–966.

128. A. Windisch, E. Timin, T. Schwarz, D. Stork-Riedler, T. Erker, G. Ecker and S. Hering, *Br. J. Pharmacol.*, 2011, **162**, 1542–1552.

129. D. Angus, M. Bingham, D. Buchanan, N. Dunbar, L. Gibson, R. Goodwin, A. Haunso, A. Houghton, M. Huggett, R. Morphy, S. Napier, O. Nimz, J. Passmore and G. Walker, *Bioorg. Med. Chem. Lett.*, 2011, **21**, 271–275.

130. M. T. Bilodeau, A. E. Balitza, T. J. Koester, P. J. Manley, L. D. Rodman, C. Buser-Doepner, K. E. Coll, C. Fernandes, J. B. Gibbs, D. C. Heimbrook, W. R. Huckle, N. Kohl, J. J. Lynch, X. Mao, R. C. McFall, D. McLoughlin, C. M. Miller-Stein, K. W. Rickert, L. Sepp-Lorenzino, J. M. Shipman, R. Subramanian, K. A. Thomas, B. K. Wong, S. Yu and G. D. Hartman, *J. Med. Chem.*, 2004, **47**, 6363–6372.

131. C. Blackburn, M. J. LaMarche, J. Brown, J. L. Che, C. A. Cullis, S. Lai, M. Maguire, T. Marsilje, B. Geddes, E. Govek, V. Kadambi, C. Doherty, B. Dayton, S. Brodjian, K. C. Marsh, C. A. Collins and P. R. Kym, *Bioorg. Med. Chem. Lett.*, 2006, **16**, 2621–2627.

132. M. Rowley, D. J. Hallett, S. Goodacre, C. Moyes, J. Crawforth, T. J. Sparey, S. Patel, R. Marwood, S. Thomas, L. Hitzel, D. O'Connor, N. Szeto, J. L. Castro, P. H. Hutson and A. M. MacLeod, *J. Med. Chem.*, 2001, **44**, 1603–1614.

133. H. C. Zhang, C. K. Derian, D. F. McComsey, K. B. White, H. Ye, L. R. Hecker, J. Li, M. F. Addo, D. Croll, A. J. Eckardt, C. E. Smith, Q. Li, W. M. Cheung, B. R. Conway, S. Emanuel, K. T. Demarest, P. Andrade-Gordon, B. P. Damiano and B. E. Maryanoff, *J. Med. Chem.*, 2005, **48**, 1725–1728.

134. I. T. Huscroft, E. J. Carlson, G. G. Chicchi, M. M. Kurtz, C. London, P. Raubo, A. Wheeldon and J. J. Kulagowski, *Bioorg. Med. Chem. Lett.*, 2006, **16**, 2008–2012.

135. J. T. Sisko, T. J. Tucker, M. T. Bilodeau, C. A. Buser, P. A. Ciecko, K. E. Coll, C. Fernandes, J. B. Gibbs, T. J. Koester, N. Kohl, J. J. Lynch, X. Mao, D. McLoughlin, C. M. Miller-Stein, L. D. Rodman, K. W. Rickert, L. Sepp-Lorenzino, J. M. Shipman, K. A. Thomas, B. K. Wong and G. D. Hartman, *Bioorg. Med. Chem. Lett.*, 2006, **16**, 1146–1150.

136. D. Rampe, B. Wible, A. M. Brown and R. C. Dage, *Mol. Pharmacol.*, 1993, **44**, 1240–1245.

137. B. Y. Zhu, Z. J. Jia, P. Zhang, T. Su, W. Huang, E. Goldman, D. Tumas, V. Kadambi, P. Eddy, U. Sinha, R. M. Scarborough and Y. Song, *Bioorg. Med. Chem. Lett.*, 2006, **16**, 5507–5512.

138. R. A. Pearlstein, R. J. Vaz, J. Kang, X. L. Chen, M. Preobrazhenskaya, A. E. Shchekotikhin, A. M. Korolev, L. N. Lysenkova, O. V. Miroshnikova, J. Hendrix and D. Rampe, *Bioorg. Med. Chem. Lett.*, 2003, **13**, 1829–1835.

139. C. Jamieson, E. M. Moir, Z. Rankovic and G. Wishart, *J. Med. Chem.*, 2006, **49**, 5029–5046.

140. A. Cavalli, E. Poluzzi, F. De Ponti and M. Recanatini, *J. Med. Chem.*, 2002, **45**, 3844–3853.

141. M. Masetti, A. Cavalli and M. Recanatini, *J. Comput. Chem.*, 2008, **29**, 795–808.

142. E. Raschi, L. Ceccarini, F. De Ponti and M. Recanatini, *Expert Opin. Drug Metab. Toxicol.*, 2009, **5**, 1005–1021.

143. R. Farid, T. Day, R. A. Friesner and R. A. Pearlstein, *Bioorg. Med. Chem.*, 2006, **14**, 3160–3173.

144. S. Ekins and S. A. Wrighton, *J. Pharmacol. Toxicol. Methods*, 2001, **45**, 65–69.

145. S. Ekins, W. J. Crumb, R. D. Sarazan, J. H. Wikel and S. A. Wrighton, *J. Pharmacol. Exp. Ther.*, 2002, **301**, 427–434.

146. A. Stary, S. J. Wacker, L. Boukharta, U. Zachariae, Y. Karimi-Nejad, J. Aqvist, G. Vriend and B. L. de Groot, *ChemMedChem*, 2010, **5**, 455–467.

147. D. A. Price, D. Armour, M. de Groot, D. Leishman, C. Napier, M. Perros, B. L. Stammen and A. Wood, *Bioorg. Med. Chem. Lett.*, 2006, **16**, 4633–4637.

148. D. A. Price, D. Armour, M. de Groot, D. Leishman, C. Napier, M. Perros, B. L. Stammen and A. Wood, *Curr. Top. Med. Chem.*, 2008, **8**, 1140–1151.

149. J. A. McCauley, C. R. Theberge, J. J. Romano, S. B. Billings, K. D. Anderson, D. A. Claremon, R. M. Freidinger, R. A. Bednar, S. D. Mosser, S. L. Gaul, T. M. Connolly, C. L. Condra, M. Xia, M. E. Cunningham, B. Bednar, G. L. Stump, J. J. Lynch, A. Macaulay, K. A. Wafford, K. S. Koblan and N. J. Liverton, *J. Med. Chem.*, 2004, **47**, 2089–2096.

150. L. R. Fish, M. T. Gilligan, A. C. Humphries, M. Ivarsson, T. Ladduwahetty, K. J. Merchant, D. O'Connor, S. Patel, E. Philipps, H. M. Vargas, P. H. Hutson and A. M. MacLeod, *Bioorg. Med. Chem. Lett.*, 2005, **15**, 3665–3669.

151. M. Perry, M. Sanguinetti and J. Mitcheson, *J. Physiol.*, 2010, **588**, 3157–3167.

152. D. Hanahan and R. A. Weinberg, *Cell*, 2011, **144**, 646–674.

153. S. P. Fraser and L. A. Pardo, *EMBO Rep.*, 2008, **9**, 512–515.

154. A. Becchetti and A. Arcangeli, *J. Gen. Physiol.*, 2008, **132**, 313–314.

155. N. Prevarskaya, R. Skryma and Y. Shuba, *Trends Mol. Med.*, 2010, **16**, 107–121.

156. J. Jehle, P. A. Schweizer, H. A. Katus and D. Thomas, *Cell Death Dis.*, 2011, **2**, e193.

157. A. Arcangeli, L. Bianchi, A. Becchetti, L. Faravelli, M. Coronnello, E. Mini, M. Olivotto and E. Wanke, *J. Physiol.*, 1995, **489**(Pt 2), 455–471.

158. L. Bianchi, B. Wible, A. Arcangeli, M. Taglialatela, F. Morra, P. Castaldo, O. Crociani, B. Rosati, L. Faravelli, M. Olivotto and E. Wanke, *Cancer Res.*, 1998, **58**, 815–822.

159. A. Cherubini, G. L. Taddei, O. Crociani, M. Paglierani, A. M. Buccoliero, L. Fontana, I. Noci, P. Borri, E. Borrani, M. Giachi, A. Becchetti, B. Rosati, E. Wanke, M. Olivotto and A. Arcangeli, *Br. J. Cancer*, 2000, **83**, 1722–1729.

160. A. Masi, A. Becchetti, R. Restano-Cassulini, S. Polvani, G. Hofmann, A. M. Buccoliero, M. Paglierani, B. Pollo, G. L. Taddei, P. Gallina, N. Di Lorenzo, S. Franceschetti, E. Wanke and A. Arcangeli, *Br. J. Cancer*, 2005, **93**, 781–792.

161. E. Lastraioli, L. Guasti, O. Crociani, S. Polvani, G. Hofmann, H. Witchel, L. Bencini, M. Calistri, L. Messerini, M. Scatizzi, R. Moretti, E. Wanke, M. Olivotto, G. Mugnai and A. Arcangeli, *Cancer Res.*, 2004, **64**, 606–611.

162. E. Lastraioli, L. Bencini, E. Bianchini, M. R. Romoli, O. Crociani, E. Giommoni, L. Messerini, S. Gasperoni, R. Moretti, F. Costanzo, L. Boni and A. Arcangeli, *Transl. Oncol.*, 2012, **5**(2), 105–112.

163. G. Hofmann, P. A. Bernabei, O. Crociani, A. Cherubini, L. Guasti, S. Pillozzi, E. Lastraioli, S. Polvani, B. Bartolozzi, V. Solazzo,

L. Gragnani, P. Defilippi, B. Rosati, E. Wanke, M. Olivotto and A. Arcangeli, *J. Biol. Chem.*, 2001, **276**, 4923–4931.

164. S. Pillozzi, M. F. Brizzi, M. Balzi, O. Crociani, A. Cherubini, L. Guasti, B. Bartolozzi, A. Becchetti, E. Wanke, P. A. Bernabei, M. Olivotto, L. Pegoraro and A. Arcangeli, *Leukemia*, 2002, **16**, 1791–1798.

165. S. Pillozzi, M. F. Brizzi, P. A. Bernabei, B. Bartolozzi, R. Caporale, V. Basile, V. Boddi, L. Pegoraro, A. Becchetti and A. Arcangeli, *Blood*, 2007, **110**, 1238–1250.

166. S. Pillozzi, M. Masselli, E. De Lorenzo, B. Accordi, E. Cilia, O. Crociani, A. Amedei, M. Veltroni, M. D'Amico, G. Basso, A. Becchetti, D. Campana and A. Arcangeli, *Blood*, 2011, **117**, 902–914.

167. S. Wadhwa, P. Wadhwa, A. K. Dinda and N. P. Gupta, *Int. Urol. Nephrol.*, 2009, **41**, 251–257.

168. E. Afrasiabi, M. Hietamaki, T. Viitanen, P. Sukumaran, N. Bergelin and K. Tornquist, *Cell. Signalling*, 2010, **22**, 57–64.

169. E. Lastraioli, A. Taddei, L. Messerini, C. E. Comin, M. Festini, M. Giannelli, A. Tomezzoli, M. Paglierani, G. Mugnai, G. De Manzoni, P. Bechi and A. Arcangeli, *J. Cell. Physiol.*, 2006, **209**, 398–404.

170. D. T. Jones, W. R. Taylor and J. M. Thornton, *Febs Letters*, 1994, **339**, 269–275.

171. H. Wang, Y. Zhang, L. Cao, H. Han, J. Wang, B. Yang, S. Nattel and Z. Wang, *Cancer Res.*, 2002, **62**, 4843–4848.

172. H. Han, J. Wang, Y. Zhang, H. Long, H. Wang, D. Xu and Z. Wang, *Cell. Physiol. Biochem.*, 2004, **14**, 121–134.

173. X. D. Shao, K. C. Wu, Z. M. Hao, L. Hong, J. Zhang and D. M. Fan, *Cancer Biol. Ther.*, 2005, **4**, 295–301.

174. D. Thomas, R. Bloehs, R. Koschny, E. Ficker, J. Sykora, J. Kiehn, K. Schlomer, J. Gierten, S. Kathofer, E. Zitron, E. P. Scholz, C. Kiesecker, H. A. Katus and C. A. Karle, *Eur. J. Pharmacol.*, 2008, **579**, 98–103.

175. A. Becchetti, *Am. J. Physiol.: Cell Physiol.*, 2011, **301**, C255–265.

176. A. Becchetti, S. Pillozzi, R. Morini, E. Nesti and A. Arcangeli, *Int. Rev. Cel. Mol. Bio.*, 2010, **279**, 135–190.

177. A. Becchetti and A. Arcangeli, *Adv. Exp. Med. Biol.*, 2010, **674**, v–vii.

178. A. Becchetti and A. Arcangeli, *Adv. Exp. Med. Biol.*, 2010, **674**, 107–123.

179. A. Cherubini, G. Hofmann, S. Pillozzi, L. Guasti, O. Crociani, E. Cilia, P. Di Stefano, S. Degani, M. Balzi, M. Olivotto, E. Wanke, A. Becchetti, P. Defilippi, R. Wymore and A. Arcangeli, *Mol. Biol. Cell*, 2005, **16**, 2972–2983.

180. A. Masi, R. Cicchi, A. Carloni, F. S. Pavone and A. Arcangeli, *Adv. Exp. Med. Biol.*, 2010, **674**, 33–42.

181. L. W. Runnels, L. Yue and D. E. Clapham, *Science*, 2001, **291**, 1043–1047.

182. A. P. Hegle, D. D. Marble and G. F. Wilson, *Proc. Natl. Acad. Sci. U. S. A.*, 2006, **103**, 2886–2891.

183. D. Pier, G. Shehatou, S. Giblett, C. Pullar, D. Tresize, C. Pritchard, J. Challiss and J. Mitcheson, *Mol. Pharm.*, 2014, **86**(2), 211–221.

184. J. H. Dolderer, H. Schuldes, H. Bockhorn, M. Altmannsberger, C. Lambers, D. von Zabern, D. Jonas, H. Schwegler, R. Linke and U. H. Schroder, *Eur. J. Surg. Onc.*, 2010, **36**, 72–77.

185. O. Shimokawa, H. Matsui, Y. Nagano, T. Kaneko, T. Shibahara, A. Nakahara, I. Hyodo, A. Yanaka, H. J. Majima, Y. Nakamura and Y. Matsuzaki, *In Vitro Cell. Dev. Biol.: Anim.*, 2008, **44**, 26–30.

186. H. J. Witchel and J. C. Hancox, *Clin. Exp. Pharmacol. Physiol.*, 2000, **27**, 753–766.

187. W. S. Redfern, L. Carlsson, A. S. Davis, W. G. Lynch, I. MacKenzie, S. Palethorpe, P. K. Siegl, I. Strang, A. T. Sullivan, R. Wallis, A. J. Camm and T. G. Hammond, *Cardiovasc. Res.*, 2003, **58**, 32–45.

188. D. Rampe, M. K. Murawsky, J. Grau and E. W. Lewis, *J. Pharmacol. Exp. Ther.*, 1998, **286**, 788–793.

189. S. B. Ganapathi, M. Kester and K. S. Elmslie, *Am. J. Physiol.: Cell Physiol.*, 2009, **296**, C701–710.

190. E. Redaelli, R. R. Cassulini, D. F. Silva, H. Clement, E. Schiavon, F. Z. Zamudio, G. Odell, A. Arcangeli, J. J. Clare, A. Alagon, R. C. de la Vega, L. D. Possani and E. Wanke, *J. Biol. Chem.*, 2010, **285**, 4130–4142.

191. A. Arcangeli and A. Becchetti, *Pharmaceuticals*, 2010, **3**, 1202–1224.

192. J. Kang, X. L. Chen, H. Wang, J. Ji, H. Cheng, J. Incardona, W. Reynolds, F. Viviani, M. Tabart and D. Rampe, *Mol. Pharmacol.*, 2005, **67**, 827–836.

193. J. Zhou, C. E. Augelli-Szafran, J. A. Bradley, X. Chen, B. J. Koci, W. A. Volberg, Z. Sun and J. S. Cordes, *Mol. Pharmacol.*, 2005, **68**, 876–884.

194. O. Casis, S. P. Olesen and M. C. Sanguinetti, *Mol. Pharmacol.*, 2006, **69**, 658–665.

195. Z. Su, J. Limberis, A. Souers, P. Kym, A. Mikhail, K. Houseman, G. Diaz, X. Liu, R. L. Martin, B. F. Cox and G. A. Gintant, *Biochem. Pharmacol.*, 2009, **77**, 1383–1390.

196. A. C. Gerlach, S. J. Stoehr and N. A. Castle, *Mol. Pharmacol.*, 77, 58–68.

197. G. Gessner, R. Macianskiene, J. G. Starkus, R. Schonherr and S. H. Heinemann, *Eur. J. Pharmacol.*, 2010, **632**, 52–59.

198. M. Perry, F. B. Sachse and M. C. Sanguinetti, *Proc. Natl. Acad. Sci. U. S. A.*, 2007, **104**, 13827–13832.

199. X. Xu, M. Recanatini, M. Roberti and G. N. Tseng, *Mol. Pharmacol.*, 2008, **73**, 1709–1721.

200. M. Perry, F. B. Sachse, J. Abbruzzese and M. C. Sanguinetti, *Proc. Natl. Acad. Sci. U. S. A.*, 2009, **106**, 20075–20080.

201. V. Garg, A. Stary-Weinzinger, F. Sachse and M. Sanguinetti, *Mol. Pharmacol.*, 2011.

202. H. Zeng, I. M. Lozinskaya, Z. Lin, R. N. Willette, D. P. Brooks and X. Xu, *J. Pharmacol. Exp. Ther.*, 2006, **319**, 957–962.

203. M. Grunnet, R. S. Hansen and S. P. Olesen, *Prog. Biophys. Mol. Biol.*, 2008, **98**, 347–362.

204. M. Grunnet, *Acta Physiol. (Oxf)*, 2010, **198**(Suppl. 676), 1–48.

205. G. Szabo, V. Farkas, M. Grunnet, A. Mohacsi and P. P. Nanasi, *Curr. Med. Chem.*, 2011, **18**, 3607–3621.

206. C. Patel and C. Antzelevitch, *Heart Rhythm*, 2008, **5**, 585–590.
207. B. H. Bentzen, S. Bahrke, K. Wu, A. P. Larsen, K. E. Odening, G. Franke, K. S. vans Gravesande, J. Biermann, X. Peng, G. Koren, M. Zehender, C. Bode, M. Grunnet and M. Brunner, *J. Cardiovasc. Pharmacol.*, 2011, **57**, 223–230.
208. A. L. Pond, B. K. Scheve, A. T. Benedict, K. Petrecca, D. R. Van Wagoner, A. Shrier and J. M. Nerbonne, *J. Biol. Chem.*, 2000, **275**, 5997–6006.
209. M. Mewe, M. Mauerhofer, I. Wulfsen, K. Szlachta, X. B. Zhou, J. R. Schwarz and C. K. Bauer, *J. Mol. Cell. Cardiol.*, 2010, **49**, 48–57.
210. F. Charpentier, J. Merot, D. Riochet, H. Le Marec and D. Escande, *Biochem. Biophys. Res. Commun.*, 1998, **251**, 806–810.
211. R. S. Wymore, G. A. Gintant, R. T. Wymore, J. E. Dixon, D. McKinnon and I. S. Cohen, *Circ. Res.*, 1997, **80**, 261–268.
212. A. Emmi, H. J. Wenzel, P. A. Schwartzkroin, M. Taglialatela, P. Castaldo, L. Bianchi, J. Nerbonne, G. A. Robertson and D. Janigro, *J. Neurosci.*, 2000, **20**, 3915–3925.
213. S. Nedergaard, *Neuroscience*, 2004, **125**, 841–852.
214. W. Hirdes, N. Napp, I. Wulfsen, M. Schweizer, J. R. Schwarz and C. K. Bauer, *Pflugers Arch., EJP*, 2009, **459**, 55–70.
215. W. Shi, R. S. Wymore, H. S. Wang, Z. Pan, I. S. Cohen, D. McKinnon and J. E. Dixon, *J. Neurosci.*, 1997, **17**, 9423–9432.
216. A. B. Hardy, J. E. Fox, P. R. Giglou, N. Wijesekara, A. Bhattacharjee, S. Sultan, A. V. Gyulkhandanyan, H. Y. Gaisano, P. E. MacDonald and M. B. Wheeler, *J. Biol. Chem.*, 2009, **284**, 30441–30452.
217. F. Barros, D. del Camino, L. A. Pardo, T. Palomero, T. Giraldez and P. de la Pena, *Pflugers Arch., EJP*, 1997, **435**, 119–129.
218. C. K. Bauer, R. Schafer, D. Schiemann, G. Reid, I. Hanganu and J. R. Schwarz, *Mol. Cell. Endocrinol.*, 1999, **148**, 37–45.
219. E. Muhlbauer, I. Bazwinsky, S. Wolgast, A. Klemenz and E. Peschke, *Cell. Mol. Life Sci.*, 2007, **64**, 768–780.
220. M. Yamashita, Y. Oki, K. Iino, C. Hayashi, F. Matsushita, A. Faje and H. Nakamura, *Regul. Pept.*, 2009, **152**, 73–78.
221. W. Hirdes, C. Dinu, C. K. Bauer, U. Boehm and J. R. Schwarz, *Endocrinology*, 2010, **151**, 1079–1088.
222. A. M. Farrelly, S. Ro, B. P. Callaghan, M. A. Khoyi, N. Fleming, B. Horowitz, K. M. Sanders and K. D. Keef, *Am. J. Physiol.: Gastrointest. Liver Physiol.*, 2003, **284**, G883–895.
223. F. Shoeb, A. P. Malykhina and H. I. Akbarali, *J. Biol. Chem.*, 2003, **278**, 2503–2514.
224. Y. Zhu, C. M. Golden, J. Ye, X. Y. Wang, H. I. Akbarali and J. D. Huizinga, *Am. J. Physiol.: Gastrointest. Liver Physiol.*, 2003, **285**, G1249–1258.
225. E. J. White, S. J. Park, J. A. Foster and J. D. Huizinga, *Am. J. Physiol.: Gastrointest. Liver Physiol.*, 2008, **295**, G691–699.
226. H. Li, L. Liu, L. Guo, J. Zhang, W. Du, X. Li, W. Liu, X. Chen and S. Huang, *Int. J. Hematol.*, 2008, **87**, 387–392.

227. G. A. M. Smith, H. W. Tsui, E. W. Newell, X. P. Jiang, X. P. Zhu, F. W. L. Tsui and L. C. Schlichter, *J. Biol. Chem.*, 2002, **277**, 18528–18534.
228. J. R. Agarwal, F. Griesinger, W. Stuhmer and L. A. Pardo, *Mol. Cancer*, 2010, **9**, 18.
229. S. Z. Chen, M. Jiang and Y. S. Zhen, *Cancer Chemother. Pharmacol.*, 2005, **56**, 212–220.
230. J. H. Gong, X. J. Liu, B. Y. Shang, S. Z. Chen and Y. S. Zhen, *Oncology reports*, 2010, **23**, 1747–1756.
231. X. D. Shao, K. C. Wu, X. Z. Guo, M. J. Xie, J. Zhang and D. M. Fan, *Cancer Biol. Ther.*, 2008, **7**, 45–50.
232. J. Roy, B. Vantol, E. A. Cowley, J. Blay and P. Linsdell, *Oncol. Rep.*, 2008, **19**, 1511–1516.
233. P. Fortunato, S. Pillozzi, A. Tamburini, L. Pollazzi, A. Franchi, A. La Torre and A. Arcangeli, *BMC Cancer*, 2010, **10**, 504.
234. H. Lin, J. Xiao, X. Luo, H. Wang, H. Gao, B. Yang and Z. Wang, *J. Cell. Physiol.*, 2007, **212**, 137–147.
235. O. Crociani, F. Zanieri, S. Pillozzi, E. Lastraioli, M. Stefanini, A. Fiore, A. Fortunato, M. D'Amico, M. Masselli, E. De Lorenzo, L. Gasparoli, M. Chiu, O. Bussolati, A. Becchetti and A. Arcangeli, *Sci. Rep.*, 2013, **3**, 3308.
236. O. Crociani, E. Lastraioli, L. Boni, S. Pillozzi, M. R. Romoli, M. D'Amico, M. Stefanini, S. Crescioli, A. Taddei, L. Bencini, M. Bernini, M. Farsi, S. Beghelli, A. Scarpa, L. Messerini, A. Tomezzoli, C. Vindigni, P. Morgagni, L. Saragoni, E. Giommoni, S. Gasperoni, F. Di Costanzo, F. Roviello, G. De Manzoni, P. Bechi and A. Arcangeli, *Clin. Cancer Res.*, 2014, **20**, 1502–1512.
237. S. Pillozzi, B. Accordi, P. Rebora, V. Serafin, M. G. Valsecchi, G. Basso and A. Arcangeli, *Leukemia*, 2014, DOI: 10.1038/leu.2014.26.
238. A. Arcangeli and A. Becchetti, *Trends Cell Biol.*, 2006, **16**, 631–369.

CHAPTER 12

Does Nature do Ion Channel Drug Discovery Better than Us?

RICHARD J. LEWIS,* IRINA VETTER, FERNANDA C. CARDOSO, MARCO INSERRA AND GLENN KING

Institute for Molecular Bioscience, The University of Queensland, Brisbane 4072, Australia
*Email: r.lewis@imb.uq.edu.au

Ion channels are important drug targets for a range of diseases including pain, epilepsy and addiction. However, progress towards the development of more selective inhibitors that generate fewer dose-limiting side effects, or open up new therapeutic opportunities, has been slow. Consequently, only a few new ion channel drugs have emerged in recent years. Historically, the most potent and selective ion channel modulators have been naturally occurring peptide toxins.[1] Today, ion channel modulatory toxins are entering clinical use for indications such as pain, with the venom-derived peptide Prialt® marketed as an intrathecal analgesic.[2] Due to the potentially higher selectivity offered by venom peptides, many pharmaceutical companies are embracing biological-based approaches to the identification of novel ion channel modulators to overcome some of the limitations of low molecular weight modulators, whose affinity is often driven by factors such as lipid solubility and interactions with more conserved transmembrane domains. Table 12.1 compares the underlying strengths and weaknesses of small

RSC Drug Discovery Series No. 39
Ion Channel Drug Discovery
Edited by Brian Cox and Martin Gosling
© The Royal Society of Chemistry 2015
Published by the Royal Society of Chemistry, www.rsc.org

Table 12.1 Advantages and disadvantages of peptide and small molecule ion channel modulators.

	Venom peptides		Small molecules	
	pros	cons	pros	cons
Delivery	Typically injectable into relevant compartment (i.t., s.c., i.v.) reducing on- and off-target side effects	Injectable nature limits range of clinical uses, especially for chronic diseases	Includes oral delivery allowing a broader range of clinical uses	Broad drug distribution increasing on- and off-target side effects
Pharmacokinetics	Short-half life and limited bioavailability (mostly kidney clearance) allowing dynamic dosing in restricted distribution	Short half-life can increase dosing rate and frequency limiting longer term use and bioavailability limits CNS applications	Short to long half-life broadens potential clinical applications and bioavailability and allows CNS penetration	Very long half-life and different half-life in different compartments can be problematic
Pharmacodynamics	Larger and complex contact surface area enhances selectivity and specificity	Advantages of mixed pharmacology or functional selectivity often missed and CNS opportunities are limited	Smaller and less complex contact surface area allows CNS actions and opportunities for mixed on-disease pharmacology	Smaller and less complex contact surface area required for broad bioavailability reduces specificity and selectivity
Cost	Typically higher cost of manufacture offset by potency and smaller volume of distribution		Typically lower cost of manufacture offset by larger volume of distribution	
Preclinical	Fewer off-target liabilities		More off-target liabilities	
Phase I	Fewer on- and off-target liabilities		Significant on- and off-target liabilities	
Phase II	Fewer on- and off-target liabilities but significant efficacy failures		Significant on- and off-target liabilities and efficacy failures	
Phase III	Efficacy failures significant		Efficacy and safety failures significant	
Market	Typically niche and personalised medicines		Typically broader market penetration	

molecule and peptide approaches that need to be considered during the development of ion channel drugs. This chapter will cover this rapidly emerging field, providing examples of venom peptide and small molecule approaches towards the development of $Ca_v2.2$, $Na_v1.7$ and $K_v1.3$ inhibitors for the treatment of pain and autoimmune diseases.

12.1 Voltage-gated Calcium Channels

Voltage-gated calcium (Ca_v) channels are attractive drug targets for a number of pathological conditions (Table 12.2) due to their critical role in translating membrane potential changes in excitable cells into regulated secretion of neurotransmitters and hormones, muscle contraction and gene transcription. Based on their distinct pharmacological and biophysical properties, at least five families of Ca_v channels, specifically L-, N-, P/Q-, R- and T-type channels, have been defined (for review see[3,4]). These channels are expressed in a spatially and temporally restricted manner and can be targeted by both small molecules and venom peptides with high potency and selectivity.

Detailed biochemical characterisation has identified voltage-gated calcium channels as multi-subunit complexes composed of different pore-forming $\alpha1$ subunit types, and several $\alpha2\delta$, β and γ regulatory subunit isoforms.[5,6] The $\alpha1$ subunit forms the main channel pore and determines voltage-dependence characteristics as well as ion selectivity. The ~190 kDa pore-forming transmembrane Ca_v $\alpha1$ subunits (~2000 amino acids) are structurally analogous to the related voltage-gated sodium (Na_v) channels and comprise four homologous domains (I-IV) each consisting of six transmembrane helices termed S1–S6. The ion pore is formed by the P-loop between S5 and S6, while the S4 transmembrane domain is a crucial component of the voltage sensor module which undergoes a conformational change in response to changes of the membrane electric field and ultimately results in pore opening.

The mammalian genome encodes ten distinct $\alpha1$ subunits. $Ca_v1.1$ ($\alpha1S$), $Ca_v1.2$ ($\alpha1C$), $Ca_v1.3$ ($\alpha1D$) and $Ca_v1.4$ ($\alpha1F$) give rise to distinct L-type calcium channels, $Ca_v2.1$ ($\alpha1A$), $Ca_v2.2$ ($\alpha1B$) and $Ca_v2.3$ ($\alpha1E$) encode for the P/Q-type, N-type and R-type channels, respectively, while $Ca_v3.1$ ($\alpha1G$), Ca_v 3.2 ($\alpha1H$) and $Ca_v3.3$ ($\alpha1I$) encode for distinct T-type channel subtypes (reviewed in[7]). In contrast to the $\alpha1$ subunit, auxiliary subunits modulate the functional properties of the pore-forming $\alpha1$ subunit and are involved in regulation of Ca_v channels by second messenger pathways, channel trafficking and modulation of biophysical properties.[8,9] To date, four auxiliary $\alpha2\delta$ subunits as well as four auxiliary β subunits and one γ subunit comprised of four transmembrane segments have been identified.[4,10,11] These auxiliary subunits form heteromeric complexes with the $\alpha1$ subunit and play an important functional role, modifying and regulating the kinetic as well as pharmacological properties of Ca_v channels.[8,11,12]

Table 12.2 Therapeutic indications for voltage-gated calcium channels and small molecule/peptide modulators.

Calcium channel subtype	Tissue distribution	Therapeutic indication(s)/ channelopathies	Small molecules	Peptides
Ca$_v$1.1	Skeletal muscle	Hypokalemic periodic paralysis (channelopathy);[108] malignant hyperthermia susceptibility (channelopathy)[109]		
Ca$_v$1.2	Heart CNS Smooth muscle	Hypertension, Angina Supra-ventricular arrhythmias Timothy syndrome (channelopathy)[113]	Dihydropyridines (*e.g.* nifedipine) Benzothiazepines (diltiazem) Phenylalkylamines (verapamil)	Calciseptine[110] ω-ctenitoxin-Pn1a[111] ω-agatoxin-Aa1a[112]
Ca$_v$1.3	Heart CNS PNS Cochlea	Depression, anxiety[114] Deafness (channelopathy)[115]		
Ca$_v$1.4	Retina	Congenital stationary night blindness (channelopathy);[116] X-linked cone rod dystrophy (channelopathy)[117]		
Ca$_v$2.1	CNS	Migraine Familial Hemiplegic Migraine (FHM1, channelopathy); Episodic Ataxia Type 2 (EA2, channelopathy); Spinocerebellar Ataxia Type 6 (SCA6, channelopathy); Lambert-Eaton Myasthenic Syndrome (LEMS, channelopathy)[118]	Roscovitine[119] Dodecylamine[120] A-1048400[34]	ω-agatoxin IVA[121] ω-conotoxin MVIIC[20]

Channel	Tissue distribution	Indication	Small molecule	Toxin/peptide
Ca$_v$2.2	CNS, PNS	Pain[14,15]	NMED-160[27], Cilnidipine[28], ZC88[29], TROX-1[30], A-1048400[34]	ω-conotoxin MVIIA[20], ω-conotoxin GVIA[122], ω-conotoxin CVID[24], ω-segestritoxin-Sf1a[123], ω-agatoxin-Aa2a[124], ω-theraphotoxin-Hh1a[125], SNX-482[126]
Ca$_v$2.3	CNS, PNS	Pain[126], Seizures[127]		
Ca$_v$3.1	CNS, Ovaries, Placenta, Heart	Seizures[128]		β/ω-theraphotoxin-Tp2a (ProTxII, non-specific)[134]
Ca$_v$3.2	CNS, Heart, Liver, Kidney, Lung, Skeletal muscle, Pancreas	Pain[135], Idiopathic Generalized Epilepsy (channelopathy)[136]	A-1048400[34], Pimozide[129], Mibefradil[130], TTA-P2[131], TTA-A1[132], Z123212[133]	
Ca$_v$3.3	CNS			

12.2 Inhibition of Ca$_v$2.2 by Small Molecules and Natural Products

The N-type voltage-gated Ca^{2+} channel Ca$_v$2.2 is expressed throughout the central (CNS) and peripheral (PNS) nervous systems where it is crucial for neurotransmitter release from presynaptic nerve terminals. Ca$_v$2.2 appears particularly important in nociceptive pathways including the ventral and dorsal spinal horn as well as dorsal root ganglion neurons.[13] Indeed, several lines of evidence validate Ca$_v$2.2 as an important analgesic target. In animal models of spinal nerve ligation-induced neuropathy and formalin-induced pain, loss of Ca$_v$2.2 resulted in decreased nociceptive responses while CNS and motor function were unaffected.[14,15] In addition, Ca$_v$2.2 is the pharmacological target for a number of analgesic drugs including opioids, which modulate Ca$_v$2.2 through Gβγ-mediated signalling. Gabapentin and pregabalin inhibit association of the α2δ subunit with Ca$_v$2.2 channels, while the peptidic Ca$_v$2.2 inhibitor ziconotide, a synthetic analogue of the venom peptide ω-conotoxin MVIIA, binds to residues in the external pore vestibule of Ca$_v$2.2 and directly blocks channel conductance.[13,16] Selected venom peptide and small molecule inhibitors of Ca$_v$ channels and their therapeutic potential are outlined in Table 12.2.

12.2.1 Natural Ca$_v$2.2 Inhibitors from Cone Snail Venoms

The ω-conotoxins, a group of small peptides isolated from the venom of fish-hunting cone snails, represent some of the most subtype-selective Ca$_v$2.2 inhibitors known (for review see[17]). These peptides have evolved as part of the 'motor cabal' prey capture strategy that leads to rapid paralysis of fish prey. The ω-conotoxins are generally considered to act as simple pore blockers, with the binding site of ω-conotoxin GVIA mapped primarily to a group of residues in the extracellular loop of the domain IIIS5-S6 region.[18] However, the presence of auxiliary subunits, in particular α2δ and β subunits, can affect both ω-conotoxin affinity as well as reversibility of block. The pharmacophore for the high-affinity interaction of ω-conotoxins with Ca$_v$2.2 comprises several key residues, including a charged residue (lysine) in position 2 and a tryptophan in position 13, while additional residues in loop 2 and 4 appear to determine Ca$_v$ subtype selectivity.[19–21]

The ω-conotoxins are stabilised by three intramolecular disulfide bonds, which define the triple-stranded β-sheet/inhibitor cystine-knot framework[22] and convey great structural stability as well as resistance to degradation. These characteristics, together with the exquisite selectivity for Ca$_v$2.2, which remains unmatched by small molecules developed to date, give rise to the therapeutic potential of ω-conotoxins. Indeed, ziconotide (Prialt) is the synthetic equivalent of ω-conotoxin MVIIA and, despite extensive efforts, no additional structure–activity benefits could be engineered into the native sequence, although the unstable methionine does cause issues related to stability in formulation. Due to its peptidic nature, ziconotide is administered by the intrathecal route, which currently limits its use to the management of severe refractory pain associated with cancer, AIDS and neuropathies.[23] Importantly, ziconotide can be opioid-sparing and provides analgesia without inducing tolerance or addiction. ω-Conotoxin CVID[24] has also produced indications of pain reversal in a Phase II study in cancer pain patients. However, significant dose-limiting neurological side effects, which most likely arise from inhibition of $Ca_v2.2$ in supraspinal brain centres,[23,25] combined with the requirement for spinal delivery currently restrict the clinical applications for ω-conotoxins.

12.2.2 Small Molecule $Ca_v2.2$ Inhibitors

Due to the limitations associated with clinical use of ω-conotoxins, orally bioavailable small molecule $Ca_v2.2$ blockers have been suggested as an alternative approach to the development of efficacious analgesics that provide therapeutic benefit to a broader patient population. Indeed, significant efforts have been made towards the development of small molecule and peptidomimetic $Ca_v2.2$ inhibitors.[16] For example, NMED-160, an orally available $Ca_v2.2$ inhibitor, initially showed promise for the treatment of pain and advanced to clinical trials.[26] A particularly attractive attribute of this molecule was its use-dependence of $Ca_v2.2$ inhibition,[27] which could decrease on-target side effects and provide an improved safety profile. Unfortunately, this compound was withdrawn from Phase II trials due to less than 'ideal pharmaceutical characteristics'. Other $Ca_v2.2$ inhibitors showing efficacy in animal models of pain include cilnidipine, ZC88 and TROX-1, although the selectivity profile of these compounds remains inferior to ω-conotoxins.[28-30]

More recently, a novel aryl indole with modest selectivity for $Ca_v2.2$ was found to be analgesic in a range of animal models; however, cardiovascular side effects attributed to inhibition of $Ca_v1.2$ were identified.[31] It is likely that inhibition of peripheral $Ca_v2.2$ also contributed to this effect, as systemically administered ω-conotoxins also produce hypotension.[32] A novel pyrazolylpiperidine with good selectivity over $Ca_v1.2$ has also been reported to be effective in models of inflammatory and neuropathic pain.[33] However, the selectivity of this series of compounds over $Ca_v2.1$ has not been reported. Surprisingly, despite relatively poor selectivity over $Ca_v2.1$, the modestly potent $Ca_v2.2$ and Ca_v3 inhibitor A-1048400 was reported to dose-dependently decrease nociceptive, neuropathic and inflammatory pain without affecting psychomotor or hemodynamic function.[34]

An alternative approach to the development of $Ca_v2.2$-selective small molecule inhibitors is to design peptidomimetics, based on the premise that nature has already provided the best possible scaffolds for potent inhibition of $Ca_v2.2$. Indeed, drug development efforts based on the pharmacophore scaffolds of ω-GVIA and ω-MVIIA[35-37] have resulted in molecules with activity at $Ca_v2.2$, which retained some of the notable selectivity characteristics of the ω-conotoxins. However, the therapeutic potential of such compounds remains to be determined.

12.3   Voltage-gated Sodium Channels

Na_v channels are key to the initiation and propagation of action potentials in electrically excitable cells. They are heteromeric membrane proteins comprising a large α subunit and smaller β subunits ($\beta1-\beta4$).[38–40] The larger subunit forms the pore, while the auxiliary subunits are important for modulating channel functions such as the voltage dependence and kinetics of channel activation and inactivation, as well as localization at the cell surface.[41] The Na_v channel family contains nine members differentiated according to their α subunits ($Na_v1.1-Na_v1.9$) (Table 12.3). These subtypes have been classified according to their sensitivity to tetrodotoxin (TTX), where $Na_v1.1$, $Na_v1.2$, $Na_v1.3$, $Na_v1.4$, $Na_v1.6$ and $Na_v1.7$ are TTX-sensitive, and $Na_v1.5$, $Na_v1.8$ and $Na_v1.9$ are TTX-resistant. Most of these isoforms are found in the CNS and/or PNS,[42] except for $Na_v1.4$ that is found in adult skeletal muscle, and $Na_v1.5$ that is responsible for the rising phase of the action potential in cardiac muscle.

Recent human molecular genetics studies have convincingly proven that inherited disorders such as cardiac arrhythmias,[43,44] epileptic disorders,[45,46] and loss and gain of pain sensation, including congenital insensitivity to pain (CIP),[47,48] erythromelalgia (IEM)[49,50] and paroxysmal extreme pain disorder (PEPD),[51,52] can be directly linked to mutations of genes encoding specific Na_v channel subtypes. Furthermore, the expression patterns and localization of diverse Na_v channel isoforms in CNS, PNS and cardiac tissue reinforce the roles of these channels as important mediators in such disorders. Table 12.3 summarizes the correlation between mutations in Na_v channel subtypes and the aforementioned disorders. Among these subtypes, $Na_v1.7$ is predominantly expressed in dorsal root ganglia (DRG), consistent with a key role in nociception.[53,54] This ion channel is at the vanguard of drug discovery efforts for delivery of innovative analgesics due to its target validation in humans and strategic peripheral distribution. It is now well established that gain-of-function mutations in $Na_v1.7$ lead to IEM (slowed deactivation and shifted activation to more hyperpolarized voltage)[55,56] and PEP (destabilized fast inactivation by depolarizing shift in steady-state inactivation),[52,57,58] whereas loss-of-function mutations in $Na_v1.7$ lead to CIP.[59] Patients with CIP have a complete insensitivity to pain without impairment of any other sensory modalities except a loss of smell (anosmia).[60] Specific knockout of $Na_v1.7$ in mouse DRG abolished inflammatory pain responses, while mechanical and cold-evoked pain responses were retained.[61,62] Thus, molecules able to modulate the activity of $Na_v1.7$, as well as other Na_v channel subtypes, are prime candidates for the next generation of analgesic drugs with an improved therapeutic index.

Some other Na_v channel subtypes are also being studied as potential therapeutic targets. Knockout of $Na_v1.8$ and $Na_v1.9$ in mice have revealed a critical role for these channels in pain sensation.[63,64] $Na_v1.8$ knockout mice are unable to sense cold pain or mechanical pressure and show deficits in inflammatory pain sensation.[63,65] Chronic inflammation is also associated

Table 12.3 Therapeutic indications for voltage-gated sodium channels and small molecule/peptide modulators.

Sodium channel subtype	Tissue distribution	Therapeutic indication(s)/channelopathies	Small molecules	Peptides
Na$_V$1.1	CNS, PNS	Epilepsy[137–139]	None known	None known
Na$_V$1.2	CNS, PNS	Epilepsy[137,139,140]	Carisbamate (Johnson & Johnson; phase II)	KIIIA[141]
Na$_V$1.3	CNS	Pain[70,71]	None known	None known
Na$_V$1.4	Skeletal muscle	Hyperkalemic periodic paralysis (PP), paramyotonia congenital[75,76]	1-(2,4-xylyl) guanidinium[142]	GIIIA/B[143]
Na$_V$1.5	Cardiac muscle	Long QT syndrome (LQT-3), Brugada syndrome (BrS)[144–146]	F15845[147–150] (phase II)	None known
Na$_V$1.6	CNS, PNS	Multiple sclerosis, autoimmune encephalomyelitis[73,74]	4,9-anhydro-tetrodotoxin[151]	None known
Na$_V$1.7	PNS	Erythromelalgia (IEM), paroxysma extreme pain disorder (PEPD),[52,57,58] congenital inability to pain sensation (CIP),[59] inflammatory pain[61]	CNV1014802 (Convergence; phase II) PF05089771 (Pfizer; phase II) XEN402 and XEN403 (Xenon; phase II and I, respectively)	ProTxII[82] Huwentoxin-IV[152]
Na$_V$1.8	PNS	Pain[65,67]	A803467	MrVIB, MfVIA[77,78]
Na$_V$1.9	PNS	Pain[68,69]	None known	None known

with increased levels of $Na_v1.8$ expression,[66] and intrathecal delivery of $Na_v1.8$ antisense oligo-siRNA decreases mechanical allodynia and thermal hyperalgesia.[67] Decreased hypersensitivity to inflammatory hyperalgesia in $Na_v1.9$ knockout mice[68] and up-regulation of this channel in inflammation have also been described.[69] $Na_v1.3$ has been reported to be important in certain pain conditions[70,71] but mice with a knockout of this channel show normal acute, inflammatory and neuropathic pain behaviours.[72] $Na_v1.6$ is highly expressed in activated microglia and macrophages in animal models of autoimmune encephalomyelitis and multiple sclerosis (MS),[73] and also in human MS patients.[74] Finally, gain-of-function mutations in $Na_v1.4$ cause myotonia and familial hyperkalemic periodic paralysis.[75,76] Despite their therapeutic potential, few peptides or small molecules have so far been reported that are highly selective for these Na_v channel subtypes. Exceptions include the $Na_v1.8$ selective µO-conotoxin MrVIB and MfVIA[77,78] and the small molecule A803467[159] produced efficacy with relatively few side effect in animal models of pain.

12.4 Inhibition of Na_v by Small Molecules and Natural Products

The first generation of small-molecule Na_v channel modulators, including phenytoin and carbamazepine, proved efficacious in treating ion-channel-related disorders, but limited subtype selectivity caused dose-limiting side effects. Since small molecule Na_v channel blockers bind *via* more conserved transmembrane regions of the α-subunit, their selectivity across the nine Na_v channel subtypes ($Na_v1.1$–$Na_v1.9$) is typically limited and Na_v channel modulators with improved subtype selectivity and therapeutic index are required for more effective treatment. A diverse range of venom peptides have been reported to modulate the activity of Na_v channels, but very few are selective for desired subtypes such as $Na_v1.7$ (reviewed in[17,79,80]). Selected venom peptide and small molecule inhibitors of Nav channels and their therapeutic potential are outlined in Table 12.3.

12.4.1 Natural $Na_v1.7$ Inhibitors from Venoms

Peptidic Na_v channel blockers have been isolated from the venoms of numerous animals, including centipedes, cone snails, scorpions, and spiders. However, spiders appear to have evolved a significantly larger repertoire of Na_v channel toxins than any other venomous animal, with at least 12 different classes of Na_v channel modulators.[80] The vast majority of these peptides are gating modifiers, but they allosterically modulate channel activity in quite different ways. For example, some spider toxins inhibit Na_v channel *inactivation* while others alter the voltage-dependence of *activation*.[80]

Although a number of spider toxins have been isolated that potently block the analgesic target $Na_v1.7$, they generally have poor selectivity. For example,

ProTx-II is a 30-residue peptide from Green velvet tarantula that potently blocks human $Na_V1.7$ ($IC_{50} = 0.3$ nM) as well as human $Na_V1.2$, $Na_V1.5$, and $Na_V1.6$ subtypes with $IC_{50} < 80$ nM.[81,82] As a result of these off-target effects, ProTx-II was found to be lethal to rats when administered intravenously at doses ≥ 1.0 mg/kg or intrathecally at doses ≥ 0.1 mg/kg.[82] This highlights the fact that for therapeutic uses, $Na_V1.7$ blockers need to be highly selective, a major challenge in progressing them to the clinic. On a positive note, spider-venom peptides have already been isolated that potently block $Na_V1.7$ and have significant selectivity over other Na_V subtypes.[80] No small molecules with such features are reported to date.

12.4.2 Small Molecule $Na_V1.7$ Inhibitors

A few small-molecule $Na_V1.7$ blockers are currently in phase I and II clinical trials (Table 12.3). PF05089771 from Pfizer is in clinical trials for treating postoperative dental pain (http://clinicaltrials.gov/ct2/show/NCT01529346, accessed on 11 October 2012). XEN 402 from Xenon is being tested for treatment of erythromelalgia and postherpetic neuralgia (http://clinicaltrials.gov/ct2/show?term = XEN402&rank = 2 and http://clinicaltrials.gov/ct2/ show/NCT01195636?term = XEN402&rank = 1, accessed on 10 October 2012). CNV1014802 from Convergence Pharmaceuticals has been tested for neuropathic pain (http://clinicaltrials.gov/ct2/show/NCT01561027?term = CNV1014802&rank = 2, accessed on 10 October 2012) and is entering a second phase II trial for trigeminal neuralgia (http://clinicaltrials.gov/ct2/show/NCT01540630?term = CNV1014802&rank = 1, accessed on 10 October 2012). Although these companies have claimed that these compounds are subtype selective, extensive characterisation remains to be reported.

12.5 Voltage-gated K^+ Channels

Voltage-gated potassium (K_V) channels are expressed throughout the mammalian nervous system where they play critical roles in excitable tissues by modulating neurotransmitter release, neuronal excitability and cardiac rhythm.[83–87] As the membranes of excitable nerve tissue becomes depolarized, K_V channels open to allow the passage of K^+ ions out of the cell to facilitate membrane repolarisation, thereby allowing the cells to enter a resting state. K_V channels are also found in non-excitable tissues such as the pancreas and immune cells where they regulate insulin secretion and inflammatory responses. K_V channels are comprised of four membrane-spanning subunits arranged concentrically around a central pore through which K^+ ions can pass. These channels often form heterotetrameric complexes made up of α and β subunits. Classes of α subunits are based on the homology of the sequence of their transmembrane domain and are named K_V1–K_V12 (Table 12.4). To date, 40 α-subunits have been identified which form a voltage-sensitive pore. These α-subunits are commonly associated with a number of auxiliary β-subunits that regulate channel activity and

Table 12.4 Therapeutic indications for voltage-gated potassium channels and small molecule/peptide modulators.

Voltage-gated K$^+$ channel subtype	Tissue distribution	Therapeutic indication(s)	Small molecules	Peptides
Kv1.1	CNS, PNS, DRG, Heart, Pancreas	Ataxia, Epilepsy[88]	5-bromo-2,2-dimethyl-5-nitro-1,3-dioxane[153]	Bgk[96]
Kv1.2	CNS, PNS, DRG, Heart, Pancreas	Ataxia, Epilepsy[88]		κM-RIIIJ[154]
Kv1.3	Skeletal Muscle, Immune Cells CNS, PNS, DRG	Autoimmune diseases[155]	5-(4-phenoxybutoxy) psoralen[103]	Shk[96]
Kv1.4	Glial Cells, CNS, PNS, DRG	Myasthenia gravis[156]		
Kv1.5	CNS, Kidney, Heart	Cardiac diseases[157]	Diphenyl phosphine oxide-1[158]	
Kv1.6	CNS, DRG			Margatoxin[95]

expression. β-subunits do not form functional K_V channels in the absence of the α-subunit.

From a therapeutic perspective, the K_v1 family of channels, also known as "shaker-related" channels due to the phenotype induced by mutations in this channel family in *Drosophila melanogaster*, are of particular interest. These channels are expressed throughout the CNS and PNS, and the ubiquitous nature of their neuronal distribution renders them important pharmacological targets for the study of neurological disorders including ataxia, epilepsy and neuropathic pain.[88,89] However, due to the lack of highly selective K_v1 ligands, the search remains for an effective therapeutic for these disorders of the nervous system.

$K_V1.3$ is one of eight subtypes in the K_V1 family. $K_V1.3$ channels are strongly upregulated during activation of terminally differentiated human effector memory T cells, which play a crucial role in autoimmune diseases such as MS, type-1 diabetes, and rheumatoid arthritis. The $K_V1.3$ channel has consequently become a target for drugs to treat autoimmune diseases.[90–92] However, in order to be therapeutically useful, $K_V1.3$ blockers need to be highly selective in order to avoid deleterious effects due to off-target activity on other K_V subtypes. Selected venom peptide and small molecule inhibitors of K_v channels and their therapeutic potential are outlined in Table 12.4.

12.6 Peptide Modulators of $K_v1.3$

Numerous $K_V1.3$ blockers have been isolated from scorpion venom, with certain members of the α-KTx subfamily of scorpion toxins displaying an extraordinary ability to distinguish between K_V1 channels and the structurally related large conductance calcium-activated potassium channels (BK channels).[93] For example, margatoxin, a 39-residue α-KTx from the venom of the scorpion *Centruroides margaritatus*, inhibits $K_V1.3$ with an IC_{50} of 30 pM.[94,95] ShK-192, an analogue of the 35-residue sea anemone toxin ShK, inhibits $K_V1.3$ with an IC_{50} of 140 pM and, most importantly, displays over 100-fold greater selectivity than other Kv1 channel subtypes.[96] ShK prevents the symptoms of experimental autoimmune encephalomyelitis[90,97,98] and arthritis in animal models.[92] On the other hand, inhibition of $K_V1.3$ by margatoxin and ShK could have beneficial effects in cases of MS where T-cell activation reduces neural progenitor cell activity and neurogeneration.[99,100]

Peptidic blockers of $K_V1.3$ are typically more potent and selective that their small-molecule counterparts.[88,89] However, these peptides are often susceptible to enzymatic degradation and hence the unmodified native peptides are generally not suitable for oral or topical application for therapeutic purposes. In some cases, peptide stability can be significantly improved by simple modifications to the amino acid sequence. For example, the protease resistance of ShK was enhanced by C-terminal amidation and problems with methionine oxidation were obviated by replacing this residue with norleucine.[2] Amgen have also developed a PEGylated analogue of ShK, ShK-Q16K[PEG20K], that has increased half-life and binds $K_V1.3$ with an affinity 500-fold higher than at any other K_V subtype.[101,102] ShK-186, a synthetic analogue of ShK, will soon enter Phase I clinical trials for treatment of MS.[2]

12.7 Small Molecule Modulators of $K_V1.3$

Currently, the most potent and selective small-molecule inhibitor of $K_V1.3$ is PAP-1 (5-(4-phenoxybutoxy)psoralen).[103] PAP-1 exhibits moderate selectivity across K_V channels with the IC_{50} of 2 nM for inhibition of $K_V1.3$ being 23-fold more potent than on any other K_V channel subtype.[104] PAP-1 delays the onset of type 1 diabetes in autoimmune diabetes-prone rats[92] and inhibited oxazolone-induced allergic contact dermatitis in an animal model of psoriasis.[105] The therapeutic potential of this small molecule is enhanced by the fact that it is not only a potent antagonist of $K_V1.3$ but it is also active following oral and topical application with no signs of acute toxicity in animal models of disease.[104,106]

12.8 Outlook

Small molecules are typically preferred in drug development as they are often orally available, achieve good tissue penetration after systemic administration, and possess more favourable pharmacokinetics. Analgesic peptides that block $Ca_V2.2$ in particular suffer from the need for intrathecal administration, a route that requires highly skilled staff and is associated with significant risks. However, while removing the need for intrathecal administration seems intuitively attractive, administration of peptides directly to this compartment may provide surprising benefits. Re-distribution of intrathecally administered ω-conotoxins is limited due to their high polarity and large size, which may avoid some of the side-effects associated with inhibition of $Ca_V2.2$ in the periphery, such as orthostatic hypotension, dizziness, sinus bradycardia, rhinitis, and nausea.[32] Inhibitors of sodium channels have also proven efficacious for treating disorders such as epilepsy and neuropathic pain. However, current therapies utilise small molecules with low therapeutic index caused by a lack of selectivity for Na_V channels subtypes. Natural peptide toxins can be significantly more selective than small molecules, and have revealed key characteristics for selectivity over important sodium channels subtypes such as $Na_V1.7$. Despite efforts to improve the efficacy of small molecules, venom peptides currently remain the richest source of molecules to selectively modulate sodium channels subtypes. However, challenges relating to the delivery of these peptides need to be overcome to expand their use as therapeutic drugs. K_V channels are another promising therapeutic target, particularly in the area of inflammatory and autoimmune diseases. While small-molecule K_V channel blockers have desirable pharmacokinetic attributes, their lack of selectivity and potency detracts from their therapeutic potential. Peptide modulators of these channels on the other hand possess superior selectivity and efficacy compared to small molecules. If issues surrounding their degradation and bioavailability can be addressed, peptidic blockers of $K_V1.3$ such as ShK might become valuable therapeutics for treating a range of autoimmune diseases.

From this overview it is evident that venom peptides which have evolved to modulate ion channels are a promising alternative for ion channel drug development. Disulfide-bonded venom peptides can be surprisingly stable,[2,107] suggesting that formulation in a slow-release or depot form might provide prolonged efficacy. Despite their clinical potential and ability to validate new therapeutic targets, only a few venom peptides such as ziconotide and exenatide have reached the clinic.[1] However, our enhanced appreciation of the chemical and pharmacological diversity of venom peptides in recent years has led to an explosion of interest in the therapeutic potential of venom peptides. As a result, there are currently six venom-derived peptides in clinical trials and many more in various stages of preclinical development.

References

1. R. J. Lewis and M. L. Garcia, *Nat. Rev. Drug Discov.*, 2003, **2**(10), 790–802.
2. G. F. King, *Expert Opin. Biol. Ther.*, 2011, **11**(11), 1469–1484.
3. W. A. Catterall, J. Striessnig, T. P. Snutch and E. Perez-Reyes, *Pharmacol. Rev.*, 2003, **55**(4), 579–581.
4. W. A. Catterall, *Cold Spring Harb. Perspect. Biol.*, 2011, **3**(8), a003947.
5. W. A. Catterall, *Annu. Rev. Cell Dev. Biol.*, 2000, **16**, 521–555.

6. M. E. Williams, P. F. Brust, D. H. Feldman, S. Patthi, S. Simerson, A. Maroufi, A. F. McCue, G. Velicelebi, S. B. Ellis and M. M. Harpold, *Science*, 1992, **257**(5068), 389–395.

7. W. A. Catterall, E. Perez-Reyes, T. P. Snutch and J. Striessnig, *Pharmacol. Rev.*, 2005, **57**(4), 411–425.

8. J. Mould, T. Yasuda, C. I. Schroeder, A. M. Beedle, C. J. Doering, G. W. Zamponi, D. J. Adams and R. J. Lewis, *J. Biol. Chem.*, 2004, **279**(33), 34705–34714.

9. A. C. Dolphin, *Curr. Opin. Neurobiol.*, 2009, **19**(3), 237–244.

10. Z. Buraei and J. Yang, *Physiol. Rev.*, 2010, **90**(4), 1461–506.

11. J. Arikkath and K. P. Campbell, *Curr. Opin. Neurobiol.*, 2003, **13**(3), 298–307.

12. G. Berecki, L. Motin, A. Haythornthwaite, S. Vink, P. Bansal, R. Drinkwater, C. I. Wang, M. Moretta, R. J. Lewis, P. F. Alewood, M. J. Christie and D. J. Adams, *Mol. Pharmacol.*, 2010, **77**(2), 139–148.

13. G. W. Zamponi, R. J. Lewis, S. M. Todorovic, S. P. Arneric and T. P. Snutch, *Brain Res. Rev.*, 2009, **60**(1), 84–89.

14. H. Saegusa, T. Kurihara, S. Zong, A. Kazuno, Y. Matsuda, T. Nonaka, W. Han, H. Toriyama and T. Tanabe, *EMBO J.*, 2001, **20**(10), 2349–2356.

15. S. Hatakeyama, M. Wakamori, M. Ino, N. Miyamoto, E. Takahashi, T. Yoshinaga, K. Sawada, K. Imoto, I. Tanaka, T. Yoshizawa, Y. Nishizawa, Y. Mori, T. Niidome and S. Shoji, *Neuroreport*, 2001, **12**(11), 2423–2427.

16. T. Pexton, T. Moeller-Bertram, J. M. Schilling and M. S. Wallace, *Expert Opin. Investig. Drugs*, 2011, **20**(9), 1277–1284.

17. R. J. Lewis, S. Dutertre, I. Vetter and M. J. Christie, *Pharmacol. Rev.*, 2012, **64**(2), 259–298.

18. P. T. Ellinor, J. F. Zhang, W. A. Horne and R. W. Tsien, *Nature*, 1994, **372**(6503), 272–275.

19. M. J. Lew, J. P. Flinn, P. K. Pallaghy, R. Murphy, S. L. Whorlow, C. E. Wright, R. S. Norton and J. A. Angus, *J. Biol. Chem.*, 1997, **272**(18), 12014–12023.

20. K. J. Nielsen, D. Adams, L. Thomas, T. Bond, P. F. Alewood, D. J. Craik and R. J. Lewis, *J. Mol. Biol*, 1999, **289**(5), 1405–1421.

21. J. P. Flinn, P. K. Pallaghy, M. J. Lew, R. Murphy, J. A. Angus and R. S. Norton, *Eur. J. Biochem.*, 1999, **262**(2), 447–455.

22. R. S. Norton and P. K. Pallaghy, *Toxicon*, 1998, **36**(11), 1573–1583.

23. A. Schmidtko, J. Lotsch, R. Freynhagen and G. Geisslinger, *Lancet*, 2010, **375**(9725), 1569–1577.

24. R. J. Lewis, K. J. Nielsen, D. J. Craik, M. L. Loughnan, D. A. Adams, I. A. Sharpe, T. Luchian, D. J. Adams, T. Bond, L. Thomas, A. Jones, J. L. Matheson, R. Drinkwater, P. R. Andrews and P. F. Alewood, *J. Biol. Chem.*, 2000, **275**(45), 35335–35344.

25. R. D. Penn and J. A. Paice, *Pain*, 2000, **85**(1–2), 291–296.

26. T. Yamamoto and A. Takahara, *Curr. Top. Med. Chem.*, 2009, **9**(4), 377–395.

27. N. C. L. McNaughton, E. Horridge, R. Gleave, P. J. Beswick, Y. H. Chen and A. J. Powell, Piperazine amide calcium canne blockers such as NMED-160 block Cav2.2, Cav3.2 and Cav1.2 human recombinant calcium channels in both a tonic and use-dependent manner., in *FENS Abstr.*, 2008, p. 4, 124.24.

28. H. Koganei, M. Shoji and S. Iwata, *Biol. Pharm. Bull.*, 2009, **32**(10), 1695–1700.

29. G. Meng, N. Wu, C. Zhang, R. B. Su, X. Q. Lu, Y. Liu, L. H. Yun, J. Q. Zheng and J. Li, Analgesic activity of ZC88, a novel N-type voltage-dependent calcium channel blocker, and its modulation of morphine analgesia, tolerance and dependence, *Eur. J. Pharmacol.*, 2008, **586**(1–3), 130–138.

30. C. Abbadie, O. B. McManus, S. Y. Sun, R. M. Bugianesi, G. Dai, R. J. Haedo, J. B. Herrington, G. J. Kaczorowski, M. M. Smith, A. M. Swensen, V. A. Warren, B. Williams, S. P. Arneric, C. Eduljee, T. P. Snutch, E. W. Tringham, N. Jochnowitz, A. Liang, D. Euan MacIntyre, E. McGowan, S. Mistry, V. V. White, S. B. Hoyt, C. London, K. A. Lyons, P. B. Bunting, S. Volksdorf and J. L. Duffy, *J. Pharmacol. Exp. Ther.*, 2010, **334**(2), 545–555.

31. S. Tyagarajan, P. K. Chakravarty, M. Park, B. Zhou, J. B. Herrington, K. Ratliff, R. M. Bugianesi, B. Williams, R. J. Haedo, A. M. Swensen, V. A. Warren, M. Smith, M. Garcia, G. J. Kaczorowski, O. B. McManus, K. A. Lyons, X. Li, M. Madeira, B. Karanam, M. Green, M. J. Forrest, C. Abbadie, E. McGowan, S. Mistry, N. Jochnowitz and J. L. Duffy, *Bioorg. Med. Chem. Lett.*, 2011, **21**(2), 869–873.

32. D. McGuire, S. Bowersox, J. D. Fellmann and R. R. Luther, *J. Cardiovasc. Pharmacol.*, 1997, **30**(3), 400–403.

33. N. L. Subasinghe, M. J. Wall, M. P. Winters, N. Qin, M. L. Lubin, M. F. Finley, M. R. Brandt, M. P. Neeper, C. R. Schneider, R. W. Colburn, C. M. Flores and Z. Sui, *Bioorg. Med. Chem. Lett.*, 2012, **22**(12), 4080–4083.

34. V. E. Scott, T. A. Vortherms, W. Niforatos, A. M. Swensen, T. Neelands, I. Milicic, P. N. Banfor, A. King, C. Zhong, G. Simler, C. Zhan, N. Bratcher, J. M. Boyce-Rustay, C. Z. Zhu, P. Bhatia, G. Doherty, H. Mack, A. O. Stewart and M. F. Jarvis, *Biochem. Pharmacol.*, 2012, **83**(3), 406–418.

35. J. B. Bael, P. J. Duggan, S. A. Forsyth, R. J. Lewis, Y. P. Lok and C. I. Schroeder, *Bioorg. Med. Chem.*, 2004, **12**(15), 4025–4037.

36. S. Menzler, J. A. Bikker, N. Suman-Chauhan and D. C. Horwell, *Bioorg. Med. Chem. Lett.*, 2000, **10**(4), 345–347.

37. C. E. Tranberg, A. Yang, I. Vetter, J. R. McArthur, J. B. Baell, R. J. Lewis, K. L. Tuck and P. J. Duggan, *ω-Conotoxin GVIA mimetics that bind and inhibit neuronal Ca(v)2.2 ion channels. Mar Drugs.* 2012 Oct; 10(10):2349–68.

38. J. Patlak, *Physiol. Rev.*, 1991, **71**(4), 1047–1080.

39. W. A. Catterall, *Physiol. Rev.*, 1992, 72(4 Suppl), S15–S48.

40. T. T. Tseng, A. M. McMahon, V. T. Johnson, E. Z. Mangubat, R. J. Zahm, M. E. Pacold and E. Jakobsson, *J. Mol. Microbiol. Biotechnol.*, 2007, **12**(3–4), 249–262.

41. W. A. Catterall, A. L. Goldin and S. G. Waxman, *Pharmacol. Rev.*, 2005, **57**(4), 397–409.
42. P. A. Felts, S. Yokoyama, S. Dib-Hajj, J. A. Black and S. G. Waxman, *Brain Res. Mol. Brain Res.*, 1997, **45**(1), 71–82.
43. Q. Wang, J. Shen, Z. Li, K. Timothy, G. M. Vincent, S. G. Priori, P. J. Schwartz and M. T. Keating, *Hum. Mol. Genet.*, 1995, **4**(9), 1603–1607.
44. C. M. Kotta, A. Anastasakis, K. Gatzoulis, J. Papagiannis, P. Geleris and C. Stefanadis, *J. Appl. Genet.*, 2010, **51**(4), 515–518.
45. A. Escayg and A. L. Goldin, *Epilepsia*, 2010, **51**(9), 1650–1658.
46. R. H. Wallace, I. E. Scheffer, S. Barnett, M. Richards, L. Dibbens, R. R. Desai, T. Lerman-Sagie, D. Lev, A. Mazarib, N. Brand, B. Ben-Zeev, I. Goikhman, R. Singh, G. Kremmidiotis, A. Gardner, G. R. Sutherland, A. L. George, Jr., J. C. Mulley and S. F. Berkovic, *Am. J. Hum. Genet.*, 2001, **68**(4), 859–865.
47. N. Danziger and J. C. Willer, *Rev. Neurol. (Paris)*, 2009, **165**(2), 129–136.
48. R. Staud, D. D. Price, D. Janicke, E. Andrade, A. G. Hadjipanayis, W. T. Eaton, L. Kaplan and M. R. Wallace, *Eur. J. Pain.*, 2011, **15**(3), 223–230.
49. P. L. Sheets, J. O. Jackson, 2nd, S. G. Waxman, S. D. Dib-Hajj and T. R. Cummins, *J. Physiol.*, 2007, **581**(Pt 3), 1019–1031.
50. N. Skeik, T. W. Rooke, M. D. Davis, D. M. Davis, H. Kalsi, I. Kurth and R. C. Richardson, *Vasc. Med.*, 2012, **17**(1), 44–49.
51. J. S. Choi, F. Boralevi, O. Brissaud, J. Sanchez-Martin, R. H. Te Morsche, S. D. Dib-Hajj, J. P. Drenth and S. G. Waxman, *Nat. Rev. Neurol.*, 2011, **7**(1), 51–55.
52. C. R. Fertleman, C. D. Ferrie, J. Aicardi, N. A. Bednarek, O. Eeg-Olofsson, F. V. Elmslie, D. A. Griesemer, F. Goutieres, M. Kirkpatrick, I. N. Malmros, M. Pollitzer, M. Rossiter, E. Roulet-Perez, R. Schubert, V. V. Smith, H. Testard, V. Wong and J. B. Stephenson, *Neurology*, 2007, **69**(6), 586–595.
53. J. A. Black, S. Dib-Hajj, K. McNabola, S. Jeste, M. A. Rizzo, J. D. Kocsis and S. G. Waxman, *Brain Res. Mol. Brain Res.*, 1996, **43**(1–2), 117–131.
54. J. J. Toledo-Aral, B. L. Moss, Z. J. He, A. G. Koszowski, T. Whisenand, S. R. Levinson, J. J. Wolf, I. Silos-Santiago, S. Halegoua and G. Mandel, *Proc. Natl. Acad. Sci. U.S.A.*, 1997, **94**(4), 1527–1532.
55. X. Cheng, S. D. Dib-Hajj, L. Tyrrell, R. H. Te Morsche, J. P. Drenth and S. G. Waxman, *Brain*, 2011, **134**(Pt 7), 1972–1986.
56. S. G. Waxman and S. Dib-Hajj, *Trends Mol. Med.*, 2005, **11**(12), 555–562.
57. J. P. Drenth and S. G. Waxman, *J. Clin. Invest.*, 2007, **117**(12), 3603–3609.
58. C. R. Fertleman, M. D. Baker, K. A. Parker, S. Moffatt, F. V. Elmslie, B. Abrahamsen, J. Ostman, N. Klugbauer, J. N. Wood, R. M. Gardiner and M. Rees, *Neuron*, 2006, **52**(5), 767–774.
59. J. J. Cox, F. Reimann, A. K. Nicholas, G. Thornton, E. Roberts, K. Springell, G. Karbani, H. Jafri, J. Mannan, Y. Raashid, L. Al-Gazali, H. Hamamy, E. M. Valente, S. Gorman, R. Williams, D. P. McHale, J. N. Wood, F. M. Gribble and C. G. Woods, *Nature*, 2006, **444**(7121), 894–898.

60. J. Weiss, M. Pyrski, E. Jacobi, B. Bufe, V. Willnecker, B. Schick, P. Zizzari, S. J. Gossage, C. A. Greer, T. Leinders-Zufall, C. G. Woods, J. N. Wood and F. Zufall, *Nature*, 2011, **472**(7342), 186–190.

61. M. A. Nassar, L. C. Stirling, G. Forlani, M. D. Baker, E. A. Matthews, A. H. Dickenson and J. N. Wood, *Proc. Natl. Acad. Sci. U.S.A.*, 2004, **101**(34), 12706–12711.

62. M. S. Minett, M. A. Nassar, A. K. Clark, G. Passmore, A. H. Dickenson, F. Wang, M. Malcangio and J. N. Wood, *Nat. Commun.*, 2012, 3, 791.

63. K. Zimmermann, A. Leffler, A. Babes, C. M. Cendan, R. W. Carr, J. Kobayashi, C. Nau, J. N. Wood and P. W. Reeh, *Nature*, 2007, **447**(7146), 855–858.

64. S. Leo, R. D'Hooge and T. Meert, *Behav. Brain Res.*, 2010, **208**(1), 149–157.

65. A. N. Akopian, V. Souslova, S. England, K. Okuse, N. Ogata, J. Ure, A. Smith, B. J. Kerr, S. B. McMahon, S. Boyce, R. Hill, L. C. Stanfa, A. H. Dickenson and J. N. Wood, *Nat. Neurosci.*, 1999, 2(6), 541–548.

66. M. Tanaka, T. R. Cummins, K. Ishikawa, S. D. Dib-Hajj, J. A. Black and S. G. Waxman, *Neuroreport*, 1998, 9(6), 967–972.

67. X. W. Dong, S. Goregoaker, H. Engler, X. Zhou, L. Mark, J. Crona, R. Terry, J. Hunter and T. Priestley, *Neuroscience*, 2007, **146**(2), 812–821.

68. B. T. Priest, B. A. Murphy, J. A. Lindia, C. Diaz, C. Abbadie, A. M. Ritter, P. Liberator, L. M. Iyer, S. F. Kash, M. G. Kohler, G. J. Kaczorowski, D. E. MacIntyre and W. J. Martin, *Proc. Natl. Acad. Sci. U.S.A.*, 2005, **102**(26), 9382–9387.

69. S. Tate, S. Benn, C. Hick, D. Trezise, V. John, R. J. Mannion, M. Costigan, C. Plumpton, D. Grose, Z. Gladwell, G. Kendall, K. Dale, C. Bountra and C. J. Woolf, *Nat. Neurosci.*, 1998, 1(8), 653–655.

70. T. J. Boucher, K. Okuse, D. L. Bennett, J. B. Munson, J. N. Wood and S. B. McMahon, *Science*, 2000, **290**(5489), 124–127.

71. B. C. Hains, C. Y. Saab, J. P. Klein, M. J. Craner and S. G. Waxman, *J. Neurosci.*, 2004, **24**(20), 4832–4839.

72. M. A. Nassar, M. D. Baker, A. Levato, R. Ingram, G. Mallucci, S. B. McMahon and J. N. Wood, *Mol. Pain*, 2006, 2, 33.

73. M. J. Craner, J. Newcombe, J. A. Black, C. Hartle, M. L. Cuzner and S. G. Waxman, *Proc. Natl. Acad. Sci. U.S.A.*, 2004, **101**(21), 8168–8173.

74. M. J. Craner, T. G. Damarjian, S. Liu, B. C. Hains, A. C. Lo, J. A. Black, J. Newcombe, M. L. Cuzner and S. G. Waxman, *Glia*, 2005, **49**(2), 220–229.

75. K. Jurkat-Rott and F. Lehmann-Horn, *Neurotherapeutics*, 2007, **4**(2), 216–224.

76. J. Zhao, N. Dupre, J. Puymirat and M. Chahine, *J. Physiol.*, 2012, **590**(Pt 11), 2629–2644.

77. J. Ekberg, A. Jayamanne, C. W. Vaughan, S. Aslan, L. Thomas, J. Mould, R. Drinkwater, M. D. Baker, B. Abrahamsen, J. N. Wood, D. J. Adams, M. J. Christie and R. J. Lewis, *Proc. Nalt. Acad. Sci. U.S.A.*, 2006, **103**(45), 17030–17035.

78. I. Vetter, Z. Dekan, O. Knapp, D. J. Adams, P. F. Alewood and R. J. Lewis, *Biochem. Pharmacol.*, 2012, **84**(4), 540–548.

79. S. England and M. J. de Groot, *Br. J. Pharmacol.*, 2009, **158**, 1413–1425.
80. J. K. Klint, S. Senff, D. B. Rupasinghe, S. Y. Er, V. Herzig, G. M. Nicholson and G. F. King, *Toxicon*, 2012, **60**(4), 478–491.
81. R. E. Middleton, V. A. Warren, R. L. Kraus, J. C. Hwang, C. J. Liu, G. Dai, R. M. Brochu, M. G. Kohler, Y. D. Gao, V. M. Garsky, M. J. Bogusky, J. T. Mehl, C. J. Cohen and M. M. Smith, *Biochemistry*, 2002, **41**(50), 14734–14747.
82. W. A. Schmalhofer, J. Calhoun, R. Burrows, T. Bailey, M. G. Kohler, A. B. Weinglass, G. J. Kaczorowski, M. L. Garcia, M. Koltzenburg and B. T. Priest, *Mol. Pharmacol.*, 2008, **74**(5), 1476–1484.
83. K. Takimoto, R. Gealy, A. F. Fomina, J. S. Trimmer and E. S. Levitan, *J. Neurosci.*, 1995, **15**(1 Pt 1), 449–457.
84. S. M. Schumacher and J. R. Martens, *Heart Rhythm*, 2010, **7**(9), 1309–1315.
85. Z. J. Xu and D. J. Adams, *J. Physiol.*, 1992, **456**, 405–424.
86. R. E. Brooke, T. S. Moores, N. P. Morris, S. H. Parson and J. Deuchars, *Eur. J. Neurosci.*, 2004, **20**(12), 3313–3321.
87. D. Singer-Lahat, D. Chikvashvili and I. Lotan, *PLoS One*, 2008, **3**, 1.
88. C. A. Robbins and B. L. Tempel, *Epilepsia*, 2012, **53**(Suppl. 1), 134–41.
89. M. Takeda, Y. Tsuboi, J. Kitagawa, K. Nakagawa, K. Iwata and S. Matsumoto, *Mol. Pain*, 2011, **7**, 5.
90. C. Beeton, H. Wulff, J. Barbaria, O. Clot-Faybesse, M. Pennington, D. Bernard, M. D. Cahalan, K. G. Chandy and E. Beraud, *Proc. Natl. Acad. Sci. U.S.A.*, 2001, **98**(24), 13942–13947.
91. H. Wulff, C. Beeton and K. G. Chandy, *Curr. Opin. Drug Discov. Dev.*, 2003, **6**(5), 640–647.
92. C. Beeton, H. Wulff, N. E. Standifer, P. Azam, K. M. Mullen, M. W. Pennington, A. Kolski-Andreaco, E. Wei, A. Grino, D. R. Counts, P. H. Wang, C. J. LeeHealey, S. A. B. A. Sankaranarayanan, D. Homerick, W. W. Roeck, J. Tehranzadeh, K. L. Stanhope, P. Zimin, P. J. Havel, S. Griffey, H. G. Knaus, G. T. Nepom, G. A. Gutman, P. A. Calabresi and K. G. Chandy, *Proc. Natl. Acad. Sci. U.S.A.*, 2006, **103**(46), 17414–17419.
93. J. Tytgat, K. G. Chandy, M. L. Garcia, G. A. Gutman, M. F. Martin-Eauclaire, J. J. van der Walt and L. D. Possani, *Trends Pharmacol. Sci.*, 1999, **20**(11), 444–447.
94. N. Abbas, M. Belghazi, Y. Abdel-Mottaleb, J. Tytgat, P. E. Bougis and M. F. Martin-Eauclaire, *Biochem. Biophys. Res. Commun.*, 2008, **376**(3), 525–530.
95. M. Garciacalvo, R. J. Leonard, J. Novick, S. P. Stevens, W. Schmalhofer, G. J. Kaczorowski and M. L. Garcia, *J. Biol. Chem.*, 1993, **268**(25), 18866–18874.
96. J. Cotton, M. Crest, F. Bouet, N. Alessandri, M. Gola, E. Forest, E. Karlsson, O. Castaneda, A. L. Harvey, C. Vita and A. Menez, *Eur. J. Biochem.*, 1997, **244**(1), 192–202.
97. C. Beeton, J. Barbaria, P. Giraud, J. Devaux, A. M. Benoliel, M. Gola, J. M. Sabatier, D. Bernard, M. Crest and E. Beraud, *J. Immunol.*, 2001, **166**(2), 936–944.

98. C. Beeton, M. W. Pennington, H. Wulff, S. Singh, D. Nugent, G. Crossley, I. Khaytin, P. A. Calabresi, C. Y. Chen, G. A. Gutman and K. G. Chandy, *Mol. Pharmacol.*, 2005, **67**(4), 1369–1381.

99. H. Peng and D. J. Huss, *J. Neurosci.*, 2010, **30**(32), 10609–10611.

100. T. Wang, M. H. Lee, T. Johnson, R. Allie, L. Hu, P. A. Calabresi and A. Nath, *J. Neurosci.*, 2010, **30**(14), 5020–2027.

101. V. Chi, M. W. Pennington, R. S. Norton, E. J. Tarcha, L. M. Londono, B. Sims-Fahey, S. K. Upadhyay, J. T. Lakey, S. Iadonato, H. Wulff, C. Beeton and K. G. Chandy, *Toxicon*, 2012, **59**(4), 529–546.

102. J. K. Sullivan *et al.*, Selective and Potent Peptide Inhibitors of Kv1.3, 2010, *United States of America Pat.*, US20120121591 A1, 2010.

103. B. Hao, Z. W. Chen, X. J. Zhou, P. I. Zimin, G. P. Miljanich, H. Wulff and Y. X. Wang, *Xenobiotica*, 2011, **41**(3), 198–211.

104. A. Schmitz, A. Sankaranarayanan, P. Azam, K. Schmidt-Lassen, D. Homerick, W. Hansel and H. Wulff, *Mol. Pharmacol.*, 2005, **68**(5), 1254–1270.

105. P. Azam, A. Sankaranarayanan, D. Homerick, S. Griffey and H. Wulff, *J. Invest. Dermatol.*, 2007, **127**(6), 1419–1429.

106. L. E. Pereira, F. Villinger, H. Wulff, A. Sankaranarayanan, G. Raman and A. A. Ansari, *Exp. Biol. Med. (Maywood)*, 2007, **232**(10), 1338–1354.

107. N. J. Saez, S. Senff, J. E. Jensen, S. Y. Er, V. Herzig, L. D. Rash and G. F. King, *Toxins*, 2010, **2**, 2851–2871.

108. K. Jurkat-Rott, M. A. Weber, M. Fauler, X. H. Guo, B. D. Holzherr, A. Paczulla, N. Nordsborg, W. Joechle and F. Lehmann-Horn, *Proc. Natl. Acad. Sci. U.S.A.*, 2009, **106**(10), 4036–4041.

109. N. Monnier, V. Procaccio, P. Stieglitz and J. Lunardi, *Am. J. Hum. Genet.*, 1997, **60**(6), 1316–1325.

110. J. R. de Weille, H. Schweitz, P. Maes, A. Tartar and M. Lazdunski, *Proc. Natl. Acad. Sci. U.S.A.*, 1991, **88**(6), 2437–2440.

111. E. Kalapothakis, C. L. Penaforte, R. M. Leao, J. S. Cruz, V. F. Prado, M. N. Cordeiro, C. R. Diniz, M. A. Romano-Silva, M. A. Prado, M. V. Gomez and P. S. Beirao, *Toxicon*, 1998, **36**(12), 1971–1980.

112. R. H. Scott, A. C. Dolphin, V. P. Bindokas and M. E. Adams, *Mol. Pharmacol.*, 1990, **38**(5), 711–718.

113. J. C. Hoda, F. Zaghetto, A. Koschak and J. Striessnig, *J. Neurosci.*, 2005, **25**(1), 252–259.

114. P. Busquet, N. K. Nguyen, E. Schmid, N. Tanimoto, M. W. Seeliger, T. Ben-Yosef, F. Mizuno, A. Akopian, J. Striessnig and N. Singewald, *Int. J. Neuropsychopharmacol*, 2010, **13**(4), 499–513.

115. S. M. Baig, A. Koschak, A. Lieb, M. Gebhart, C. Dafinger, G. Nurnberg, A. Ali, I. Ahmad, M. J. Sinnegger-Brauns, N. Brandt, J. Engel, M. E. Mangoni, M. Farooq, H. U. Khan, P. Nurnberg, J. Striessnig and H. J. Bolz, *Nat. Neurosci.*, 2011, **14**(1), 77–84.

116. A. Hemara-Wahanui, S. Berjukow, C. I. Hope, P. K. Dearden, S. B. Wu, J. Wilson-Wheeler, D. M. Sharp, P. Lundon-Treweek, G. M. Clover,

J. C. Hoda, J. Striessnig, R. Marksteiner, S. Hering and M. A. Maw, *Proc. Natl. Acad. Sci. U.S.A.*, 2005, **102**(21), 7553–7558.

117. R. Jalkanen, M. Mantyjarvi, R. Tobias, J. Isosomppi, E. M. Sankila, T. Alitalo and N. T. Bech-Hansen, *Pflugers Arch.*, 2010, **460**(2), 375–393.

118. D. Pietrobon, *Pflugers Arch.: Eur. J. Physiol.*, 2010, **460**, 375–393.

119. Z. Buraei, G. Schofield and K. S. Elmslie, *Neuropharmacology*, 2007, **52**(3), 883–894.

120. A. M. Beedle and G. W. Zamponi, *Biophys. J.*, 2000, **79**(1), 260–270.

121. I. M. Mintz, V. J. Venema, K. M. Swiderek, T. D. Lee, B. P. Bean and M. E. Adams, *Nature*, 1992, **355**(6363), 827–829.

122. B. M. Olivera, W. R. Gray, R. Zeikus, J. M. McIntosh, J. Varga, J. Rivier, V. de Santos and L. J. Cruz, *Science*, 1985, **230**(4732), 1338–1343.

123. R. Newcomb, A. Palma, J. Fox, S. Gaur, K. Lau, D. Chung, R. Cong, J. R. Bell, B. Horne, L. Nadasdi and J. Ramachandran, *Biochemistry*, 1995, **34**(26), 8341–8347.

124. M. E. Adams, V. P. Bindokas, L. Hasegawa and V. J. Venema, *J. Biol. Chem.*, 1990, **265**(2), 861–867.

125. Z. Liu, J. Dai, L. Dai, M. Deng, Z. Hu, W. Hu and S. Liang, *J. Biol. Chem.*, 2006, **281**(13), 8628–86235.

126. E. A. Matthews, L. A. Bee, G. J. Stephens and A. H. Dickenson, *Eur J. Neurosci.*, 2007, **25**(12), 3561–3569.

127. M. Weiergraber, M. Henry, K. Radhakrishnan, J. Hescheler and T. Schneider, *J. Neurophysiol.*, 2007, **97**(5), 3660–3669.

128. S. M. Cain and T. P. Snutch, Voltage-Gated Calcium Channels in *Epilepsy, in Jasper's Basic Mechanisms of the Epilepsies*, ed. J. L. Noebels, M. Avoli, M. A. Rogawski, R. W. Olsen and A. V. Delgado-Escueta, National Center for Biotechnology Information, Bethesda, MD, USA, 2012, 4th edn.

129. J. J. Enyeart, B. A. Biagi, R. N. Day, S. S. Sheu and R. A. Maurer, *J. Biol. Chem.*, 1990, **265**(27), 16373–16379.

130. S. K. Mishra and K. Hermsmeyer, *Circ. Res.*, 1994, **75**(1), 144–148.

131. W. Choe, R. B. Messinger, E. Leach, V. S. Eckle, A. Obradovic, R. Salajegheh, V. Jevtovic-Todorovic and S. M. Todorovic, *Mol. Pharmacol.*, 2011, **80**(5), 900–910.

132. V. N. Uebele, C. E. Nuss, S. V. Fox, S. L. Garson, R. Cristescu, S. M. Doran, R. L. Kraus, V. P. Santarelli, Y. Li, J. C. Barrow, Z. Q. Yang, K. A. Schlegel, K. E. Rittle, T. S. Reger, R. A. Bednar, W. Lemaire, F. A. Mullen, J. E. Ballard, C. Tang, G. Dai, O. B. McManus, K. S. Koblan and J. J. Renger, *Cell Biochem. Biophys.*, 2009, **55**(2), 81–93.

133. M. E. Hildebrand, P. L. Smith, C. Bladen, C. Eduljee, J. Y. Xie, L. Chen, M. Fee-Maki, C. J. Doering, J. Mezeyova, Y. Zhu, F. Belardetti, H. Pajouhesh, D. Parker, S. P. Arneric, M. Parmar, F. Porreca, E. Tringham, G. W. Zamponi and T. P. Snutch, *Pain*, 2011, **152**(4), 833–843.

134. G. B. Edgerton, K. M. Blumenthal and D. A. Hanck, *Toxicon*, 2010, **56**(4), 624–636.

135. X. J. Wen, S. Y. Xu, Z. X. Chen, C. X. Yang, H. Liang and H. Li, *Pharmacology*, 2010, **85**(5), 295–300.
136. H. Khosravani, C. Bladen, D. B. Parker, T. P. Snutch, J. E. McRory and G. W. Zamponi, *Ann. Neurol.*, 2005, **57**(5), 745–749.
137. I. Helbig, I. E. Scheffer, J. C. Mulley and S. F. Berkovic, *Lancet Neurol.*, 2008, **7**(3), 231–245.
138. W. A. Catterall, F. Kalume and J. C. Oakley, *J. Physiol.*, 2010, **588**(Pt 11), 1849–1859.
139. M. H. Meisler and J. A. Kearney, *J. Clin. Invest.*, 2005, **115**(8), 2010–2017.
140. J. A. Kearney, N. W. Plummer, M. R. Smith, J. Kapur, T. R. Cummins, S. G. Waxman, A. L. Goldin and M. H. Meisler, *Neuroscience*, 2001, **102**(2), 307–317.
141. M. M. Zhang, B. R. Green, P. Catlin, B. Fiedler, L. Azam, A. Chadwick, H. Terlau, J. R. McArthur, R. J. French, J. Gulyas, J. E. Rivier, B. J. Smith, R. S. Norton, B. M. Olivera, D. Yoshikami and G. Bulaj, *J. Biol. Chem.*, 2007, **282**(42), 30699–30706.
142. S. Sokolov, T. Scheuer and W. A. Catterall, *J. Gen. Physiol.*, 2010, **136**(2), 225–236.
143. T. R. Cummins, F. Aglieco and S. D. Dib-Hajj, *Mol. Pharmacol.*, 2002, **61**(5), 1192–1201.
144. A. S. Amin, A. Asghari-Roodsari and H. L. Tan, *Pflugers Arch.*, 2010, **460**(2), 223–237.
145. A. S. Amin, C. A. Klemens, A. O. Verkerk, P. G. Meregalli, A. Asghari-Roodsari, J. M. de Bakker, C. T. January, A. A. Wilde and H. L. Tan, *Neth. Heart J.*, 2010, **18**(3), 165–169.
146. C. A. Remme and C. R. Bezzina, *Cardiovasc. Ther.*, 2010, **28**(5), 287–294.
147. A. Bocquet, S. Sablayrolles, B. Vacher and B. Le Grand, *Br. J. Pharmacol.*, 2010, **161**(2), 405–415.
148. B. Le Grand, C. Pignier, R. Letienne, F. Cuisiat, F. Rolland, A. Mas and B. Vacher, *J. Med. Chem.*, 2008, **51**(13), 3856–3866.
149. R. Letienne, L. Bel, A. M. Bessac, B. Vacher and B. Le Grand, *Eur. J. Pharmacol.*, 2009, **624**(1–3), 16–22.
150. B. Vie, S. Sablayrolles, R. Letienne, B. Vacher, A. Darmellah, M. Bernard, D. Feuvray and B. Le Grand, *J. Pharmacol. Exp. Ther.*, 2009, **330**(3), 696–703.
151. C. Rosker, B. Lohberger, D. Hofer, B. Steinecker, S. Quasthoff and W. Schreibmayer, *Am. J. Physiol. Cell. Physiol.*, 2007, **293**(2), C783–C789.
152. Y. Xiao, J. P. Bingham, W. Zhu, E. Moczydlowski, S. Liang and T. R. Cummins, *J. Biol. Chem.*, 2008, **283**(40), 27300–27313.
153. Q. Lu, J. Peevey, F. Jow, M. M. Monaghan, G. Mendoza, H. Zhang, J. Wu, C. Y. Kim, J. Bicksler, L. Greenblatt, S. S. Lin, W. Childers and M. R. Bowlby, *Bioorg. Med. Chem.*, 2008, **16**(6), 3067–3075.
154. P. Chen, A. Dendorfer, R. K. Finol-Urdaneta, H. Terlau and B. M. Olivera, *J. Biol. Chem.*, 2010, **285**(20), 14882–14889.
155. R. Chen and S. H. Chung, *Biochemistry*, 2012, **51**(9), 1976–1982.

156. S. Suzuki, K. Utsugisawa, H. Yoshikawa, M. Motomura, S. Matsubara, K. Yokoyama, Y. Nagane, T. Maruta, T. Satoh, H. Sato, M. Kuwana and N. Suzuki, *Arch. Neurol.*, 2009, **66**(11), 1334–1338.

157. C. Birner, O. Husser, A. Jeron, M. Rihm, S. Fredersdorf, M. Resch, P. Schmid, D. Endemann, G. Riegger and A. Luchner, *Arch. Pharmacol.*, 2012, **385**(5), 473–480.

158. J. Karczewski, L. Kiss, S. A. Kane, K. S. Koblan, R. J. Lynch and R. H. Spencer, *Biochem. Pharmacol.*, 2009, 77(2), 177–185.

159. M. F. Jarvis, P. Honore, C. C. Shieh, M. Chapman, S. Joshi, X. F. Zhang, M. Kort, W. Carroll, B. Marron, R. Atkinson, J. Thomas, D. Liu, M. Krambis, Y. Liu, S. McGaraughty, K. Chu, R. Roeloffs, C. Zhong, J. P. Mikusa, G. Hernandez, D. Gauvin, C. Wade, C. Zhu, M. Pai, M. Scanio, L. Shi, I. Drizin, R. Gregg, M. Matulenko, A. Hakeem, M. Gross, M. Johnson, K. Marsh, P. K. Wagoner, J. P. Sullivan, C. R. Faltynek and D. S. Krafte, *Proc. Natl. Acad. Sci. U. S. A.*, 2007, **104**(20), 8520–8525.

CHAPTER 13

Antibodies as Ion Channel Modulators

WILSON EDWARDS[a] AND ALAN D. WICKENDEN*[b]

[a] Biotechnology Center of Excellence, Janssen Research & Development, L. L. C., 3210 Merryfield Row, San Diego, CA 92121, USA; [b] Neuroscience Discovery, Janssen Research & Development, L. L. C., 3210 Merryfield Row, San Diego, CA 92121, USA
*Email: awickend@its.jnj.com

13.1 Introduction

Ion channels allow the movement of ions across cell membranes. In this role, ion channels are responsible for all electrical signaling. In addition, ion channels either directly by allowing calcium flux or indirectly by modulating membrane potential, affect the intracellular calcium concentration, which in turn regulates a variety of functions including muscle contraction, hormone secretion and gene transcription. Based on the central role ion channels play in so many physiological processes, they represent a target class with enormous potential for intervention in a wide range of disease states. Indeed, the successful introduction of numerous ion channel drug classes (*e.g.*, calcium antagonists, sodium channel blockers, gamma-aminobutyric acid [GABA] receptor potentiators, sulfonylureas *etc.*) onto the market has firmly established ion channels as valid drug targets. However, most marketed ion channel drugs are relatively non-selective, often exhibit dose limiting side effects and are rarely 100% effective and there remains a clear need for improved ion channel drugs (*i.e.*, second generation ion channel

RSC Drug Discovery Series No. 39
Ion Channel Drug Discovery
Edited by Brian Cox and Martin Gosling
© The Royal Society of Chemistry 2015
Published by the Royal Society of Chemistry, www.rsc.org

drugs) that are either more selective or that target novel ion channels with fewer side effects and improved efficacy. Thankfully, the availability of the near complete ion channel gene set along with a growing appreciation of the functional roles of many of these proteins has provided a multitude of new ion channel drug targets and numerous opportunities for the discovery of optimized second generation ion channel drugs. Since the cloning of the first ion channel (the nicotinic acetylcholine receptor α-subunit) in 1982[1–3] there has been an explosion in the number of cloned ion channels culminating in the publication of the human genome and the availability of the complete ion channel gene set in 2000. The result of this major effort was the identification of approximately 400 ion channel pore forming genes and accessory sub-units, which together form a bewildering set of potential ion channel drug targets with diverse properties, expression profiles and function (http://www.iuphar-db.org/DATABASE/ReceptorFamiliesForward? type = IC).[4]

The search for optimized, second generation ion channel drugs has relied almost exclusively on high throughput screening of large collections of small molecules against recombinant ion channels in heterologous expression systems.[5] As a result of this effort over the last two decades a large number of small molecule drug candidates have been discovered, many have advanced into clinical trials and a few have received marketing approval, including Varenicline (Chantix), a selective α4β2 nicotinic acetylcholine receptor partial agonist for smoking cessation,[6] Retigabine (Trobalt in Europe and Potiga in the USA), a Kv7.2 activator for treatment resistant epilepsy,[7] VX-770 (Kalydeco), a cystic fibrosis transmembrane conductance regulator (CFTR) potentiator for treatment of a specific type of cystic fibrosis[8] and Perampanel (Fycompa), a non-competitive 2-amino-3-(5-methyl-3-oxo-1,2-oxazol-4-yl) propanoic acid (AMPA) antagonist for the treatment of epilepsy.[9] However, ion channels remain under-exploited and despite a large investment in ion channel screening, the promise of improved second generation drugs is yet to be fulfilled. There are a multitude of reasons for this state of affairs and many are almost certainly not unique to ion channel drug discovery. At the same time, ion channel drug discovery presents its own unique challenges that may contribute to the somewhat limited progress in developing new drugs.[10] Obtaining selectivity between closely related members of the ion channel target family to avoid undesirable off-target effects remains a particularly tough challenge that may have significantly hampered progress in small molecule ion channel drug discovery. In cases where ion channels have proven intractable from a small molecule drug discovery perspective alternative or complimentary, non-small molecule approaches are required. One approach that seems to offer some promise is targeting ion channels with antibodies.

Antibodies are large molecular weight (∼150 000 Da) glycoproteins secreted by B lymphocytes as part of the immune system to bind foreign antigens. Antibodies are a dimer of paired heavy and light chains connected through disulfide bonds. Selective recognition of antigen epitopes is

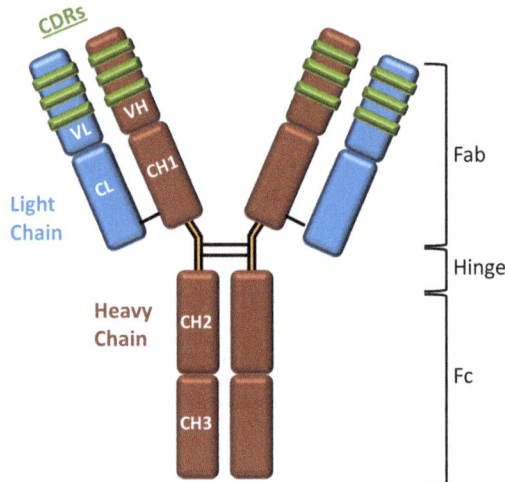

Figure 13.1   Structure of an IgG Antibody. The IgG antibody is composed of heavy
and light chain pairs stabilized through disulfide bonds (—). Light
chains consist of a variable (VL) and constant (CL) domain, and heavy
chains consist of one variable domain(VH) and three constant domains (CH1, CH2, and CH3). The VL and CL domains paired with the
VH and CH1 domains form the Fab fragment of the antibody, which
contain the complementary determining residues (CDRs) responsible
for antigen recognition and binding. The CH2 and CH3 domains form
the Fc fragment of the antibody, which along with the hinge region
are responsible for binding to complement C1q and Fc receptors
resulting in effector functions and long circulating half-life.

mediated through N-terminal variable domains containing the antigen
compliment determining regions (CDRs) (Figure 13.1). Monoclonal antibodies have become a well-established therapeutic platform for many disease areas. As therapeutics, they have several theoretical advantages over
small molecules, including their long half-life and high affinity. However,
perhaps the most significant advantage of antibody drugs over small
molecules is their potential for true pharmacological selectivity. Indeed,
antibodies can exhibit exquisite selectivity, and are often capable of differen-
tiating between highly homologous proteins. By virtue of their larger interaction surfaces, which allow multiple contact sites along the antigen epitope,
antibodies offer the potential for pharmacological potency and selectivity that
far exceeds that achievable with most small molecules. Given the potential for
meaningful selectivity, antibody based drug discovery may represent an
attractive approach for targeting ion channels, especially where sub-type
selectivity is essential and has proven challenging with small molecules.

In the following sections, we will summarize the evidence suggesting that
targeting ion channels with function modifying antibodies (functional
antibodies) is feasible, we will describe the current status of antibody based
ion channel drug-discovery and discuss some possible technical challenges

that may need to be addressed before the promise of ion channel antibody therapeutics can be fulfilled (also reviewed in [11]).

13.2 Modifying Ion Channel Function with Antibodies

For ion channel targeted antibodies (and related approaches—see below) to be viable as therapeutics, it is essential that they not only bind to the target channel with high affinity and selectivity, but that they are also capable of modifying channel, and ultimately cellular function (*i.e.*, they are functional). The feasibility of modifying channel function with antibodies is firmly established in the scientific literature. Ion channel function (biophysical or physiological) can be modified by at least three mechanisms. Antibodies can modulate channel gating or permeation directly by binding to either extracellular or intracellular epitopes (Figure 13.2A). Since the focus of this chapter is the therapeutic potential of ion channel antibodies, we will only discuss antibodies that bind to epitopes accessible from the extracellular space. Indeed, since the extracellular loops and domains of ion channels are often less conserved between related ion channels, these extracellular epitopes make excellent targets for sub-type selective

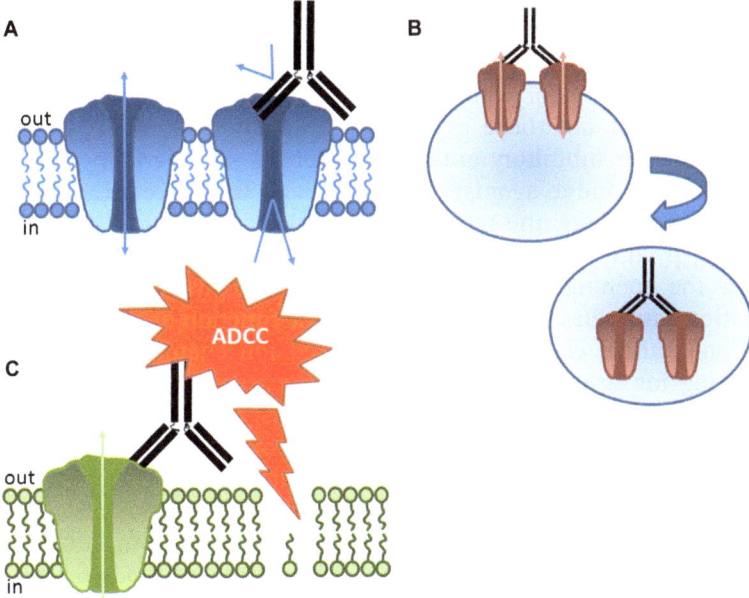

Figure 13.2 Antibodies can modulate the biophysical or physiological function of ion channels in at least three ways: (A) direct modulation of permeation or gating; (B) cross-linking adjacent channels resulting in channel internalization; or (C) *via* antibody-dependent cell-mediated cytotoxicity (ADCC) to ion channel expressing cells.

antibodies. Readers are referred to a recent article by Dallas and colleagues for a brief introduction to antibodies that act against intracellular epitopes.[11] A second mechanism that has been described is antibody mediated channel cross-linking, channel internalization and a reduction in channel surface expression and there is some evidence to suggest that such a mechanism may operate in some forms of ion channel autoimmune diseases (Figure 13.2B). Finally, instead of directly modifying channel function or expression, ion channel targeted antibodies may be able to alter cellular function or initiate complete removal of channel expressing cells through antibody-mediated complement activation and antibody-dependent cell-mediated cytotoxicity (ADCC, Figure 13.2C).

13.2.1   Direct Modulation of Channel Function by Antibodies

There is a substantial body of evidence to support the feasibility of developing antibodies that directly modify channel function. For ligand-gated channels, antibodies have been described that modify or mimic agonist-induced channel activation by binding to extracellular ligand recognition domains. The first example of an antibody that inhibits function by binding to a ligand recognition domain was targeted at the acetylcholine receptor from *Torpedo Californica*.[12,13] In a second, more recent example, direct inhibition of channel function was demonstrated for a monoclonal antibody targeting the purinergic ion channel, P2X7.[14] Direct inhibition of P2X7 channel function was demonstrated by acute (<10 min), concentration-dependent inhibition of whole cell currents in HEK293 cells expressing human P2X7 but not those expressing P2X4 or P2X3. While both these examples are for inhibitory antibodies, there is some evidence that antibodies can also bind to agonist recognition sites to induce activation rather than inhibition. Thus, the monoclonal antibody B6B21 was shown to displace [3H]-glycine that was bound specifically to an NMDA receptor, and enhance the opening of the cation channel in a glycine-like fashion.[15,16] Generating antibodies that directly modify the function of other, non-ligand gated ion channel classes, *e.g.* voltage-gated ion channels, may be more challenging for several reasons. Firstly, the extracellular regions of these channels tend to be smaller and possibly less accessible than the extracellular regions of ligand-gated channels. Secondly, in the absence of an obvious ligand binding site, it is often difficult to know what region of the channel to target for functional effect. Nevertheless, modulation of non-ligand gated channels with antibodies does seem feasible. Antibodies raised against mammalian brain sodium channels are able to block rat brain sodium channels in planar lipid bilayers in a voltage-independent manner,[17] and antibodies targeting the pore-forming SS1–SS2 loop in domain IV of alpha 1D calcium channels are able to block channel function when incubated under depolarizing conditions.[18] Recently, several other groups have successfully targeted the pore domain (so-called "E3 targeting") to generate polyclonal antibodies that are capable of blocking voltage-gated channels and TRP channels, including Nav1.5, Kv1.2, Kv3.1, short transient receptor potential channel 5 (TRPC5), transient receptor potential cation channel subfamily M member 3 (TRPM3), TRPV1 and Kv1.3.[19-24] From a therapeutic perspective, however, polyclonal antibodies are not ideal. Although the studies described above clearly show that function blocking is possible with polyclonal antibodies, much less evidence is available to show that this can be recapitulated with monoclonals. Indeed, in the study by Klionsky and colleagues rabbit polyclonal antiserum blocked human TRPV1 activation but rabbit monoclonal antibodies (identified on the basis of selective binding to Chinese hamster ovary cells expressing human TRPV1) did not.[19] Encouragingly though, one recent report has described the generation of a function blocking monoclonal antibody raised against the E3 loop of human Ether-á-go-go potassium channel 1(EAG1).[25] Although not particularly potent or efficacious, this antibody was capable of inhibiting tumor cell growth *in vitro* and *in vivo*. Interestingly, other E3 binding antibodies did not exhibit functional effects in this study, suggesting that E3 binding alone is not sufficient for channel inhibition. It will be important to understand the additional requirements for functional activity in order to optimize channel blocking antibodies for use as therapeutic and diagnostic agents in the future. Targeting other extracellular domains that have important functional roles may also be sufficient to generate functional antibodies. For example, antibodies raised against synthetic peptides corresponding to a part of the voltage sensor of eel electroplax sodium channels exhibit voltage-dependent binding to the external side of rat brain synaptosomal vesicles, and in some cases are able to depress the action potential of rat sciatic nerve.[26] Similarly, polyclonal antibodies raised against the S3–S4 region of D1 of the neonatal variant of Nav1.5 blocked ion conductance when applied extracellularly.[27] Although not proven, it seems possible that antibody binding to the voltage sensor domain can impair its movement and inhibit voltage-dependent gating, in much the same way as spider toxins inhibit voltage-dependent channels.[28,29]

Further evidence for direct modification of channel function by antibodies comes from the literature on autoimmune ion channelopathies. These autoimmune diseases represent a growing list of neurological disorders involving highly specific autoantibodies to ion channels of the central and peripheral nervous system (for recent reviews see [30,31]). These autoimmune channelopathies are often paraneoplastic, presumably due to an autoimmune response to cancer cells expressing target ion channels, and usually respond to immunotherapies that reduce the levels of the pathogenic autoantibodies. Myasthenia gravis, perhaps the best known ion channel autoimmune disorder, is associated with the presence of antibodies to nicotinic acetylcholine receptors of the neuromuscular junction. Other ion channel autoimmune disorders include Lambert-Eaton Myasthenic Syndrome (LEMS) and Paraneoplastic Cerebellar Ataxia (PCA) involving anti-voltage-gated calcium channel antibodies; Neuromyotonia, Morvan's syndrome and Limbic Encephalitis involving anti-voltage-gated potassium channel antibodies; Neuromyelitis Optica involving anti-aquaporin antibodies; Autoimmune Dysautonomias involving antibodies to the α3 nicotinic acetylcholine receptor of autonomic ganglia; Encephalitis (*e.g.*, Rasmussen's encephalitis) associated with NMDA and AMPA receptor antibodies; and epilepsy and hyperexcitability associated with antibodies to GABA and glycine receptors. The pathogenicity of many of these disease associated auto-antibodies has been proven through induction of a disease phenotype following passive immunization of mice with purified IgG (for example see [30–34]). In many cases, disease associated autoantibodies can directly inhibit ion channel function, similar to the experimentally generated antibodies described above. Direct inhibition of nicotinic acetylcholine receptors was first demonstrated with purified antibodies from patients with Myasthenia Gravis, where IgG from patients was shown to inhibit native nicotinic ACh receptors in mouse myotubes and recombinant receptors in HEK293 cells.[35,36] Block was reported to be rapid, concentration dependent and reversible.[35] In some cases, block became partially irreversible with increased duration of exposure.[37] In a second example, direct block of voltage-gated calcium channels has been demonstrated for PCA associated auto-antibodies that target a major epitope in the S5–S6 extracellular loop of domain III of the P/Q type calcium channel. This antibody inhibited native and recombinant P/Q channels and inhibited synaptic transmission in cerebellar slices.[38] Pharmacological studies indicated that this antibody bound to the outer pore region of calcium channels and probably (partially) occluded the conduction pathway.

13.2.2   Other Mechanisms of Antibody-mediated Channel Modulation

The autoimmune channelopathy literature also provides evidence to support the feasibility of generating antibodies that modify channel function through indirect mechanisms. In some cases of autoimmune channelopathy, direct channel inhibition does not appear to underlie the pathogenicity of the auto-antibodies. Often, auto-antibodies act by cross linking adjacent channels, inducing channel internalization and a reduction in the surface expression of ion channels. For example, using freeze-fracture electron microscopy, IgG from LEMS patients was shown to cause clustering and a reduction in the number of active zone particles (putative voltage-sensitive calcium channels) in mouse diaphragm. This effect was observed with IgG and bivalent F(ab) but not monovalent F(ab), indicating that cross-linking of adjacent channels was a critical step in the IgG mediated depletion of calcium channels from the neuromuscular junction.[39] Similarly, CSF or purified IgG from patients with anti-NMDA receptor encephalitis has been shown to reduce surface expression and synaptic localization of NMDA receptors in cultured hippocampal neurons. This effect was only seen with intact IgG and not with monovalent Fab fragments, again suggesting that the mechanism underlying the loss of surface expression involves capping and cross-linking NMDA receptors.[40] Antibody mediated channel internalization and down regulation need not necessarily require direct antibody binding to the ion channel protein itself. In some cases, antibody binding to proteins intimately associated with ion channel protein complexes may be sufficient to induce disruption of channel expression or localization. For example, it has recently been determined that many of the voltage-gated potassium channel antibodies from patients with Neuromyotonia, Morvan's syndrome and Limbic Encephalitis (*i.e.*, antibodies that immunoprecipitate iodinated α-dendrotoxin (125I-α-DTX)-labeled voltage-gated potassium channel from digitonin-solubilized mammalian brain homogenates[41–43]) are not directed towards the Kv1 subunits themselves but to two proteins, Lgi1 and Caspr2, that are closely associated with VGKCs in the peripheral and central nervous system.[44,45]

Ion channel targeted antibodies can also alter cellular function in-dependent of any direct effect on ion channel function, through antibody-mediated complement activation and antibody-dependent cell-mediated cytotoxicity (ADCC). Such a mechanism may be at least partially responsible for the symptoms of Neuromyelitis optica (NMO), an inflammatory demye-linating disease of the central nervous system often leading to blindness. Nearly all NMO patients are seropositive for autoantibodies against extra-cellular epitope(s) on aquaporin-4 (AQP4), a water-selective channel expressed strongly at the plasma membrane of astrocyte foot processes throughout the CNS. NMO-IgG binding to AQP4 in astrocytes in the CNS is thought to initiate a series of inflammatory events, including antibody dependent complement and cell-mediated astrocyte damage, leukocyte recruitment, cytokine release, and demyelination.[46]

Again, these examples of antibody mediated channelopathies lend further support to the concept and feasibility for developing therapeutic antibodies targeting disease relevant ion channels, as well as the multiple mechanisms that can ultimately lead to modulation of cellular function or removal from the system.

13.3   Current Status

As discussed in earlier sections, antibody therapeutics offer several theore-tical advantages over small molecule drugs and ample proof of principal for functional targeting of ion channels with antibodies exists in the scientific literature. In theory therefore, it should be possible to identify antibodies against therapeutically relevant ion channels for use in the treatment of human disease. Despite the obvious promise, however, a review of recent patent literature suggests that there has been only limited pharmaceutical/ biotech company activity in the discovery and development anti-ion channel antibodies as therapeutics. At the time of writing, only a small number of patent applications have been published on the subject of therapeutic ion channel antibodies. US 2009/0028851, assigned to the Max Plank Society and U3 Pharma AG, describes antibodies that bind to mammalian EAG1 channels for the treatment of hyper-proliferative, inflammatory and neurodegenerative diseases.[47] Data is presented showing that anti-EAG1 antibodies directly inhibit EAG1 function and inhibit proliferation of melanoma cells. WO 2011/063277 from Amgen describes the discovery of functional anti-human Orai1 antibodies for the treatment of various immune disorders.[48] These antibodies were raised in mice immunized with either CHO or U2OS cells expressing recombinant hOrai1 and hSTIM1, with or without thapsigargin pretreatment to activate these store-operated channels. The patent application describes a series of antibodies that bind with high affinity to the putative second extracellular loop of hOrai1. Several antibodies are described that inhibit cytokine generation from thapsigargin-stimulated human whole blood at low- or sub-nanomolar concentrations. Selected antibodies also inhibit calcium entry, CRAC currents and NFAT activation in HEK cells expressing hOrai1/hStim1. One antibody, mAb2C1.1, prevented weight loss, attenuated the production of inflammatory cytokines and reduced engraftment of human T-cells in the spleen in a model of human graft *versus* host disease in mice. WO 2011/051350 from UCB Pharma describes function modifying Nav1.7 antibodies for the treatment or prophylaxis of pain.[49] The antibodies described were raised in rabbits immunized with peptide antigens from loops E1 or E3 of each of the four sodium channel domains. Numerous antibodies are described in the application, with the best producing up to 68% inhibition of Nav1.7 in patch clamp studies at 25 µg/ml, and reducing electrically-induced action potential generation in isolated DRGs. In one example, inhibition appeared to be use-dependent. US 2012/0083000 from Regeneron also describes a selection of function blocking anti-Nav1.7 antibodies.[50] In some cases, block of human Nav1.7 was greater than 30% at sub-micromolar concentrations in electrophysiological assays. Some antibodies appeared to exhibit cross-reactivity with human, rat and/or mouse Nav1.7, and could mimic the effects of TTX in a CGRP-release assay. No *in vivo* efficacy data were presented in either the UCB or the Regeneron applications. A review of business literature and company websites also suggests that there may be some limited interest in anti-ion channel antibodies. For example, Ablynx, a biotechnology company based in Belgium, has pioneered the development of single chain Llama antibodies (nanobodies). In company press releases from 2010 and 2012, Ablynx claimed to have successfully generated nanobodies capable of inhibiting unspecified ion channels (http://www.ablynx.com). All in all though, it would seem that despite the obvious promise and clear feasibility from the literature, activity around antibody-based ion channel drug discovery is minimal and progress has been limited. Although the reasons for the limited progress are not entirely clear, it would seem that there may be some significant technical challenges facing ion channel antibody-based drug discovery that will have to be addressed before the this area can reach its potential. Some of these potential challenges are discussed in more detail in the following sections.

13.4 Challenges

13.4.1 Discovery of Functional Antibodies

The requirements for reagent antibodies *versus* therapeutic antibodies differ with respect to poly- or mono-clonality, selectivity, and the need for humanization and functional activity. Since the focus of this article is on ion channel drug discovery, we will discuss only approaches and challenges relating to the discovery of therapeutic antibodies targeting and modulating ion channel activity. Many of the challenges and issues facing antibody-based ion channel drug discovery are also relevant to antibody-based approaches to other transmembrane proteins (see [51] for an excellent recent review on the subject of therapeutic antibodies to GPCRs).

13.4.1.1 Selection of Ion Channel Antigens

Ion channels present a significant challenge to therapeutic antibody development, because they are large, multi-transmembrane spanning proteins that generally require membrane insertion for proper conformation. Furthermore, only a subset of pore forming protein classes, such as the ligand-gated channels, possess large globular extracellular domains conducive to antibody interactions. Many ion channel proteins, such as voltage-gated ion channels, possess only small extracellular loops and very little of the ion channel protein is exposed to the extracellular environment. For channel proteins with only small extracellular loops, the source and conformation of the antigen is a significant consideration, and may have a profound impact on determining the quantity of any antibody hits derived and their quality in terms of specificity, affinity, functionality, and biophysical characteristics. The vast majority of attempts to generate functional ion channel antibodies described to date have used extracellular loop peptide fragments as immunogens, especially the third extracellular loop, and several function blocking polyclonal antibodies have been identified.[19,21–24] However, peptides only provide a linear epitope and may not take advantage of more complex non-contiguous conformational epitopes which provide for greater numbers of contacts along the binding surfaces, possibly explaining why many of the antipeptide antibodies described to date are of only modest potency/efficacy. More sophisticated antigens may be required to improve the potency and functionality of ion channel antibodies. As an example, the previously mentioned monoclonal antibody to the EAG1 potassium channel was generated utilizing the third extracellular loop of EAG1 fused to a tetramerization domain of the C terminal tail region as the antigen.[25] This approach displayed the E3 loop as a tetramer, ostensibly mimicking the more natural state of the channel pore region, but whether this antigen exists in a true, "native" conformation is not clear. Presenting an antigen in a "native" and functional conformation is preferable and will likely increase the probability of producing antibodies capable of modulating channel activity. Expression of full length channels or isolated channel domains in a lipid bilayer (*i.e.*, in micelles, liposomes, membrane preparations, or whole cells) is one means for achieving proper antigen conformation. Expression levels need to be sufficiently high to elicit an adequate immune response, which may present a challenge for some ion channels with poor surface expression. As an alternative to the expression in heterologous systems, DNA immunizations may provide a means of expressing target channels in host cells. Purified, functional ion channel protein provides another potential source of functional ion channel antigen. Significant advances in ion channel structural biology have been made over the last decade and isolated proteins for numerous channels and channel domains have now been described, including isolated extracellular domains of various glutamate receptors,[52–56] water soluble tKcSA variants[57] and full length voltage-gated sodium and potassium channel proteins, ASIC channels, purinergic channels and AMPA receptors.[58–62] These or related proteins, if available in high enough quantities could be ideal antigens for generating functional ion channel antibodies. Unfortunately, the use of "native" and functional antigens has not been widely explored for ion channel antibody drug discovery. Only a small number of examples of the use of non-peptide antigens to generate anti-ion channel antibodies can be found in the literature. The monoclonal antibody to the P2X7 ligand gated channel was generated by immunizing mice with a mouse myeloma cell line stably transfected with DNA encoding the human P2X7 channel.[14] Other examples include the use of a purified voltage dependent potassium channel from Aeropyrum pernix (KvAP) to generate function blocking monoclonal antibody Fabs[92] (Jiang *et al.*, 2003) and the use of either CHO or U2OS cells expressing recombinant hOrai1 and hSTIM1, to raise antibodies to store-operated channels. Clearly this is an area that warrants further attention. The use of non-optimal antigens likely underlies some of the difficulties encountered to date in generating high affinity, high efficacy functional ion channel antibodies. Careful attention to the quality of conformational antigens may be the key to advancing the field toward the identification and development of ion channel modulating therapeutic antibodies.

13.4.1.2   *Antibody Generation*

Having identified a suitable antigen, antibodies can be discovered using either *in vivo* immunization or *in vitro* antibody display techniques. The species of choice for generation of monoclonal antibodies through immunizations are mice or rats followed by rabbits, since these are the only species for which there are suitable myeloma fusion partners to generate stable hybridomas.[63] This strategy takes advantage of the diversity of the host immunoglobulin repertoire for selecting antibodies, and *in vivo* maturation of the immune response to yield high affinity IgG monoclonal antibodies. The immunization approach, however, has several limitations including some that may be particularly pertinent to ion channels. For instance, the process of generating and growing hybridoma fusions and screening for anti-target antibodies can be slow and time consuming. While this is not a problem for many targets, it could be a significant consideration when attempting to raise functional antibodies to non-traditional targets, where the number of suitable antibody hits might be quite low or when the optimal strategy is not clear, necessitating iterative adjustment and re-implementation of new strategies. The degree of homology between the target species and the host species also needs to be taken under consideration, especially for ion channels that often exhibit a high degree of sequence homology across species. Generating an adequate antibody immune response to these antigens may require strategies to break immune tolerance or immunization of an appropriate genetically modified knockout animal strain.

Alternatively, *in vitro* antibody display and selection platform approaches, such as phage, ribosomal, and yeast display, are options for deriving monoclonal antibodies.[64] Antibody libraries for display and selection can be derived from immunized *in vivo* sources as well as from completely synthetic *de novo* generation *in vitro*. It has been suggested that the diversity of a *de novo* library can equal or often exceed that of the endogenous immune system *in vivo*.[65] *In vitro* antibody display methodologies can be advantageous for targets that are highly conserved across species (as are many ion channels), circumventing the need to break tolerance or acquire specialized genetic knockout animal strains. Utilizing human libraries has the added benefit of generating fully human antibodies, which removes the need for humanization engineering of rodent monoclonal antibodies. As a potential limitation, antibodies from *de novo* libraries often have lower affinity, and require additional engineering for affinity maturation. While *in vitro* display technology seems to offer some advantages, there are no reports describing the use of this approach to identify ion channel antibodies in the scientific literature. The reason(s) for this is/are not clear.

13.4.1.3 Screening Assays and Library Panning

For immunization or *in vitro* display methodologies to be successful, robust assays and panning strategies must be available to identify high affinity binders with functional activity. The large numbers of hybridomas following fusion of B cells from immunized animals will require primary screening assays with relatively high throughput. Panning *in vitro* display libraries incorporates selection into the process and reduces but does not eliminate the need for primary screening assays with sufficient throughput. Theoretically, panning strategies can be quite sophisticated for targeting specific regions, blocking immuno-dominant epitopes to diversify hits, and incorporating subtype counter selection or absorption for enrichment of channel selective hits. For simple antigens, detection of binders typically utilizes affinity capture or panning against purified immobilized antigen. For more complex antigens, however, expression of the target in lipoparticles, membranes or whole cells may be required. For binding or panning with membrane-inserted antigens to be effective, however, expression levels need to be sufficiently high to enable detection of specific binding above non-specific background and enrichment of target specific binders. Non-specific binding is a particular problem with phage display, since phage coats are highly charged, which often leads to a high degree of non-specific interactions, potentially obscuring hits of interest. Since binding alone will not necessarily translate into functional activity, secondary assays are required to select functional antibodies. Many cell-based assays have been developed to support small molecule ion channel screening, including radioligand binding, fluorescent techniques, ion flux techniques and automated electrophysiology (for recent reviews see [10,66]). These assays are particularly well suited to the detection of direct modulators of ion channel permeation and/or gating and have been used successfully to identify functional ion channel antibodies.[19] However, as described above, direct modulation of channel permeation or gating is not the only mechanism by which antibodies can alter channel function. Antibodies can also exhibit functional effects by altering ion channel membrane expression or location. Alternative screening strategies may have to be developed to identify such antibodies, possibly involving high content analysis of primary cell cultures.

13.4.2 Biodistribution

Since most FDA approved antibody drugs target circulating factors (such as cytokines, complement system proteins and growth factors) or surface antigens on blood cells (such as T-cells or vascular endothelial cells), extensive extra vascular distribution is not necessarily required for their therapeutic activity. In contrast, many therapeutically relevant ion channel targets are found outside the vascular system. Indeed, ion channels are often best recognized as targets for CNS disorders and pain, and extravascular distribution is therefore a pre-requisite for antibodies for these indications. Considerable qualitative evidence exists to indicate that antibodies can access a variety of extravascular sites. The development of neuropathies following passive immunization of animals clearly indicates that antibodies can access peripheral nerves and/or sites at the neuromuscular junction.[30-34] Using immunocytochemical techniques, high levels of rabbit IgG and anti-NGF antibodies were found in dermis (but not epidermis) of 15 day old rats following four daily sub-cutaneous doses[67] and fluorescently labeled anti-herpes simplex virus (HSV) antibodies were found to co-localize with neuronal markers in the cornea of mice infected with HSV type 1.[68] However, these qualitative studies do not provide any information on the relative efficiency of distribution or tissue concentrations, and quantitative analyses paint a somewhat different picture. Where studied, the volume of distribution of therapeutic monoclonal antibodies is typically quite low, suggesting only limited extravascular distribution.[69-71] In keeping with these observations, cerebrospinal fluid (CSF) levels of rituximab are only 0.1% of serum levels following intravenous administration to patients with primary central nervous system lymphoma, indicating that distribution into the CNS is likely to be minimal.[72] However, it is possible that CNS penetration may increase under circumstances where blood–brain barrier integrity is compromised, such as in Alzheimer's disease, for example.[73] Similarly, bronchoalveolar lavage fluid (BALF) concentrations of mepolizumab, a humanized anti-IL-5 monoclonal antibody, were reported to be between 500- and 1,000-fold lower than the steady-state plasma concentrations in cynomolgus monkeys[74] and synovial fluid concentrations of the humanized anti-CD4 monoclonal antibody, 4162W94, were found to be between 17% and 28% of plasma concentrations in rheumatoid arthritis patients after 5 successive days of dosing with 100 mg and 300 mg 4162W94, respectively.[75] Thus, while there are exceptions, it would appear that the distribution of monoclonal antibodies into organs and tissues is generally rather limited, potentially posing a major hurdle to the successful development of anti-ion channel antibodies for many traditional ion channel indications. In order to advance the field, a better understanding of the factors that influence antibody penetration into disease relevant organs and tissues will be required. In the event that antibodies are partially or totally excluded from critical compartments, strategies will have to be developed to optimize distribution before antibody based ion channel therapeutics become a realistic option.

Antibody penetration into the CNS is particularly relevant to ion channel research and there are numerous potential strategies being explored for delivering large molecules into the CNS, including intracranial or intrathecal administration, nasal delivery, blood–brain barrier disruption, nano-particles, receptor mediated transcytosis (RMT) *etc.* (see [76]). The majority of these will require the development of drug delivery technologies and a detailed discussion of these technologies is probably beyond the scope of this chapter. RMT, however, is a strategy that seems to offer some promise as a general strategy for improving brain penetration of antibodies and as such merits brief discussion here. Receptor-mediated trancytosis describes a mechanism by which ligands are endocytosed following interaction with a receptor on the apical side of an endothelial cell, packaged into intracellular transport vesicles, sorted and sent to the basolateral side of the polarized endothelium where they are released. Brain uptake of nutrients such as iron and insulin occurs *via* receptor-mediated transcytosis following binding of these ligands with their respective receptors (transferrin receptors or insulin receptors). In order to take advantage of endogenous RMT systems, it is necessary to make bi-specific antibodies or antibody fusion proteins comprising both a transport domain that recognizes a receptor on the BBB and a therapeutic domain to modulate the therapeutic target. There are several reports in the literature that appear to show improved delivery of antibody-based therapeutics into the CNS using this strategy. Most examples described to date are fusion proteins in which the transport domain is a monoclonal antibody directed against an exofacial epitope of either the human insulin receptor or the rat transferrin receptor, and the therapeutic domain consists of a peptide or protein cargo fused to the carboxyl terminus of the antibody (so-called Molecular Trojan Horses, or MTHs). Examples of MTHs described in the literature include anti-transferrin antibodies fused to either nerve growth factor,[77] brain derived neurotrophic factor,[78] basic fibroblast growth factor,[79] or tumor necrosis factor decoy receptors,[80] and anti-insulin antibodies fused to either glial derived neurotrophic factor,[81] iduronate-2-sulfatase,[82] erythropoietin,[83] and tumor necrosis factor decoy receptors.[84] In a recent development, Yu and colleagues demonstrated that lower affinity anti-transferrin variants were more efficiently transported across the BBB into brain parenchyma.[85] Presumably, suitably optimized MTHs could be employed to deliver anti-ion channel antibodies to the central nervous system, thus providing a general solution to the problem of limited biodistribution.

13.5   Fusion Proteins

Although there are many potential advantages to antibodies, particularly potency, selectivity and long half-life, it is worth noting that the majority of ion channel antibodies acting *via* block of permeation or gating are neither potent nor efficacious, possibly suggesting that antibodies may have difficulty accessing the grooves and clefts in functional channel domains close to the cell membrane that are required for potent/efficacious channel modulation. While it is likely that potency/efficacy can be improved through advances in antigen design, screening technology/capacity, affinity maturation *etc.* (see above), it is also worth considering other related strategies, such as fusion proteins that combine the antibody domains responsible for long half-life (*i.e.*, Fc) with ion channel targeting domains comprising molecules with proven ability to modulate ion channels with a high degree of potency, efficacy and selectivity, such as venom-derived peptides for example. The venoms of various venomous creatures, such as snakes, snails, scorpions and spiders are a particularly rich source of ion channel modulating peptides. Venom peptides are typically 1–10 kDa in size, most of them tightly folded and stabilized by several disulfide bridges and they have evolved to modify ion channels in the nervous systems of prey species with extremely high potency and selectivity. In fact, these attributes have made venom peptides interesting as drug candidates in their own right[86–88] and the cone snail venom-derived peptide, ziconotide, a highly potent and selective inhibitor of N-type calcium channels, is FDA approved under the trade name Prialt to treat severe, treatment-refractory chronic pain.[89] Likewise, ShK-186, a potent and selective peptide blocker of Kv1.3 derived from a sea anemone toxin, is reported to be in early clinical development for the treatment of multiple sclerosis, type 1 diabetes and other autoimmune diseases.[90] Obviously, one major limitation of these peptides is their very short *in vivo* half-lives. However, it may be possible to combine the potency/selectivity of the peptides with the long half-life of the antibody by making peptide/antibody fusion proteins or peptide/Fc fusion proteins. Similar half-life extension could also be achieved by fusing the active peptide to other, non-antibody based, half-life extending moieties such as PEG or albumin. The peptide-fusion protein approach has been highly successful for an array of non-ion channel peptide drugs, such as Etanercept (Enbrel), a soluble TNF receptor 2/Fc fusion protein used for the treatment of auto-immune diseases, and recently this approach seems to have been successfully applied to ion channels. In a series of patent applications, Amgen describe the properties of several Kv1.3 blocking fusion proteins.[91] The potassium channel targeting domain of these fusion proteins comprises either ShK, a highly potent peptide toxin from the sea anemone, *Stichodactyla Helianthus*, OsK1, another potent peptide toxin from the scorpion, *Orthochirus scrobiculosus*, or variants thereof. The peptide toxins are either fused to Fc domains or conjugated to PEG. Encouragingly, several examples are provided to show that potent Kv1.3 blocking activity can be retained following fusion of the peptides to the half-life extending moieties. Not surprisingly, these fusion constructs also exhibit drastically improved pharmacokinetic profiles compared to the parent peptides. It is also a distinct possibility that some hybrid molecules (*e.g.*, PEG, albumin, *etc.*,) may distribute more effectively than monoclonal antibodies owing to their smaller size or mechanism of action. Based on these data, it would seem that generation of potent, highly selective, long half-life ion channel modulating peptide fusion proteins is technically feasible and may represent a viable alternative or compliment to antibodies for targeting some ion channels with large molecules.

13.6 Conclusion

As discussed above, ion channel modulators hold considerable promise for the treatment of a variety of serious human diseases. Unfortunately, despite a concerted effort in the pharmaceutical industry over the last two decades, the identification of potent, selective and drug-like small molecule ion channel modulators has proven to be more challenging than most people anticipated. In spite of the historical challenges, traditional small molecule-based drug discovery should remain an important component of this effort and it is hoped that recent advances in screening technologies and structural biology will accelerate progress. For some highly validated but technically challenging targets alternatives to non-small molecule approaches should also be considered, in order to increase the chance of success with these important targets. Based on emerging literature, targeting therapeutically relevant ion channels with functional monoclonal antibodies or other related large molecules appears to be a feasible alternative to small molecules, offering the potential for long half-life, high potency and exquisite selectivity. However, this field is very much still in its infancy and turning the promising concept into approved drugs will likely require significant advances in our understanding of immunization strategies, phage (or other) display methodologies, antigen conformation/presentation and in the factors that influence tissue distribution. For the channel modulating antibody approach to gain momentum and become sustainable, the efficiency and frequency of hits per target needs to increase beyond an occasional antibody per effort and will require input from disciplines that are not traditionally linked to ion channel drug discovery. In many ways, the challenge appears daunting. However, we strongly believe that ion channel drugs have the potential to make a significant impact on the lives of patients who are poorly served by existing medications and that continued investment in this area is justified.

References

1. K. Sumikawa, M. Houghton, J. C. Smith, L. Bell, B. M. Richards and E. A. Barnard, *Nucleic Acids Res.*, 1982, **10**(19), 5809–5822.
2. M. Noda, H. Takahashi, T. Tanabe, M. Toyosato, Y. Furutani, T. Hirose, M. Asai, S. Inayama, T. Miyata and S. Numa, *Nature*, 1982, **299**(5886), 793–797.
3. M. Noda, H. Takahashi, T. Tanabe, M. Toyosato, S. Kikyotani, T. Hirose, M. Asai, H. Takashima, S. Inayama, T. Miyata and S. Numa, *Nature*, 1983, **301**(5897), 251–255.
4. T. J. Jegla, C. M. Zmasek, S. Batalov and S. K. Nayak, *Comb. Chem. High Throughput Screen.*, 2009, **12**(1), 2–23.
5. J. Dunlop, M. Bowlby, R. Peri, G. Tawa, J. LaRocque, V. Soloveva and J. Morin, *Comb. Chem. High Throughput Screen.*, 2008, **11**(7), 514–522.
6. S. Tonstad and H. Rollema, *Expert Rev. Respir. Med.*, 2010, **4**(3), 291–299.
7. J. L. Weisenberg and M. Wong, *Neuropsychiatr. Dis. Treat.*, 2011, 7, 409–414.
8. A. Opar, *Nat. Rev. Drug Discov.*, 2011, **10**(7), 479–480.
9. G. L. Krauss, M. Bar, V. Biton, J. A. Klapper, I. Rektor, N. Vaiciene-Magistris, D. Squillacote and D. Kumar, *Acta Neurol. Scand.*, 2011, **125**(1), 8–15.
10. A. Wickenden, B. Priest and G. Erdemli, *Future Med. Chem.*, 2012, **4**(5), 661–679.
11. M. L. Dallas, S. A. Deuchars and J. Deuchars, *Expert Rev. Clin. Pharmacol.*, 2010, **3**(3), 281–289.
12. A. Maelicke, D. Watters, G. Fels and R. Plumer, *J. Recept. Res.*, 1984, **4**(1–6), 671–679.
13. D. Mochly-Rosen and S. Fuchs, *Biochemistry*, 1981, **20**(20), 5920–5924.
14. G. Buell, I. P. Chessell, A. D. Michel, G. Collo, M. Salazzo, S. Herren, D. Gretener, C. Grahames, R. Kaur, M. H. Kosco-Vilbois and P. P. Humphrey, *Blood*, 1998, **92**(10), 3521–3528.
15. R. Haring, P. K. Stanton, M. A. Scheideler and J. R. Moskal, *J. Neurochem.*, 1991, **57**(1), 323–332.

16. L. T. Thompson, J. R. Moskal and J. F. Disterhoft, *Nature*, 1992, **359**(6396), 638–641.
17. M. D. Jarnot and A. M. Corbett, *Brain Res.*, 1995, **674**(1), 159–162.
18. C. N. Wyatt, V. Campbell, J. Brodbeck, N. L. Brice, K. M. Page, N. S. Berrow, K. Brickley, C. M. Terracciano, R. U. Naqvi, K. T. MacLeod and A. C. Dolphin, *J. Physiol.*, 1997, **502**(Pt 2), 307–319.
19. L. Klionsky, R. Tamir, B. Holzinger, X. Bi, J. Talvenheimo, H. Kim, F. Martin, J. C. Louis, J. J. Treanor and N. R. Gavva, *J. Pharmacol. Exp. Ther.*, 2006, **319**(1), 192–198.
20. J. Naylor, C. J. Milligan, F. Zeng, C. Jones and D. J. Beech, *Br. J. Pharmacol.*, 2008, **155**(4), 567–573.
21. S. Z. Xu, G. Boulay, R. Flemming and D. J. Beech, *Am. J. Physiol. Heart. Circ. Physiol.*, 2006, **291**(6), H2653–2659.
22. S. Z. Xu, F. Zeng, M. Lei, J. Li, B. Gao, C. Xiong, A. Sivaprasadarao and D. J. Beech, *Nat. Biotechnol.*, 2005, **23**(10), 1289–1293.
23. B. Y. Zhou, W. Ma and X. Y. Huang, *J. Gen. Physiol.*, 1998, **111**(4), 555–563.
24. X. F. Yang, Y. Yang, Y. T. Lian, Z. H. Wang, X. W. Li, L. X. Cheng, J. P. Liu, Y. F. Wang, X. Gao, Y. H. Liao, M. Wang, Q. T. Zeng and K. Liu, *PLoS One*, 2012, **7**(4), e36379.
25. D. Gomez-Varela, E. Zwick-Wallasch, H. Knotgen, A. Sanchez, T. Hettmann, D. Ossipov, R. Weseloh, C. Contreras-Jurado, M. Rothe, W. Stuhmer and L. A. Pardo, *Cancer Res.*, 2007, **67**(15), 7343–7349.
26. M. Sammar, G. Spira and H. Meiri, *J. Membr. Biol.*, 1992, **125**(1), 1–11.
27. A. M. Chioni, S. P. Fraser, F. Pani, P. Foran, G. P. Wilkin, J. K. Diss and M. B. Djamgoz, *J. Neurosci. Methods*, 2005, **147**(2), 88–98.
28. S. Sokolov, R. L. Kraus, T. Scheuer and W. A. Catterall, *Mol. Pharmacol.*, 2008, **73**(3), 1020–1028.
29. K. J. Swartz, *Toxicon*, 2007, **49**(2), 213–230.
30. K. A. Kleopa, *Curr. Neuropharmacol.*, 2011, **9**(3), 458–467.
31. A. Vincent, C. G. Bien, S. R. Irani and P. Waters, *Lancet Neurol.*, 2011, **10**(8), 759–772.
32. M. Kinoshita, Y. Nakatsuji, T. Kimura, M. Moriya, K. Takata, T. Okuno, A. Kumanogoh, K. Kajiyama, H. Yoshikawa and S. Sakoda, *Biochem. Biophys. Res. Commun.*, 2009, **386**(4), 623–627.
33. B. Lang, J. Newsom-Davis, D. Wray, A. Vincent and N. Murray, *Lancet*, 1981, **2**(8240), 224–226.
34. S. Vernino, L. G. Ermilov, L. Sha, J. H. Szurszewski, P. A. Low and V. A. Lennon, *J. Neurosci.*, 2004, **24**(32), 7037–7042.
35. J. Bufler, R. Pitz, M. Czep, M. Wick and C. Franke, *Ann. Neurol.*, 1998, **43**(4), 458–464.
36. K. Krampfl, R. Dengler and J. Bufler, *Muscle Nerve*, 2002, **25**(3), 433–437.
37. K. Jahn, C. Franke and J. Bufler, *Neurology*, 2000, **54**(2), 474–479.
38. Y. J. Liao, P. Safa, Y. R. Chen, R. A. Sobel, E. S. Boyden and R. W. Tsien, *Proc. Natl. Acad. Sci., U. S. A.*, 2008, **105**(7), 2705–2710.
39. A. Nagel, A. G. Engel, B. Lang, J. Newsom-Davis and T. Fukuoka, *Ann. Neurol.*, 1988, **24**(4), 552–558.

40. E. G. Hughes, X. Peng, A. J. Gleichman, M. Lai, L. Zhou, R. Tsou, T. D. Parsons, D. R. Lynch, J. Dalmau and R. J. Balice-Gordon, *J Neurosci*, 2010, **30**(17), 5866–5875.

41. P. Shillito, P. C. Molenaar, A. Vincent, K. Leys, W. Zheng, R. J. van den Berg, J. J. Plomp, G. T. van Kempen, G. Chauplannaz and A. R. Wintzen, *Ann. Neurol.*, 1995, **38**(5), 714–722.

42. I. K. Hart, C. Waters, A. Vincent, C. Newland, D. Beeson, O. Pongs, C. Morris and J. Newsom-Davis, *Ann. Neurol.*, 1997, **41**(2), 238–246.

43. M. J. Thieben, V. A. Lennon, B. F. Boeve, A. J. Aksamit, M. Keegan and S. Vernino, *Neurology*, 2004, **62**(7), 1177–1182.

44. S. R. Irani, S. Alexander, P. Waters, K. A. Kleopa, P. Pettingill, L. Zuliani, E. Peles, C. Buckley, B. Lang and A. Vincent, *Brain*, 2010, **133**(9), 2734–2748.

45. E. Lancaster, M. G. Huijbers, V. Bar, A. Boronat, A. Wong, E. Martinez-Hernandez, C. Wilson, D. Jacobs, M. Lai, R. W. Walker, F. Graus, L. Bataller, I. Illa, S. Markx, K. A. Strauss, E. Peles, S. S. Scherer and J. Dalmau, *Ann. Neurol.*, 2011, **69**(2), 303–311.

46. S. R. Hinson, S. J. Pittock, C. F. Lucchinetti, S. F. Roemer, J. P. Fryer, T. J. Kryzer and V. A. Lennon, *Neurology*, 2007, **69**(24), 2221–2231.

47. Maxplank-Gesellshaft zur forderung der wissensch and U3 Pharma AG, US 2009/0028851, 2009.

48. Amgen, WO 2011/063277, 2011.

49. UCB Pharma S.A, WO 2011/051350, 2011.

50. Regeneron Pharmaceuticals, Inc., US 2012/0083000, 2012.

51. C. J. Hutchings, M. Koglin and F. H. Marshall, *MAbs*, 2010, **2**(6), 594–606.

52. R. Jin, S. K. Singh, S. Gu, H. Furukawa, A. I. Sobolevsky, J. Zhou, Y. Jin and E. Gouaux, *EMBO J.*, 2009, **28**(12), 1812–1823.

53. J. Kumar, P. Schuck, R. Jin and M. L. Mayer, *Nat. Struct. Mol. Biol.*, 2009, **16**(6), 631–638.

54. G. Yao, Y. Zong, S. Gu, J. Zhou, H. Xu, I. I. Mathews and R. Jin, *Biochem. J.*, 2011, **438**(2), 255–263.

55. J. Kumar and M. L. Mayer, *J. Mol. Biol.*, 2010, **404**(4), 680–696.

56. Y. Yao and M. L. Mayer, *J. Neurosci.*, 2006, **26**(17), 4559–4566.

57. A. M. Slovic, H. Kono, J. D. Lear, J. G. Saven and W. F. DeGrado, *Proc. Natl. Acad. Sci., U. S. A.*, 2004, **101**(7), 1828–1833.

58. S. B. Long, E. B. Campbell and R. Mackinnon, *Science*, 2005, **309**(5736), 897–903.

59. J. Payandeh, T. Scheuer, N. Zheng and W. A. Catterall, *Nature*, 2011, **475**(7356), 353–358.

60. J. Jasti, H. Furukawa, E. B. Gonzales and E. Gouaux, *Nature*, 2007, **449**(7160), 316–323.

61. T. Kawate, J. C. Michel, W. T. Birdsong and E. Gouaux, *Nature*, 2009, **460**(7255), 592–598.

62. A. I. Sobolevsky, M. P. Rosconi and E. Gouaux, *Nature*, 2009, **462**(7274), 745–756.

63. H. Spieker-Polet, P. Sethupathi, P. C. Yam and K. L. Knight, *Proc. Natl, Acad. Sci, U. S. A.*, 1995, **92**(20), 9348–9352.

64. H. R. Hoogenboom, *Nat. Biotechnol.*, 2005, **23**(9), 1105–1116.
65. S. S. Sidhu and F. A. Fellouse, *Nat. Chem. Biol.*, 2006, **2**(12), 682–688.
66. G. C. Terstappen, R. Roncarati, J. Dunlop and R. Peri, *Future Med. Chem.*, 2010, **2**(5), 715–730.
67. J. R. Tonra and L. M. Mendell, *J. Neuroimmunol.*, 1997, **80**(1–2), 97–105.
68. P. P. Sanna, T. J. Deerinck and M. H. Ellisman, *J. Virol.*, 1999, **73**(10), 8817–8823.
69. D. Tracey, L. Klareskog, E. H. Sasso, J. G. Salfeld and P. P. Tak, *Pharmacol. Ther.*, 2008, **117**(2), 244–279.
70. M. Tabrizi, G. G. Bornstein and H. Suria, *AAPS J.*, 2010, **12**(1), 33–43.
71. J. H. Lin, *Curr. Drug Metab.*, 2009, **10**(7), 661–691.
72. J. L. Rubenstein, D. Combs and J. Rosenberg *et al.*, *Blood*, 2003, **101**(2), 466–468.
73. G. L. Bowman, J. A. Kaye, M. Moore, D. Waichunas, N. E. Carlson and J. F. Quinn, *Neurology*, 2007, **68**(21), 1809–1814.
74. T. K. Hart, R. M. Cook, P. Zia-Amirhosseini, E. Minthorn, T. S. Sellers, B. E. Maleeff, S. Eustis, L. W. Schwartz, P. Tsui, E. R. Appelbaum, E. C. Martin, P. J. Bugelski and D. J. Herzyk, *J. Allergy Clin. Immunol.*, 2001, **108**(2), 250–257.
75. E. H. Choy, D. J. Connolly, N. Rapson, S. Jeal, J. C. Brown, G. H. Kingsley, G. S. Panayi and J. M. Johnston, *Rheumatology (Oxford)*, 2000, **39**(10), 1139–1146.
76. M. Rajadhyaksha, T. Boyden, J. Liras, A. El-Kattan and J. Brodfuehrer, *Curr. Drug. Discov. Technol.*, 2011, **8**(2), 87–101.
77. J. H. Kordower, V. Charles, R. Bayer, R. T. Bartus, S. Putney, L. R. Walus and P. M. Friden, *Proc. Natl. Acad. Sci., U. S. A.*, 1994, **91**(19), 9077–9080.
78. Y. Zhang and W. M. Pardridge, *Brain Res.*, 2006, **1111**(1), 227–229.
79. B. W. Song, H. V. Vinters, D. Wu and W. M. Pardridge, *J. Pharmacol. Exp. Ther.*, 2002, **301**(2), 605–610.
80. Q. H. Zhou, R. J. Boado, E. K. Hui, J. Z. Lu and W. M. Pardridge, *Drug Metab. Dispos.*, 2011, **39**(1), 71–76.
81. R. J. Boado and W. M. Pardridge, *Drug Metab. Dispos.*, 2009, **37**(12), 2299–2304.
82. J. Z. Lu, R. J. Boado, E. K. Hui, Q. H. Zhou and WM Pardridge, *Biotechnol. Bioeng.*, 2011, **108**(8), 1954–1964.
83. A. Fu, E. K. Hui, J. Z. Lu, R. J. Boado and W. M. Pardridge, *Brain Res.*, 2010, **1369**, 203–207.
84. R. J. Boado, E. K. Hui, J. Z. Lu, Q. H. Zhou and W. M. Pardridge, *J. Biotechnol*, 2010, **146**(1–2), 84–91.
85. Y. J. Yu, Y. Zhang, M. Kenrick, K. Hoyte, W. Luk, Y. Lu, J. Atwal, J. M. Elliott, S. Prabhu, R. J. Watts and M. S. Dennis, *Sci. Transl. Med.*, 2011, **3**(84), 84ra44.
86. P. Escoubas and G. F. King, *Expert. Rev. Proteomics*, 2009, **6**(3), 221–224.
87. G. F. King, *Expert Opin. Biol. Ther.*, 2011, **11**(11), 1469–1484.
88. B. M. Olivera, D. R. Hillyard, M. Marsh and D. Yoshikami, *Trends Biotechnol.*, 1995, **13**(10), 422–426.

89. G. P. Miljanich, *Curr. Med. Chem.*, 2004, **11**(23), 3029–3040.
90. V. Chi, M. W. Pennington, R. S. Norton, E. J. Tarcha, L. M. Londono, B. Sims-Fahey, S. K. Upadhyay, J. T. Lakey, S. Iadonato, H. Wulff, C. Beeton and K. G. Chandy, *Toxicon*, 2012, **59**(4), 529–546.
91. Amgen Inc., WO 2006/116156, 2006.
92. Y. Jiang, A. Lee, J. Chen, V. Ruta, M. Cadene, B. T. Chait and R. MacKinnon, *Nature*, 2003, **423**(6935), 33–41.

CHAPTER 14

Ion Channel Drug Discovery: Future Perspectives

MARTIN GOSLING

Respiratory Diseases, Novartis Institutes for Biomedical Research, Horsham, UK
Email: Martin.Gosling@Novartis.com

14.1   Introduction

The last two decades have witnessed an explosion in the level of interest the pharmaceutical and biotechnology industries have displayed in ion channels as a therapeutic target class. This interest has been fuelled by a number of factors, including:

i. Ion channels are the targets of therapeutically useful agents estimated to account for worldwide sales of >US$12 billion, highlighting their 'tractable' nature.
ii. The annotation of the human genome suggests that there approximately 400 ion channel pore-forming and accessory ion channel subunits yet current ion channel therapeutics target only a small number (<10%).
iii. The historical perspective that ion channels are only therapeutically relevant in electrically excitable tissues or cannot be selectively modulated, leading to nonselective dose limiting side effects, has been eroded.
iv. The ability to prosecute ion channel targets using 'industrial scale' high throughput biology has been dramatically enhanced by a

RSC Drug Discovery Series No. 39
Ion Channel Drug Discovery
Edited by Brian Cox and Martin Gosling
© The Royal Society of Chemistry 2015
Published by the Royal Society of Chemistry, www.rsc.org

combination of technological developments in frontline high throughput screening assays and automation of the benchmark patch clamp electrophysiology technique.

v. An increasing number of diverse disease pathologies have been shown to be the result of mutations in ion channel genes ('channelopathies') providing both insight into the physiological function of ion channels and novel therapeutic opportunities for intervention.

The result of this renewed interest by the industry has been the discovery and transition of an increased number of ion channel modulators into clinical studies in recent years. Additionally, these candidates are aimed at treating a wide spectrum of diseases and target a diverse cross-section of channel families ranging from potassium channel openers for epilepsy[1] through to ryanodine receptor stabilizers for heart failure.[2] Although not all have or will achieve success in terms of registration, several have achieved that milestone, the most recent being the CFTR potentiator VX-770 (Ivacaftor, Kalydeco™; **1**) from Vertex Pharmaceuticals which received FDA approval for the treatment of cystic fibrosis in January 2012.

1

2

In this chapter I have aimed to provide an insight into the future directions ion channel drug discovery may take highlighting some of the opportunities, approaches and challenges that these enigmatic targets pose.

14.2 Channels, Channels, Channels: How Many and What Do They Do?

The publication of the human genome was a milestone achievement but the first step in the path to understanding human development, health and disease—focus is now directed towards the larger task of annotating function to each of the component genes and understanding their interconnectivity. Although the 'channelome' is potentially defined, the exact molecular subunit/gene composition which underlies many 'native' currents is not, nor is the exact physiological/pathophysiological role of a significant number of channels. Historically the late 1980s—early 1990s saw the channel field move from the electrophysiological 'cataloguing' of currents measured in cultured native and primary cells to characterizing channel gene products in heterologous expression systems. This advanced the channel field in defining the molecular identity of the channels underlying many currents, at least from a pore-forming unit perspective. However, in many cases the role of accessory subunits and by association their importance in determining channel behavior, channel trafficking and pharmacological sensitivity remain loosely defined or controversial. This can be important highlighting that currents which look very similar from a gross electrophysiological perspective may be the result of entirely different gene products or subunit combinations. Arguably this can be viewed from polar perspectives. Firstly as an unwanted and potentially unfathomable degree of complexity raising challenges to understand which is the relevant channel complex to modulate, particularly if multiple subunits are expressed within the same cell and from a channel family that can readily heteromultimerize (*e.g.* TRPC channels). An alternative and more optimistic view is that this complexity provides opportunity. What appear to be ubiquitously expressed currents may actually be the result of the expression of a diversity of differing channel subunit combinations offering the potential for tissue specific modulation. An example of this from the author's experience is the large conductance Ca^{2+}-activated potassium channel (BK_{Ca}). Although BK_{Ca} currents are widely expressed throughout the periphery and the CNS, the channel's molecular composition differs. BK_{Ca} is composed of a pore-forming alpha (KCNMA1) and an accessory beta subunit (KCNMB). Although there is only a single pore-forming alpha subunit, albeit with multiple splice variants, there are four beta subunits, with KCNMB3 displaying four splice variants (named a–d). As outlined in Figure 14.1, the expression profile of the beta subunits can be highly tissue dependent. These subunits not only differentially modulate the channels behavior (sensitivity to voltage, calcium *etc.*) but there is suggestion that these subunits modulate pharmacological sensitivity too.[3] BK_{Ca} channels are certainly not the only example of this (*e.g.* several Kv channels, TMEM16A channels exhibit splice variants) and this may be an area for future investigation when considering selection of channel targets. This approach does of course still require selective compounds which can discriminate between these subtle variants. However, these molecules may be more readily discovered by targeting regions of structural diversity (accessory subunits/unique pore-forming accessory subunit combinations) rather than the pore-forming subunits themselves which have high degrees of conservancy, a point I will return to later in the chapter.

Returning to the 'channelome', having a catalogue of channel 'electrophysiological' signatures and the human genome sequences available has empowered definition of the majority of human ion channels. However, for

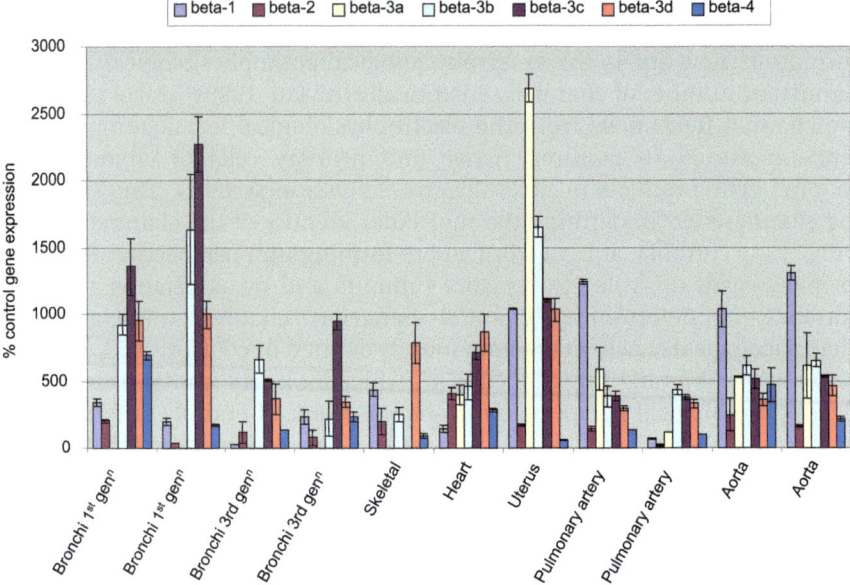

Figure 14.1   Heterogeneity of large conductance, calcium-sensitive K channel accessory subunits in selected human tissues. Expression of large conductance calcium-sensitive potassium channel (BK_{Ca}) subunit RNA in laser captured muscle. Transcripts were measured by quantitative RT-PCR (Taqman) and normalized to a reference (control) housekeeping gene to facilitate comparison.
Gosling and Jarai, unpublished.

some channels reconciling the two has been more challenging and in some cases work yet remains to be done. Relatively recent examples of ascribing gene to channel function are Orai1,[4,5] which encodes the pore-forming subunit of the calcium release activated channel (CRAC), TMEM16A, the epithelial calcium-activated chloride channel[6,7] and the Piezos, which encode stretch-activated cation channels.[8] In all cases the genes encoded proteins with novel structural motifs; 4 transmembrane domains with a unique pore architecture for Orai1, 8 transmembrane domains for TMEM16A and between 24 and 36 predicted transmembrane domains for the Piezos. This structural novelty is perhaps the explanation why these channel proteins were not predicted from the 'looking under the lamp post' approach of sequence homologies/alignments with known channels (Figure 14.2). Interestingly there are some parallels in the approaches with which these channel proteins were identified. The discovery of Orai1 was reported by the Penner/Kinet and Feske/Rao groups almost simultaneously and using almost identical approaches[4,5]—both employed *Drosophila* genome-wide knockdown screens. However the Feske/Rao group combined this with a single nucleotide polymorphism (SNP) linkage analysis of patients with a hereditary severe combined immune deficiency (SCID) characterized

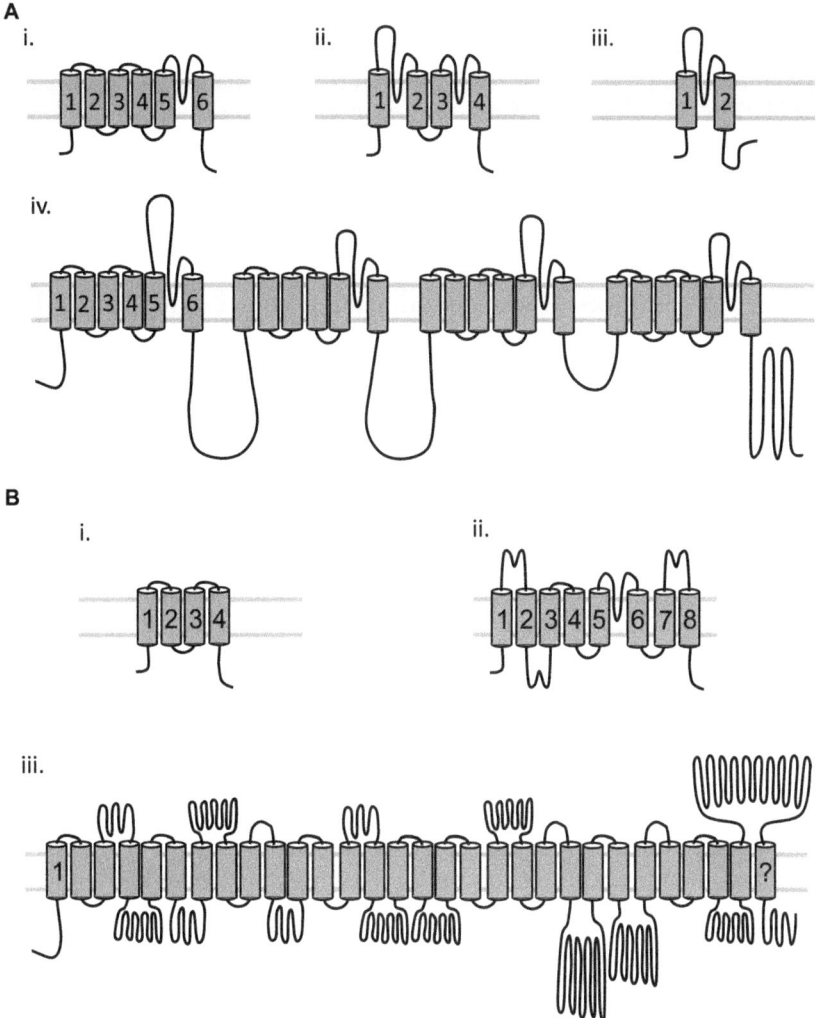

Figure 14.2 Common and prototypical ion channel pore-forming subunit motifs. A. Common generalized ion channel pore forming motifs: i. 6 TM domains with a single pore loop as exemplified in many Kv, Kca and TRP channels; ii. 4 TM domains and two pore loops as found in twin pore (K2P) channels; iii. 2 TM domains and a single pore loop motif of Kir channels; iv. 24 TM domains and 4 pore loops as found in NaV and CaV channels. B. Novel motifs of recently identified ion channels: i. Orai1 4 TM motif; ii. 8 TM motif of TMEM16A; ii. Piezo channel structural motif (redrawn from [37]) whose structure is yet to be fully resolved but proposed to contain between 24 and 36 TM domains.

by defective CRAC channel function providing clear links between channel, function and pathophysiology. The discovery of the elusive TMEM16A and Piezo channels by the Galietta[6] and Patapoutian groups,[8] respectively, also resulted from a genomics based approach. However, rather than a knock-down screen, these genes were identified based upon differential gene expression profiling. In the case of TMEM16A the Galietta group exploited the upregulation of calcium-activated chloride currents in epithelial cells by IL-4, whilst the Patapoutian group identified candidate genes by comparison of cell types with and without stretch activated channels, limiting the list to proteins with a minimum of two predicted transmembrane domains. Both groups undertook subsequent functional confirmation using gene silencing. In the case of the Piezo channels this was a significant undertaking as function was confirmed using low throughput conventional patch clamp electrophysiology with the eventual successful candidate (Fam38a) being the 73rd gene tested! Pharmacological modulation of each of the relatively new additions to the channelome may offer the opportunity to deliver novel ion channel-based therapeutics in the near future. Orai1 is an influential controller of mast cell/leukocyte activation so has the potential for next generation approaches for allergic/immune diseases;[9] recent patent and public domain company information (*e.g.* Calcimedica, www.calcimedica.com) show efforts are well advanced with this channel. TMEM16A is highly expressed by respiratory and intestinal epithelia offering potential approaches to treating cystic fibrosis and secretory diarrhea—it is also highly expressed in a number of cancers and implicated in cell proliferation, growth and invasiveness.[10] Whilst the biological implications of modulating mechanotransduction *via* the Piezos remains to be fully unraveled, several groups have reported human gain of function mutations in Piezo1 in a number of hemolytic anemias suggesting the channel plays a key role in maintaining erythrocyte volume.[11–13] Although it is widely believed that the majority of the human channelome is defined there are still a few discrete enigmas to be solved. For example what is the molecular entity responsible for the ubiquitously expressed volume-regulated anion currents? Historical and more recent suggestions have included P-glycoprotein,[14] ClC family members, bestrophins, CFTR and TMEM16 proteins.[15] Although each of these has been shown to be affected by cell swelling and may contribute to the cellular response to osmotic challenge, there are biophysical differences when compared to native currents that suggest there is more to be discovered.[†]

Selecting and validating ion channel targets is arguably no more challenging than for any other target class, and in common with every drug discovery program this is a critical determinant of its eventual therapeutic success. Obviously the highest degree of validation is a pharmacological precedent with an efficacious drug. However, this is not always available and in a current climate where a high degree of innovation and novelty are actively sought this may not always be a positive criteria for selecting a target. As already mentioned in previous paragraphs human genetic data can provide powerful insight into a channels cellular function and the potential

---

[†]Note added in proof - Two recent publications[38,39] have reported LRRC8A (SWELL1) as key component of the volume-regulated anion currents.

consequences of its pharmacological modulation. However, there are caveats to the latter as the effects of a life-long channel gene mutation, with its potential developmental implications and '24/7' duration, may not be re-capitulated by a relatively acute pharmacological approach. Additionally there are non-conducting roles of the channel protein to consider as well as the spectrum of functional consequences of the specific mutation. The biophysical consequences of even minor sequence changes can be con-siderable, ranging from full loss to full gain of function with a significant number of degrees of separation resulting from effects on level of channel protein expression, effects on channel biophysics (opening, closing, acti-vation, inactivation) and altered channel subunit composition. Despite these caveats genetic ion channel dysfunctions when associated with human dis-ease, termed 'channelopathies', present a strong rationale for their selection as targets. The poster child channelopathy, cystic fibrosis, does however tell us that this path will not necessarily be straightforward. It has taken a considerable amount of scientific dedication, time, effort and resources to translate the identification of the genetic defect in CFTR in 1989[16] into the registration of Kalydeco™ in 2012; however, by addressing the primary defect Kalydeco™ is potentially disease modifying and may represent a cure for many CF patients, an exciting advance. To date the number of reported channelopathies is far in excess of 100, with >40 associated with cardiac pathologies alone.[17,18] Perhaps one of the most engaging recent channelo-pathies are those associated with the NaV1.7 gene (SCN9A gene) which im-plicate the channel with a major role in pain sensation.[19] Both loss and gain of functions in NaV1.7 have been described in familial cohorts. Individuals with loss of function mutations have a congenital insensitivity to pain and interestingly also are reported to have an impaired sense of smell. Un-fortunate individuals with gain of function mutations suffer the debilitating chronic pain syndromes primary erythermalgia (IEM) and paroxysmal ex-treme pain disorder (PEPD). As there are currently a number of NaV1.7 blockers reported to be undergoing clinical evaluation, relief for IEM, PEPD and patients experiencing other forms of chronic pain may be close at hand.

Perhaps as important as selecting the right target for a drug discovery program is testing the candidate molecules and the clinical hypothesis in the correct patient population. This can be challenging and has led to a trend in the pharmaceutical/biotech industries to currently focus early clinical testing (*i.e.* proof of concept studies) in highly stratified and ideally genetically defined patient cohorts. Channelopathies by definition, meet these criteria, yet the number being actively pursued (as reported in public domain databases) remains relatively low—an underexplored opportunity?

14.3 Ion Channel Modulators: More, Better…Different?

As highlighted in previous chapters, channels have historically suffered from a drug discovery perspective from bad 'PR' and a view that they are highly challenging to prosecute. Although there are a number of factors (opinions?) that can be cited to support these views they were predominantly based upon the combination of low potency, poorly selective molecules available at that time and that high fidelity assays amenable to high throughput industrial scale biology were not readily available. Fortunately these interrelated issues (*i.e.* it's difficult to get good molecules when you don't have good assays/ techniques/technologies to find them) have been somewhat overcome with relatively recent additions to the ion channel drug discoverer's armament- arium. These include automated patch clamp electrophysiology (Chapter 3), structure-based molecule design (Chapter 6) and biologics based approaches (Chapters 12 and 13). Even historical 'villains' such as the hERG channel, responsible for several drug withdrawals, now offers exciting therapeutic potential (Chapter 11). So where are the opportunities for the future in finding better, more effective ion channel targeted drugs?

Starting with the technological aspects of discovering ion channel modulators the implementation of automated whole-cell patch clamping with the ability to generate thousands of datapoints per day has been a revolution. Electrophysiology assays can now be frontline in hit identifi- cation/lead optimization flowcharts or used to screen focused libraries of tens or hundreds of thousands of compounds, rather than languishing as a tertiary assay capable of providing data for only one or two compounds per week. The introduction of these systems has created an entirely new dis- cipline as automated electrophysiology, in common with all other electro- physiological techniques, has idiosyncrasies all of its own. These range from knowledge of where/how non-specific compound binding arises through to correction of voltage offsets to the challenges of analyzing large quantities of high content information accurately and ensuring that artifacts do not arise as a result of either the experimental protocol or the data handling. Cur- rently the field is experiencing the 3rd generation of systems with manu- facturers focusing on developing higher throughput systems with the presumed expectation that these be used to interrogate larger and larger compound collections as a front line hit finding tool (*e.g.* Sophion Qube, www.sophion.com; Molecular Devices Barracuda, www.moleculardevices. com). In the author's view these systems will deliver advances, particularly for channels where state-dependent modulators are sought, and the pre- ferred state requires accurate voltage and/or fast/defined ligand control as these assays are challenging to configure with other formats. But the ad- vances will be incremental rather than disruptive. Two other areas may however provide opportunity for automation to offer a more significant impact. The first relates to the challenge of defining the native channel complex and therefore optimizing molecules to engage it appropriately. Much early channel screening utilizes recombinant systems for convenience by providing optimal signal to noise ratios in assays, compatibility with cell husbandry equipment and cost. Automated systems are highly reliant on a small subset of cell types (*e.g.* HEK, CHO, U2OS) which effectively seal with the planar arrays these systems employ and can be readily genetically manipulated to overexpress the channel of interest. If primary cells, expressing native channel complex, are used with the current systems success rates drop significantly requiring high levels of redundancy in screening wells and an associated increase in cost per datapoint making this an almost intractable proposition for even a low number of compounds (several hundreds). Perhaps the most successful platform for automated patching of primary cells based on currently available data is the Nanion Patchliner system[20] (www.nanion.de); however, this is a relatively low throughput system by current benchmark standards with eight recording channels and a claimed delivery of around 500 data points per day. A high throughput system compatible with primary cells is a current gap and thus opportunity. The second area relates to the electrophysiological recording format. Current systems are almost exclusively focused on delivering whole cell data from channel populations or in some systems, from populations of cells themselves. This is undoubtedly powerful and can provide high information content on compound behavior at a certain biophysical level. However, if more detailed kinetic information is required then the electrophysiologist commonly resorts to single channel recording and studying the behavior of only one or two channels in a single recording. Single channel recording can provide detailed information about compound mechanism of action as it allows definition of transition rates between channel states in addition to kinetically determined affinity constants. Although there is not a marketed system yet available there are an increasing number of reports in the literature of groups who have successfully incorporated single channels into immobilized lipid membranes and measured channel activity.[21–24] If this approach is readily scalable then an automated system could soon be a reality. The potential advances of such a system are many fold: i. reduced need for cells and culture as cell membrane preparations can be used; ii. screening on native channel complexes which can be enriched *via* separation techniques; iii. ease of access to both intracellular and extracellular faces of the channel; iv. recording time likely to be longer lasting than for current cell systems, and perhaps most excitingly; iv. the opportunity to record from intracellular ion channels. Intracellular ion channels have been somewhat neglected as targets by the channel field, potentially due to challenges in studying them. However, this also may be a changing paradigm *e.g.* ryanodine receptor stabilizers termed 'Rycals' for cardiovascular diseases [www.armgo.com].[25]

Although automated electrophysiology has revolutionized some aspects of screening and empowered lower fidelity fluorescence based approaches by virtue of rapid triage of hit lists to remove false positives arising from assay interference (fluorescent compounds, quench agents *etc.*) the cellular reagent employed by both formats may still provide issues. As mentioned above, screening on primary cell systems versus the native channel complex is where all in the field would like to be. However, this presents challenges of its own as these cells, often terminally differentiated, do not always readily proliferate and maintain the desired channel expression profile to a level that would support even modest screening approaches (how many and which compounds should be screened is a different challenge addressed later in the chapter). One approach is to use cells in more complex phenotypic assays, ideally with a functional readout and a low cell burden. A recent and elegant example of this are intestinal organoids developed from primary biopsies from CF patients by the Clevers/Beekman groups.[26] These cells form spherical cysts whose swelling is optically assessed and dependent on CFTR function providing a novel assay to screen for CFTR modulators and potentially identify surrogate pathways as alternate targets/approaches. Another opportunity is the use of human pluripotent stem cells, either embryonic (ESC) or induced (iPSC), which would facilitate an inexhaustible supply of genetically stable and homogeneous cells to support discovery efforts. This is an exciting and emerging area with some success in relation to channels, albeit currently limited to the cardiovascular and neuronal areas.[27]

Many first generation ion channel modulators were characterized by their lack of selectivity between different channel families, with selectivity becoming even less evident within channel families, *e.g.* NaV channels. This is perhaps unsurprising if the approach to finding them is considered; commonly using heterologous expression systems with high levels of the pore forming domains expressed in isolation. For many channels these domains, and particularly the residues lining the channel pore regions, are very highly conserved and show little, if any, differences from close family member to member. However, advances in the structural biological understanding of many channels provided by combining high resolution crystallography and functional studies with structurally modified channels has enabled light to be shed upon areas of diversity in terms of their function and pharmacology. The first clear example of this type of approach and the potential it has for defining selective molecules relates to NaV channels.[28] The voltage sensing domains of NaV channels show significant sequence diversity, particularly in the paddle domains—the Bosmans/Swartz groups mapped the binding sites for NaV specific blocking spider toxins which affected channel activation/inactivation kinetics to these regions.[28] This was confirmed by generating chimeras with the voltage paddles from the NaV channels transferred to Kv channels and showing that pharmacological sensitivity followed, *i.e.* the chimeric Kv channels were inhibited by the toxins whereas the native Kv channels were not. This was an elegant approach and highlights a novel screening strategy to identify molecules that bind to a defined region with the potential to deliver subtype selectivity. A similar approach employing chimeras and point mutations has subsequently been shown by others to explain the NaV subtype selectivity of their molecules.[29] In addition to selectivity considerations, and guiding medicinal chemistry efforts (if structural information is available in parallel), early identification and delineation of pharmacological binding sites offers the potential to discover molecules with diverse mechanisms of action; *i.e.* by screening/triaging using recombinant channels with point mutations in known binding sites can reveal molecules with activity in new ones. In some cases prior knowledge of the binding site is required or alternatively a more unbiased 'shotgun' approach using high-throughput random residue mutagenesis can be considered, a strategy employed by Bandell and colleagues when delineating key residues responsible for the activation of TRPM8 by menthol.[30] With the expectation that an increasing amount of ion channel structural information will become available in the near future, these kinds of approach, and that of structure-based drug design as exemplified by Simon Ward with AMPA receptors (Chapter 6) will hopefully become more mainstream and represent an opportunity for defining more selective and efficacious ion channel modulators.

As already alluded to in this chapter, where the drug discoverer goes looking for ion channel modulators is equally as important a determinant of overall success as the how (assay) and the who (which channel). Large multimillion compound collections enriched with chemotypes known to interact with common 'druggable' families (*e.g.* GPCRs, kinases) may not be expected to yield good tractable chemical starting points for ion channel medicinal chemistry. A number of companies now offer ion channel targeted libraries but there is little data in the literature to fully validate higher success with these either. A number of innovative additional approaches to expanding ion channel chemical space have already been covered in previous chapters including peptides/toxins (Chapter 12) and antibody approaches (Chapter 13) and there are an array of other biologics based approaches that are also being explored including aptamers and siRNAs.[31] However, there are two other approaches to ion channel modulation that I would like to highlight. The first is synthetic pores. Antibiotics and antifungals such as gramicidin and amphotericin B have been in clinical use since the mid-19th century— their mechanism of action is the formation of a transmembrane channel which destroys the cytoplasmic ionic gradients leading to cell death. The pore forming capabilities of amphotericin B and related molecules are still exploited today by electrophysiologists to gain electrical access to the cell interior whilst preserving intracellular contents in the perforated patch recording technique (by virtue of the molecular weight cutoff the pore will permeate). However, the development of relatively simple low molecular weight molecules that can exhibit true ion channel behavior has lagged but are receiving renewed attention.[32,33] These molecules, although still relatively embryonic, represent a novel approach to increasing ion conductance in a cell or tissue. There are obviously a number of challenges to making these molecules widely therapeutically useful but where topical application is readily achievable (*e.g.* skin, lung) these hurdles may be significantly reduced. Returning to CF as an example, a synthetic chloride channel has been reported to restore chloride conductance in human CF epithelial cells, albeit *in vitro*,[34] offering an alternative approach to treat this devastating disease. The second approach is an alternate modality of modulation. The majority of channel drug discovery efforts have focused upon modulators which affect the gating of the channel, a panoply of inhibitors, blockers, activators, potentiators, positive/negative allosteric modulators and gating modifiers. However, affecting the level of channel expression by modulation of protein trafficking is not widely explored. This is despite there being a number of channelopathies that result from reduced/aberrant channel expression levels such as some long QT syndromes,[17,18] and additionally emerging evidence that reduced channel levels may play a role in pathologies such as pulmonary arterial hypertension.[35,36] The notable exception is again CF, as outlined by Sheppard and colleagues in Chapter 8. Here the opportunity has been aggressively pursued by the field resulting in several molecules that enhance mutant CFTR expression level showing benefit in a clinical setting (*e.g.* VX-809; **2**). As the precedent becomes set perhaps this approach will be a more realistic opportunity for other channel related diseases?

14.4   Concluding Remarks

The ion channel drug discovery field has made considerable advances in the last decade and recent analyses suggest that there are >100 novel ion channel modulators reported to be in clinical development [www.insightpharmareports.com]. However, the number that have received marketing approval still remains very low suggesting that considerable challenges, but also opportunities, remain for the future. A better understanding of channel function in health and disease, combined with the ability to discover better and diverse therapeutic molecules will be required if this enigmatic class of proteins is to fully deliver on its therapeutic potential, an aspiration for all working in the ion channel field.

Acknowledgements

My sincere thanks to Drs Pamela Tranter and Martin Verkuijl for reviewing this chapter and their constructive comments and suggestions for improvement.

References

1. L. J. Stephen and M. J. Brodie, *CNS Drugs*, 2011, **25**(2), 89–107.
2. A. R. Marks, *J. Clin. Invest.*, 2013, **123**(1), 46–52.
3. R. Lu, A. Alioua, Y. Kumar, M. Eghbali and E. Stefani, *J. Physiol.*, 2006, **570**(1), 65–72.
4. S. Feske, Y. Gwack, M. Prakriya, S. Srikanth, S. H. Puppel, B. Tanasa, P. G. Hogan, R. S. Lewis, M. Daly and A. Rao, *Nature*, 2006, **441**, 179–185.
5. M. Vig, C. Peinelt, A. Beck, D. L. Koomoa, D. Rabah, M. Koblan-Huberson, S. Kraft, H. Turner, A. Fleig, R. Penner and J. P. Kinet, *Science*, 2006, **312**, 1220–1223.
6. A. Caputo, E. Caci, L. Ferrera, N. Pedemonte, C. Barsanti, E. Sondo, U. Pfeffer, R. Ravazzolo, O. Zegarra-Moran and L. J. Galietta, *Science*, 2008, **322**, 590–594.

7. Y. D. Yang, H. Cho, J. Y. Koo, M. H. Tak, Y. Cho, W. S. Shim, S. P. Park, J. Lee, B. Lee, B. M. Kim, R. Raouf, Y. K. Shin and U. Oh, *Nature*, 2008, **455**, 1210–1215.

8. B. Coste, J. Mathur, M. Schmidt, T. J. Earley, S. Ranade, M. J. Petrus, A. E. Dubin and A. Patapoutian, *Science*, 2010, **330**, 55–60.

9. S. Feske, *Ann. NY Acad. Sci.*, 2011, **1238**, 74–90.

10. A. Britschgi, A. Bill, H. Brinkhaus, C. Rothwell, I. Clay, S. Duss, M. Rebhan, P. Raman, C. T. Guy, K. Wetzel, E. George, M. O. Popa, S. Lilley, H. Choudhury, M. Gosling, L. Wang, S. Fitzgerald, J. Borawski, J. Baffoe, M. Labow, L. A. Gaither and M. Bentires-Alj, *Proc. Natl. Acad. Sci., U. S. A.*, 2013, **110**(11), E1026–E1034.

11. R. Zarychanski, V. P. Schulz, B. L. Houston, Y. Maksimova, D. S. Houston, B. Smith, J. Rinehart and P. G. Gallagher, *Blood*, 2012, **120**(9), 1908–1915.

12. I. Andolfo, S. L. Alper, L. De Franceschi, C. Auriemma, R. Russo, L. De Falco, F. Vallefuoco, M. R. Esposito, D. H. Vandorpe, B. E. Shmukler, R. Narayan, D. Montanaro, M. D'Armiento, A. Vetro, I. Limongelli, O. Zuffardi, B. E. Glader, S. L. Schrier, C. Brugnara, G. W. Stewart, J. Delaunay and A. Iolascon, *Blood*, 2013, **121**(19), 3925–3935.

13. J. Albuisson, S. E. Murthy, M. Bandell, B. Coste, H. Louis-Dit-Picard, J. Mathur, M. Fénéant-Thibault, G. Tertian, J. P. de Jaureguiberry, P. Y. Syfuss, S. Cahalan, L. Garçon, F. Toutain, P. Simon Rohrlich, J. Delaunay, V. Picard, X. Jeunemaitre and A. Patapoutian, *Nat. Commun.*, 2013, **4**, 1844.

14. M. A. Valverde, M. Díaz, F. V. Sepúlveda, D. R. Gill, S. C. Hyde and C. F. Higgins, *Nature*, 1992, **355**, 830–833.

15. C. Duran, C. H. Thompson, Q. Xiao and H. C. Hartzell, *Annu. Rev. Physiol.*, 2010, **72**, 95–121.

16. J. R. Riordan, J. M. Rommens, B. Kerem, N. Alon, R. Rozmahel, Z. Grzelczak, J. Zielenski, S. Lok, N. Plavsic, J. L. Chou, M. L. Drumm, M. C. Iannuzzi, F. S. Cillons and L. C. Tsui, *Science*, 1989, **245**(4922), 1066–1073.

17. G. Webster and C. L. Berul, *Circulation*, 2013, **127**(1), 126–140.

18. H. Abriel and E. V. Zaklyaminskaya, *Gene*, 2013, **517**(1), 1–11.

19. S. D. Dib-Haji, Y. Yang, J. A. Black and S. G. Waxman, *Nat. Rev. Neurosci.*, 2013, **14**(1), 49–62.

20. C. J. Milligan, J. Li, P. Sukumar, Y. Majeed, M. L. Dallas, A. English, P. Emery, K. E. Porter, A. M. Smith, I. McFadzean, D. Beccano-Kelly, Y. Bahnasi, A. Cheong, J. Naylor, F. Zeng, X. Liu, N. Gamper, L. H. Jiang, H. A. Pearson, C. Peers, B. Robertson and D. J. Beech, *Nat. Protoc.*, 2009, **4**(2), 244–255.

21. A. M. El-Arabi, C. S. Salazar and J. J. Schmidt, *Lab Chip*, 2012, **12**(13), 2409–2413.

22. J. K. Rosenstein, S. Ramakrishnan, J. Roseman and K. L. Shepard, *Nano. Lett.*, 2013, **13**(6), 2682–2686.

23. R. Kawano, Y. Tsuji, K. Sato, T. Osaki, K. Kamiya, M. Hirano, T. Ide, N. Miki and S. Takeuchi, *Sci. Rep.*, 2013, **3**, 1995.
24. A. Oshima, A. Hirano-Iwata, H. Mozumi, Y. Ishinari, Y. Kimura and M. Niwano, *Anal. Chem.*, 2013, **85**(9), 4363–4369.
25. S. O. Marx and A. R. Marks, *J. Mol. Cell. Cardiol.*, 2013, **58**, 225–231.
26. J. F. Dekkers, C. L. Wiegerinck, H. R. de Jonge, I. Bronsveld, H. M. Janssens, K. M. de Winter-de Groot, A. M. Brandsma, N. W. de Jong, M. J. Bijvelds, B. J. Scholte, E. E. Nieuwenhuis, S. van den Brink, H. Clevers, C. K. van der Ent, S. Middendorp and J. M. Beekman, *Nat. Med.*, 2013, **19**(7), 939–945.
27. J. Jiao, Y. Yang, Y. Shi, J. Chen, R. Gao, Y. Fan, H. Yao, W. Liao, X. F. Sun and S. Gao, *Hum. Mol. Genet.*, 2013, **22**(21), 4241–4252.
28. F. Bosmans, M. F. Martin-Eauclaire and K. J. Swartz, *Nature*, 2008, **456**(7219), 202–208.
29. K. McCormack, S. Santos, M. L. Chapman, D. S. Krafte, B. E. Marron, C. W. West, M. J. Krambis, B. M. Antonio, S. G. Zellmer, D. Printzenhoff, K. M. Padilla, Z. Lin, P. K. Wagoner, N. A. Swain, P. A. Stupple, M. de Groot, R. P. Butt and N. A. Castle, *Proc. Natl. Acad. Sci., U. S. A.*, 2013, **110**(29), E2724–E2732.
30. M. Bandell, A. E. Dubin, M. J. Petrus, A. Orth, J. Mathur, S. W. Hwang and A. Patapoutian, *Nat. Neurosci.*, 2006, **9**(4), 493–500.
31. E. B. De Souza, S. T. Cload, P. S. Pendergrast and D. W. Y. Sah, *Neurosychopharm. Rev.*, 2009, **34**, 142–158.
32. P. Reiß and U. Koert, *Acc. Chem. Res.*, 2013, **46**(12), 2773–2780.
33. G. W. Gokel and S. Negin, *Acc. Chem. Res.*, 2013, **46**(12), 2824–2833.
34. B. Shen, X. Li, F. Wang, X. Yao and D. Yang, *PLoS One*, 2012, **7**(4), e34694.
35. L. Ma, D. Roman-Campos, E. D. Austin, M. Eyries, K. S. Sampson, F. Soubrier, M. Germain, D. A. Trégouët, A. Borczuk, E. B. Rosenzweig, B. Girerd, D. Montani, M. Humbert, J. E. Loyd, R. S. Kass and W. K. Chung, *N. Engl. J. Med.*, 2013, **369**(4), 351–361.
36. F. K. Kuhr, K. A. Smith, M. Y. Song, I. Levitan and J. X. Yuan, *Am. J. Physiol. Heart Circ. Physiol.*, 2012, **302**(8), H1546–H1562.
37. B. Nilius, *EMBO J.*, 2010, **11**, 902–903.
38. Z. Qiu, A. E. Dubin, J. Mathur, B. Tu, K. Reddy, L. J. Miraglia, J. Reinhardt, A. P. Orth and A. Patapoutian, *Cell*, 2014, **157**, 447–458.
39. F. K. Voss, F. Ullrich, J. Munch, K. Lazarow, D. Lutter, N. Mah, M. A. Andrade-Navarro, J. P. von Kries, T. Stauber and T. J. Jentsch, *Science*, 2014, **344**, 634–638.

Subject Index